Introduction to AI Techniques for Renewable Energy Systems

Introduction to AI Techniques for Renewable Energy Systems

Edited by

Suman Lata Tripathi
Mithilesh Kumar Dubey
Vinay Rishiwal
Sanjeevikumar Padmanaban

CRC Press
Taylor & Francis Group
Boca Raton London New York

CRC Press is an imprint of the
Taylor & Francis Group, an **informa** business

First edition published 2021
by CRC Press
6000 Broken Sound Parkway NW, Suite 300, Boca Raton, FL 33487-2742

and by CRC Press
2 Park Square, Milton Park, Abingdon, Oxon, OX14 4RN

CRC Press is an imprint of Taylor & Francis Group, LLC

ISBN: 978-0-367-61092-0 (hbk)
ISBN: 978-0-367-61167-5 (pbk)
ISBN: 978-1-003-10444-5 (ebk)

Typeset in Times
by KnowledgeWorks Global Ltd.

Contents

Preface

Artificial intelligence (AI) approaches are being developed to produce more accurate predictions of *renewable energy* using AI techniques, including their generation and impacts on the electric grid such as net load forecasting, line loss predictions, maintaining system reliability, integrating hybrid solar and battery storage systems. The book mainly deals with AI techniques for modeling, analysis, prediction of the performance, and control of renewable energy systems. This book will provide conceptual as well as practical knowledge about AI techniques used in renewable energy systems to the students, researchers, and academicians.

CHAPTER ORGANIZATION

This book is organized into 24 chapters in total.

Chapter 1 intends to study the brief introduction and past history relevant to artificial intelligence (AI) in fields of solar, wind, and other renewable energy sources.

Chapter 2 presents a study on issues and challenges of renewable energy and AI applications to resolve these issues.

Chapter 3 demonstrates the key challenges faced in renewable energy using machine learning and various AI applications in real-life scenarios.

Chapter 4 deals with enhanced computational power and hybrid transaction or analytical processing systems (HTAPS) enabling ML algorithms to optimize the energy and also the power sector on an outsized scale.

Chapter 5 highlights the importance of machine learning in research and development, and also focuses on its types and techniques.

Chapter 6 is designed to provide the ins and outs of hybrid renewable energy systems (HRESs) such as the configuration architecture, stability issues, maintenance, available optimization techniques, and performance predicting simulation software for HRESs.

Chapter 7 focuses mainly on the integration of solar photovoltaic (PV) and electricity board (EB) source for sustaining the electric power system to satisfy the energy demand throughout the adequate insolation.

Chapter 8 describes the various AI techniques and machine learning algorithms and how these approaches have been applied to different renewable energy (RE) types, especially solar energy.

Chapter 9 demonstrates the accuracy and effectiveness of the proposed optimizer; a 400-W domestic standalone SPV system and its specifications are used as a model for simulation in MATLAB.

Chapter 10 throws light on the various AI-based failure prediction strategies that are deployed on renewable source infrastructure.

Chapter 11 deals with the application of AI techniques in smart energy systems and is an effort to concise all the information on the optimization of solar energy systems for common-life applications.

Chapter 12 gives an idea about AI-enabled energy forecasting systems to handle fluctuations that may adversely affect the planning, designing, and operations according to the requirements.

Chapter 13 focuses on biomass resources and their products with their utilization for renewable energy based on the current trend of research and development in these sectors.

Chapter 14 discusses the integration of renewable energy-based smart grid that comes up with some new challenges which enhance the research in the field of AI.

Chapter 15 deals with the modeling, simulation, and analysis of small-scale photovoltaic energy system (PVES).

Chapter 16 describes about the deep learning-based fault identification of microgrid transformers.

Chapter 17 aims to compare and analyze different configurations of UPQC, namely, UPQC-I (interline), UPQC-MC (multiconverter), UPQC-DG (distributed generation) supplying loads, which are very sensitive toward voltage variations and critical in operation.

Chapter 18 discusses concepts of reliability and maintainability theory, which are concerned with our work for calculating the maintainability.

Chapter 19 focuses on different machine learning models and it's achievements for damage estimation capability to design and develop efficient wind turbines for wind energy systems

Chapter 20 utilizes the data from the two commercial solar cells and for the purpose of the validation of results, the IV and PV curves of each cell are presented along with the other statistical analyses, which have been performed at various temperature and irradiance levels.

Chapter 21 deals with the modern wind generation system and its monitoring of health in operating condition, recognition of smart gird (SG) subsystem's fault pattern, and control of SG based on real-time simulator.

Chapter 22 presents the estimation of parameters of thin-film solar panels using the newly developed reverse two diode model (RTDM) applying moth flame optimizer (MFO). RTDM being not a regular model used for parameter estimation, increases the complexity of the equations.

Chapter 23 describes the traditional time series forecasting to achieve effective energy resource planning that ensures the availability of the desired amount of natural gas for various manufacturing operations at appropriate point of time.

Chapter 24 addresses the security issue in the smart grid due to large data processing and proposed machine learning-based cyber security solutions for the intelligent grid.

About the Editors

Suman Lata Tripathi is working as Professor, School of Electronics and Electrical Engineering, Lovely Professional University, India. She has over 17 years of experience in academics and has published over 35 research papers in refereed journals and conferences. She has organized several workshops, summer internships, and expert lectures for students. She has worked as a session chair, conference steering committee member, editorial board member, and reviewer in international/national *IEEE Journal* and conferences. She has been nominated for the "Research Excellence Award" in 2019 at Lovely professional University. She has received the best paper at IEEE ICICS-2018. Her areas of expertise include microelectronics device modeling and characterization, low power VLSI circuit design, VLSI design of testing, and advance FET design for IoT, embedded system design, and biomedical applications, etc.

Mithilesh Kumar Dubey is working as Associate Professor, School of Computer Science and engineering, Lovely Professional University, India. He has over 12 years of teaching experience and industry experience. He has worked with Tata Consultancy Services (TCS) as Software Engineer. He has filed and published three patents in the government of India in "E-health care system." He has published 21 research papers in various reputed journals like *Scopus/UGC*. He has also published one book entitled "*Core python programming*" in Lambert Germany (LAP publisher) in the year of 2018.

Vinay Rishiwal is working as Professor, Department of Computer Science and Information Technology, Faculty of Engineering and Technology, MJP Rohilkhand University, Bareilly, Uttar Pradesh, India. He received his PhD in Computer Science and Engineering from Gautam Buddha Technical University, Lucknow, India. He has 20 years of experience in academics. He is a senior member of IEEE. He has published 85+ research papers in various journals and conferences of international repute. He is a general chair of three International Conferences, namely, ICACCA, IoT-SIU, and ICAREMIT. He has received many awards for the best paper and best orator in various conferences.

Sanjeevikumar Padmanaban (Member'12–Senior Member'15, IEEE) received a Ph.D. degree in electrical engineering from the University of Bologna, Bologna, Italy 2012. He was an Associate Professor at VIT University from 2012 to 2013. In 2013, he joined the National Institute of Technology, India, as a Faculty Member. In 2014, he was invited as a Visiting Researcher at the Department of Electrical Engineering, Qatar University, Doha, Qatar, funded by the Qatar National Research Foundation (Government of Qatar). He continued his research activities with the Dublin Institute of Technology, Dublin, Ireland, in 2014.

Further, he served as an Associate Professor with the Department of Electrical and Electronics Engineering, University of Johannesburg, Johannesburg, South Africa, from 2016 to 2018. From March 2018 to February 2021, he has been a Faculty Member with the Department of Energy Technology, Aalborg University, Esbjerg, Denmark. Since March 2021, he has been with the CTIF Global Capsule (CGC) Laboratory, Department of Business Development and Technology, Aarhus University, Herning, Denmark.

S. Padmanaban has authored over 300 scientific papers and was the recipient of the Best Paper cum Most Excellence Research Paper Award from IET-SEISCON'13, IET-CEAT'16, IEEE-EECSI'19, IEEE-CENCON'19 and five best paper awards from ETAEERE'16 sponsored Lecture Notes in Electrical Engineering, Springer book. He is a Fellow of the Institution of Engineers, India, the Institution of Electronics and Telecommunication Engineers, India, and the Institution of Engineering and Technology, U.K. He is an Editor/Associate Editor/Editorial Board for refereed journals, in particular the IEEE SYSTEMS JOURNAL, IEEE Transaction on Industry Applications, IEEE ACCESS, *IET Power Electronics*, *IET Electronics Letters*, and Wiley-*International Transactions on Electrical Energy Systems*, Subject Editorial Board Member—*Energy Sources—Energies Journal*, MDPI, and the Subject Editor for the *IET Renewable Power Generation*, *IET Generation*, *Transmission and Distribution*, and *FACETS* journal (Canada).

1 Artificial Intelligence
A New Era in Renewable Energy Systems

Kandra Prameela[1], Challa Lahari[2], Grandhi Sai Kishore[3], Kandula Venkata Nikhil[3], and Pavuluri Hemanth[3]
[1]Department of Biotechnology, Gitam Institute of Technology, GITAM (deemed to be University), Visakhapatnam, Andhra Pradesh, India
[2]Department of Computer Science and Engineering, National Institute of Technology (NIT), Tadepalligudem, Andhra Pradesh, India
[3]Department of Computer science, Gitam Institute of Technology, GITAM (deemed to be University), Visakhapatnam, Andhra Pradesh, India

CONTENTS

1.1 INTRODUCTION

The emergence of technology has helps us to generate solutions to minimize the risk of human errors in the complex systems. Most of the technologies developed by scientists are inspired by natural system. Artificial intelligence (AI) is an interdisciplinary promising approach for prompting the human brain functions and incorporating them into the machine to make it human smart. In the modern world, science and technology is a remarkably strong venture with development of quantified self to empower the challenges faced by humans. AI is the mimic of human brain and programmed to the computer to use for several computer systems which have been generated, for example, to analyze electric circuits, to solve differential equations, to design for synthesis of complex chemical molecules, and to detect and treat diseases. Due to AI's systems degree, it is possible to understand and build a neural network to challenges, for human benefits [1]. In general, AI includes fuzzy logic, expert systems, and artificial neural networks (ANNs). These are very powerful tools of renewable energy and smart grid systems for design, stimulation, estimation, control, fault diagnostics, and fault-tolerant control [2]. AI is a creation of a system with human brain intelligence to find solutions of narrow specific manually unsolved tasks. For example, in a supervised learning process, the system will learn the mapping based on their input and output programms.

The pervasiveness of technology allows spreading it across all fields in the world, including energy systems, health, agriculture economics, engineering, marine, and others. In every field, consumption of energy is increasing drastically and is causing decline to fossil fuels. In such situation, many countries have shifted their focus to renewable energy resources such as wind and solar energy sources. The main purpose of AI is to understand the problems of human cognition, analyze circuits, limit the human speech, train the system and store the information to retrieve it. It includes powerful tools such as fuzzy logic, expert systems, and ANNs that are used for designing, simulating, controlling, and diagnosing faults in renewable energy systems and others. Genetic algorithm is also one of the optimum AI techniques that imitates the evolution and is based on natural selection which utilizes some of the genetic actions such as selection, crossover, and variation. Similarly, data mining is another potential technique used to predict information from large databases. Most of these techniques are used in the field of production control, customer behavior, and stock prices. Data mining is also applicable in astronomical and medical data. They are complex and require precise analysis to set up global technological systems. AI programs require a clear-cut explanation since detailed reasoning can be drifted out so as to reach adequate solutions. AI research in the energy field is majorly focusing on solving several technical issues. The most promising development in AI is artificial neural techniques, which can be used for wind energy source assessment and apex power point tracking in photovoltaic (PV) generators. Expert systems have helped in building efficient fault-tolerant control in modern grid systems. Currently, many countries are working in R&D of smart grid to replace and develop advanced, efficient smart grid, as older versions are insufficient to resolve some of the faults.

1.2 HISTORICAL PERSPECTIVES OF ARTIFICIAL INTELLIGENCE

The great contribution of some researchers in the development of AI gains momentum toward the change of the world's scenario. AI concept can be traced back to the ancient time when people used to mention machine learning in their mythical stories [3]. This belief has made to found an expression in many forms of AI to explore the origin and the reasons for our existence and to create workable technologies for improvement of quality of human life. Initially, people used to mention this concept as "mythical thinking machines." Later, in 10 AD–70 AD, Heron of Alexandria had written in his famous work *Pneumatica* that these mythical thinking machines were built to exhibit some form of tricked intelligence [4]. However, during that time, it was not possible to develop mechanical machines having all intelligence-based functionalities. In the 17th century, scientists worked on the development of complex machines with intelligence, where this concept was compared with animals (Descartes), which is the basis of mind, body, and soul for the treatment of animals [5]. This idea of comparing the animal body with machines and mind with the intelligence for operation of machines was the novel approach for understanding of AI [6]. There was a great furor against this concept by philosophers and mathematicians. The first consequences of mathematical logic to computers came in 1642 with the invention of the first mathematical calculator [7, 8]. It took almost 30 years to make a mechanical calculator perform functions such as division and multiplication.

The most striking attributes came at the end of the 19th century with a notable contribution from George Boole who invented Boolean algebra, which is the major component of modern computers. Almost at the end of the 19th century, Samuel Butler noted that the machines might be evolved like Darwinian evolution [9]. Samuel has correctly interrelated his concepts since nowadays, researchers and scientists are building a foundation for the early 20th century. It is important to note that in 1942 Isaac Asimov published the concepts that he contrived with allusion to his laws of responsible robotics pertaining to his belief in growing the AI [10]. He proposed three concepts with two laws. They are: robots are not harmful to humans; it obeys human orders (first law); and robots must protect its own existence (second law). Another professor from the University of Connecticut focused most of his research work on machine ethics [11]. In 1950, a conference was conducted by one of the youngest renowned computer scientists John McCarty to bring many computer scientists on a single platform to discuss the creation of machine intelligence. He is considered the "father of artificial intelligence." After 2 years, there was a drastic improvement in AI. In 1961, Marvin Minsky mentioned in his published paper the steps toward AI and he summarized much of his work during the first decade of AI [12, 13]. He was able to represent in frames what he called "society of mind." In a week-long Dartmouth Summer Research Project on AI (DSRPAI) conducted by Marvin and John McCarthy in New Hampshire, the word "artificial intelligence" was coined. The main objective of this project was to reunite scientists in different fields in order to build a novel research platform on machines that are able to simulate human intelligence. Newell and Simon's work on GPS was developed based on a psychological query and empirical methods [14]. It is important to note that contributions from the end of the 19th century help in understanding computers and their languages to store and retrieve information in a massive way. In 1977, Goldstein and Papert demonstrated and published the Dendral program [15, 16]. However, the probability of expanding AI started in the 1960s, and it took a huge leap forward in the 20th century. Initially, two laboratories were established and the burden was taken by MIT (Massachusetts Institute of Technology) and Carnegie Tech University (CMU), with the Rand Corporation establishing AI laboratories later in Stanford and Edinburg. Edinburg was the place where the first conference of AI was conducted in 1965. In 1969, International Joint Conferences on Artificial Intelligence (IJCAI) started biannual series on AI. In 1980, the American Association of Artificial Intelligence (AAAI) was established [17]. It has expanded by conducting conferences and a series of workshops, sponsoring scholarships, encouraging journals, and maintaining digital libraries.

1.3 APPLICATIONS OF AI

AI has unique applications where complicated languages allow to make quicker, better, accurate predictions than manual methods. It has many branches, such as fuzzy logic, ANN, adaptive network-based fuzzy inference system (ANFIS), and data mining. ANN is inspired by the nervous system of humans, i.e., connections between elements to establish a specific function by adjusting the connections between elements [18, 19]. Mapping multiple inputs to get a single output, pattern classification, recognition, association and generation of meaningful patterns are some of the applications of AI. AI applications have also been successfully extended to other fields such as engineering, mathematics, meteorology, medicine, neurology, economics, and psychology. One of the interesting applications of AI is the recognition of sound and speech. In 1960, the adaptive linear combiner was developed, which is a very useful law [20]. Most of the neural networks depend on algorithms, mathematical models, and nonlinear operations, which include maps of noise. AI is also applicable in market trends, weather forecasting, thermal and electrical load prediction, exploration of mineral sites, and prediction of data collected from sensors [18, 19]. Analysis of medical signatures, electromyography, and identification of explosives and military targets are other applications. One of the interesting applications of AI is in PV electricity production. The main objective of AI application is to produce viable solar emission info for real-time designing and forecasting of solar radiation and steam generators. Generating PV modules, designing solar collectors, heating controllers, sun tracking and building controlled solar air conditioning systems, predicting temperature and radiation, and developing PV-diesel and solar-wind hybrid systems are other encouraged applications [21, 22]. Another interesting domain of AI is molecular biology. AI can be applied to design and analyze scientific data, solving wide problems, hierarchical patterns, and knowledge-based maintenance and obtaining knowledge acquisition technologies. AI has momentum in the field of molecular biology for the last few years. Nutritional meal planning is another important issue for cancer patients whenever they undergo chemotherapy. Through AI, the meal components with nutrition were designed in Mary to greater variety. Now, this Mary is used as a public tool; similarly, other additional nutritional elements such as carbohydrates and cholesterol are also added to the design of the foods. AI has large applications for cancer patients such as resolving drug and dosage parameters. One of the designs for drugs and doses is quadratic phenotypic optimization platform (QPOP). This model will help in reducing the toxicity of drugs with maximum output in preclinical studies. With the QPOP platform, nearly 14 chemotherapeutic drugs were screened and are used for several cancer treatments such as multiple myeloma [23]. The combination of drugs, such as decitabine and mitomycin C, markedly shown improvement in outcomes in mouse models [24]. Designing of biomarkers is another advantage of AI [25]. The step forward in treatment of cancer patients is targeted therapies genomic alterations and treatment responses [26, 27]. Therefore, AI analysis of electronic health records and patient biomarker data may affect trial outcomes [28]. Some of the applications of AI in renewable energy with AI apps are shown in Figure 1.1.

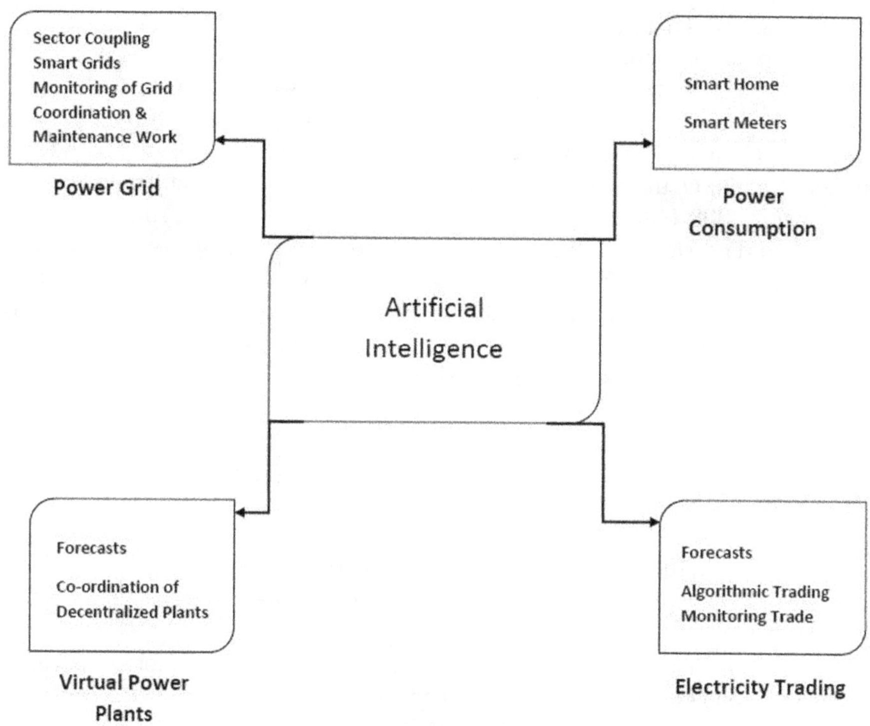

FIGURE 1.1 Some applications of AI in renewable energy with AI apps.

1.4 APPLICATIONS OF AI IN WIND RENEWABLE ENERGY SYSTEMS

Fossil fuel consumption all over the world is increasing day by day; therefore, the need for electric energy is also increasing. Effective ways of electric power generation are inherently dependent on the world economy. One of the sources for substitute renewable energy is wind. In large geographical areas such as Mexico, it was determined that the renewable energy wind is a periodical phenomenon [29]. Countries such as Mexico have a high demand for renewable energy sources because of the increase in prices of fossil fuels and high demand to minimize environmental contaminants that fossil fuels generate [30]. For better optimal designs and operations of wind energy systems, new strategies are continuously developed by many researchers [31]. Because of low-speed winds, this system does not have the same viability in all locations and is more uncertain than solar energy systems. Preferred locations of wind farms are offshore and high-altitude sites. In 2014, Norway's contribution toward renewable energy was 98%. Brazil has contributed 73.4%, New Zealand 79%, Venezuela 62.8%, and Colombia 70%. The traditional method that is used in generating wind energy is, first, installation of windmills in the farm; the wind will rotate the turbine and due to this the kinetic energy is changed to mechanical energy, which then transformed into electric energy. The first wind turbine with a capacity of 12 kW was installed in Ohio in 1887–1888. An overall growth rate of

9.96% is observed in installing wind energy scope when correlated with the previous year's scope scope in the European Union. The annual installation increased from 48 GW in 2006 to 141 GW in 2015, where the annual growth was more than 9% [32]. From 2010 to 2015, China was leading in installing the wind power units. In 2006 and 2007, Germany has greater statistics in the installation of wind power systems. In 2008 and 2009, the United States has a greater number of installed wind power units, but in 2008, both the United States and Germany contributed similar statistics in wind power systems that were installed. The efficiency of power generation using wind turbine can be increased by gathering the past wind data at turbine location. Accurate forecasting of wind speed cannot be done but this can be predicted using the past data of the wind speed. Using data analytics and AI, there are two different approaches to prediction of wind speed.

1.5 MODELING AND SIMULATION STUDIES OF AI IN WIND RENEWABLE ENERGY SYSTEM

Most of the wind designs can be recognized with the mathematical formula $V(k + l)$ of the wind vector, where $V(k)$ is based on the previous m measurements $V(k)$, $V(k - 1)$, …, $V(k - m + 1)$.

For a better wind speed forecast, l is chosen to be small; therefore, it is called short-term wind speed prediction. Wang and his co-workers have produced a general procedure to produce fuzzy rule [33, 34]. The fuzzy model will predict the wind speed and these are used to produce electrical power, especially at the wind units. ANN is a favorable technology for wind speed forecasting [35]. The proposed ANN will have fewer neuron numbers and it will have accurate wind prediction results in less time in different fields such as transient detection, approximation, pattern recognition, and time series prediction. $V(k)$, $V(k - 1)$, …, $V(k - m + 1)$ are the measured wind speeds in time series, and based on this $V(k + l)$, …, $V(k + 2)$, $V(k + 1)$ are predicted or estimated wind speeds, and some properties of time series such as standard deviation, slope, and average are calculated and these values are given as inputs to the fuzzy or neural network predictor. The fuzzy rule base will be reduced if we reduce the inputs to the fuzzy interface system. This will give all the speeds of wind without sacrificing the estimated wind speed. Accuracy of the prediction and the learning process speed will be increased by reducing the ANN size for a neural network. In the period 2002–2005, many investigations were carried out using real wind speed data in Rostamabad in northern Iran. For every 30-minute interval, these measured data were averaged and the data for the months of February, May, August, and November were averaged for the period 2002–2004 [36]. The data are used as train data for the neural network and measurements for the year 2005 are used as test data. The absolute error between predicted and real values is called RMSE – lesser the value, the better the wind speed estimation. On the other hand, the neural method gives very little error in both the COD and RMSE values, where the absolute change will be less the 20%. To determine the certainty, one's prediction from a certain model is done by COD where its value varies from 0 to 1, 0 for the menial speed forecast and 1 for best accurate speed prediction [36].

In order to see the improved performance of ANN method, a hybrid method is adopted which is a combination of neural and fuzzy methods. Data from Tasmania, Australia, are used to measure wind energy using the ANFIS method on a very short-term basis [37]. Even though it is a neural network, functionally, it is equivalent to the fuzzy inference model and thus it is called a neuro-fuzzy system [37]. The prediction is done by ANFIS at the height of 40 m, which has only 3% of the mean absolute percentage error when compared with the actual wind speed at the same height of 40 m. The RMSE between the predicted values of ANFIS and the actual values of wind is 0.230. In general, every input is presented by two fuzzy sets only, and sometimes three sets are also considered as common. With the increase of extra membership functions, the accuracy of the result will increase but it will take more time to train the model or the neural network. The architecture of the ANFIS consists of six layers, where each layer has a different application. Layer 1 is an input layer where it just sends the crisp external signals to layer 2. Layer 2 is called fuzzification layer where fuzzification is performed by each neuron. Layer 3 is called a rule layer where each neuron corresponds to a single fuzzy rule. Layer 4 is known as the normalization layer where normalized firing strength of the given rules is being calculated by taking the inputs from all neurons in the rule layer. Layer 5 is called a defuzzification layer. Every neuron in this layer is connected to the corresponding normalized neuron, and input of ×1 and ×2 is also received. Layer 6 is a single summation neuron where it sums all the defuzzification neurons and produces the output y for the ANFIS model [37].

1.6 APPLICATIONS OF AI IN SOLAR RENEWABLE ENERGY SYSTEMS

In today's world, climate change is one of the major threats currently faced by the people. As these challenges are considered by the government and various energy solution providers, it has become important for them to provide a sustainable mode of renewable energy. This made the renewable energy an alternate source to fossil fuels which we are currently using [38]. Due to climatic conditions and depletion of nonrenewable energy rapidly, research and development on sustainable energy have increased, and it affects all countries and enterprises [39]. When compared to other areas, solar energy applications are very expensive in remote areas since devices such as pyranometers which are used to measure solar radiation perfectly, are either present in only some locations or absent [40]. The increase in energy consumption from several decades has risen fright of using up reserves of resources in coming future [41]. With the tough year ahead, production and exhaustion of distinct types of fossil fuels to give energy constitute notable changes in environment [42]. According to the World Energy Forum prediction, the energy fuels such as oil, coal, and gas reserves will be drained in the coming 10 decades. Normally, the primary consumption of fossil fuel in the world is over 79% and of that over 57.7% is consumed by the transport and is drained rapidly [43]. India needs to improve energy equipping for population, and speedy improvement in economy poses a discouraging challenge.

1.7 MODELING AND SIMULATION OF SOLAR RENEWABLE ENERGY SYSTEMS

The solar energy application has become an interesting research topic. Advancement in AI has brought about more significant results in weather forecasting and henceforth in upgrading the renewable energy and more secure access. Solar energy is considered as the better common and famous source of sustainable energy, as the sunlight is abundant in nature and access to this is simple when compared to others [39]. The solar power plants convert the energy from sunlight by using either intensive solar power plants, known as concentrating solar power (CSP) plants, or solar PV plants [44]. The CSP plants generate electricity from solar without emitting any harmful gases. This results in not polluting the environment [39]. Coming to solar PV plants, PV effects are used in order to generate electricity [44]. The CSP plants use steam turbines in order to convert sunlight into electrical energy. These plants can also run with fossil fuels when there are cases like blocking of the sun or at night times [45]. Despite all these, there are some disadvantages in solar energy. Till now there is no best tech in order to transform the electrical energy into storage medium. But there are three technologies currently in use and up to some extent they provide a feasible storage solution. They are thermal energy, smart power unit depository, and hydrogen fuel cells [39]. These three provide a reliable storage medium for the generated renewable electricity. The PV storage systems are used in generating electricity from the solar energy. These systems are divided into on-grid, off-grid, and hybrid systems. The on-grid systems continuously fluctuate and modify the state of the battery according to alterations in the potential of the sun and the charge status. The on-grid systems consist of a group of inverters that convert DC power to AC and store it in the battery pack. On the other hand, the off-grid systems convert the DC power to AC power and supply to the systems or to battery units. The hybrid solar PV systems unite the on-grid systems with additional battery modules. These systems are more complex and very cost-effective [39]. PV systems encounter a problem that is nothing but the most appropriate resolution of their dimensions. The optimization problem was dealt with by using hybrid energy storage technology which was proposed by Li et al. [46].

The main component of the solar energy generation is sunlight. There are many methods and predictions in order to forecast the solar radiance of sun, by forecasting weather and other climatic conditions. To approximate the solar radiance, few of the procedures are utilized but they are efficient when there is less spatial resolution than complex ones such as mountainous areas [47]. Due to this an artificial-based technique using ANN was put forward by Bosch et al. [48] that predicts the solar radiance even on terrains. Several other algorithms to predict the solar radiance include ANN and neuro-fuzzy inference systems [21]. ANN is used for identification of optimal operating point of PV systems [49]. To teach the ANN controller in order to identify the maximal operating point, a gradient sinking algorithm is used [19]. Using ANN, the solar radiation was predicted where there is no coverage of direct measurement instrumentation [40]. This prediction has a 93% accuracy of solar radiation with a mean absolute percentage error of 7.3%. Another method, back-propagation neural network (BPNN), is also used for forecasting the solar radiation. In the years

1988–2002, global solar radiations were predicted using BPNN; for the year 2002, the RMSE value obtained by using this method was 2.823×10^4 [50]. Using BPNN, daily solar radiation was estimated with an RMSE of 5.0–7.5% [48].

Predictions of solar radiation using ANN models with distant back-propagation algorithms were carried out by Premalatha and Valan Arasu [51]. In this, they considered two ANN models, and nine inputs were given to ANN model as trained data. The first model considered four stations, and the second model consisted of five stations in India. The two models are tested using the trained data and the algorithms. Among the two models, the model with five stations (model-2) depicted greater accuracy when calculated with LM (Levenberg-Marquardt Back Propagation) algorithm since it has shown lesser RMSE with a value of 1.0416 and a maximum linear correlation (R) of 0.9545. ANFIS is a hybrid method that combines both neuro-fuzzy and fuzzy logic. It is used to suspect clarity index and daily solar emission forecasting with RMSE 0.0215–0.0235, modeling PV power supply with 98% accuracy [52]. Therefore, by using AI we can predict the factors required to generate solar energy accurately with least error, which reduces the operational cost and human error. By using solar energy, the emission of greenhouse gases can be decreased, which encourages us to produce more renewable energy. It has very less impact on producing carbon dioxide and global warming emissions. In India, it is expected that nearly 60,000 MW of power will be produced from renewable energy sources during 2031–2032. The investment required to build a solar power plant is very high. It requires more land for installing more panels. It works only during daytime (i.e., in sunlight) and thus the energy cannot be collected during the night. In order to process the equipment using solar energy at night, the energy needs to be stored in batteries and supplied to the components.

1.8 CURRENT TRENDS AND FUTURE PROSPECTS OF AI

Currently, AI is present in many areas and is much developed in the area of renewable energy. AI is being used in predicting, designing, controlling, monitoring, and optimizing renewable energy models. Compared to the first quarter of 2019, this year the global usage of renewable energy has risen to 1.5%. After the completion of more than 100 GW of solar PV and 60 GW of wind power projects in 2019, the renewable energy generation has increased to 3% [53]. As per the renewable energy policy network (REN), at least 10 GW of energy each was produced from 32 countries globally in 2019. For most of the countries, producing energy from solar and wind is cost effective than from other sources such as coal and thermal [54]. India is producing 75 W of electricity from solar and wind energy. It may increase up to 175 W by 2022 and to 500 W by 2030 [55]. Physical methods are also used for predicting the renewable energy. These methods are efficient in forecasting atmospheric movements but they require large computation resources since a lot of data are required to assess. Physical methods are suitable only for short-term forecasting [56]. The main aim in wind energy is to predict the wind power ahead and we can plan accordingly with the predicted data. ANN methods are considered as the best estimation models in to predict wind power. The major drawback in ANN methods is that their performance varies with the location. For example, BPNN methods may provide an accurate estimation of wind power in one site; in another site, the radial basis function neural network (RBFNN) can be used as the best estimation model [57. Fuzzy logic can also be used in predicting, but the major problem in this is that it should generate a large number of fuzzy rule bases. In order to generate less rule base for fuzzy logic, a proposed model is defined based on ANN and fuzzy logic, which helps in generating less rule [57]. By using some algorithms, we can decide which places are suitable to build wind farms; this may result in less operating costs [58]. The numerical weather prediction method is efficient in short-term forecasting, while ANN and statistical methods are best in very short-term forecasting [37]. Hybrid model ANFIS is also the best model in estimating the very short-term forecasting but the training of ANFIS may fail if the set consists of insufficient variation to properly model the characteristic data [37]. Germany has developed a deep learning neural network model which helps in saving the lives of birds by not being killed by the wind turbine blades. A camera is attached near the turbine and predicts the movement using image processing, so when a bird approaches the blades, the turbine is automatically switched off [59]. Although there are models to forecast the wind speed such as physical, conventional statistical, and spatial correlation models, there are some disadvantages with these models, e.g., a statistical model cannot forecast high noise, fluctuation, and irregular nonlinear trends, spatial correlational model finds it difficult to predict the wind speed since the model needs more information like wind speed values of spatial correlated sites, etc. [60]. Now, considering the solar energy, the main usage of AI is in predicting the solar radiation. ANN in line with solar energy is used for monitoring the thermal ratings and rise of temperature of overhead power lines, depending on the meteorological conditions. It is also helpful in determining the solar irradiance [19]. The hybrid model ANFIS is used for determining the frequencies from the temperature as well as the duration of sunshine [61] and the modeling of the PV power supply system with 98% accuracy [57]. BPNN is used to measure the global solar radiation. Forecasting solar radiation by BPNN consists of higher value coefficients [57]. Another method that is used for predicting the short-term solar power is the support vector machine (SVM). Compared to other methods such as RBFNN, the SVM methods perform better, having a mean absolute error of 62 W/m^2 [57]. The solar radiation is higher in India when compared to countries like Germany, where the solar radiation ranges from 800 kWh/m^2 to 1200 kWh/m^2 [62]. Some of the hybrid AI methods are used in solar energy systems, e.g., in estimating the PV power using the hybrid evolutionary optimization of ANN with particle swarm optimization and genetic algorithm [63]. Genetic swarm optimization of BPNN [64] predicts the solar radiation using a combination of autoregressive and moving average (ARMA) and time delay neural network [65]. By using AI, we can improve safety, integration of micro grids, etc., in solar energy. The world is running on fossil fuels currently and for sure they are running out and none will be left over after. This may lead the world to begin changing its course, relying on the renewable energy completely. This takes much more advancement in the area of AI being used in the renewable energy. Everything currently is being automated and we may not be surprised if the world is completely automated in the near future. All the renewable energy monitoring may be automated and checked by the machines. New methods and algorithms may predict all the conditions perfectly without any errors and make human lives easier and the environment less polluted. An estimation given by the German Advisory Council on global change states that 50% of energy will be accounted for renewable energy by 2050 [66]. The International Renewable Energy Agency suggests that renewable energy can account for 60% or more of many countries' total final energy consumption and all countries may consider the use of renewable energy in their total energy use by 2050. Hence, progress in conventional applications of AI provides the opportunities to design and develop new mechanisms for specific applications in the near future.

1.9 CONCLUSION

AI is an emerging technology in renewable energy systems. Many researchers have made efforts to integrate the applications by innovative approaches for a sustainable environment. Wind and solar energy have enabled us to assess the potentiality through neural-based approaches. Computer science and analytical techniques in the field of biology have been applied to understand basic machine learning to application challenges in medicine and agriculture. In order to advance this technology, multiple strategies were used to pace the progress for human challenges. Hence, this chapter focused on applications of AI in the fields of solar, wind, and other renewable energy sources. Moreover, it emphasizes the concepts of different AI technologies with few examples such as solar and wind renewable energy systems. It deals with a major breakthrough in acceptance, modeling, and simulation studies with current trends and future prospects of AI technologies. Hence, progress in conventional applications of AI provides opportunities to design and develop new mechanisms for specific applications in the near future.

REFERENCES

1. Nilsson N. (1980). Principles of Artificial Intelligence. Springer-Verlag, New York, NY. ISBN 3-540-11340-1.
2. Bose K. B. (2017). Artificial intelligence techniques in smart grid and renewable energy systems—Some example applications. Proceedings of the IEEE, 105(11), 2263–2273.
3. McCorduck P. (2004). Machines Who Think. https://monoskop.org/images/1/1e/McCorduck_Pamela_Machines_Who_Think_2nd_ed.pdf
4. Heron of Alexandria. Encyclopedia Britannica. Encyclopedia Britannica, 1-24-017. https://www.britannica.com/biography/Heron-of-Alexandria
5. Cottingham J. (1978). 'A Brute to the Brutes?': Descartes' treatment of animals. Philosophy, 53(206), 551–559.
6. Laland K. N. & Hoppitt W. (2003). Do animals have culture? Evolutionary Anthropology, 12(3), 150–159.
7. Karwatka P. & Dennis M. (2004). Blaise Pascal and the first calculator. Tech Directions, 64(4), 10–15. 1p. 2 Black and White Photographs. https://search.proquest.com/openview/bc3f2d92bff9d978494763814084ad0c/1.pdf?pq-origsite=gscholar&cbl=182
8. Solla Price D. D. (1984). A history of calculating machines. IEEE Micro, 4(1), 22–52.
9. Samuel B. (1873). The Book of the Machines: Chapters in *Erewhon*. The Floating Press. https://oheg.org/hyr.pdf
10. Asimov I. (2004). I Robot. (1). Spectra, New York, NY.
11. Anderson S. L. (2008). Asimov's "three laws of robotics" and machine metaethics. AI & Society, 22(4), 477–493.

12. Minsky M. (1968). Semantic Information Processing. MIT Press, Cambridge, MA.
13. Minsky M. (1961). Steps toward artificial intelligence. Proceedings of the Institute of Radio Engineers, 49(1), 8–30.
14. Newell A. & Simon H. (1972). Human Problem Solving. Prentice-Hall, Englewood Cliffs, NJ.
15. Goldstein I. & Papert S. (1977). Artificial intelligence, language and the study of knowledge. Cognitive Science, 1(1), 84–123.
16. Lindsay R. K., Buchanan B. G., Feigenbaum E. A., & Lederberg J. (1980). Applications of Artificial Intelligence for Chemical Inference: The DENDRAL Project. McGraw-Hill, New York, NY.
17. American Association for Artificial Intelligence (AAAI). (2005). Menlo Park, CA. www.aaai.org/aitopics/history
18. Kalogirou S. A. (2000). Applications of artificial neural-networks for energy systems. Applied Energy, 67(1–2), 17–35.
19. Kalogirou S. A. (2001). Artificial neural networks in renewable energy systems applications: A review. Renewable and Sustainable Energy Reviews, 5(4), 373–401.
20. Chien J. T. (2005). Predictive hidden Markov model selection for speech recognition. IEEE Transaction on Speech and Audio Processing, 13(3), 377–387.
21. Mellit A., Kalogirou S. A., Shaari S., Salhi H., & Hadj Arab A. (2008). Methodology for predicting sequences of mean monthly clearness index and daily solar radiation data in remote areas: Application for sizing a stand-alone PV system. Renewable Energy, 33(7), 1570–1590.
22. Mellit A. & Pavan A. M. (2010). A 24-h forecast of solar irradiance using artificial neural network: Application for performance prediction of a grid-connected PV plant at Trieste, Italy. Solar Energy, 84(5), 807–821.
23. Palmer A. C. & Sorger P. K. (2017). Combination cancer therapy can confer benefit via patient-to-patient variability without drug additivity or synergy. Cell, 171(7), 1678–1691.
24. Rashid M. B. M. A. et al. (2018). Optimizing drug combinations against multiple myeloma using a quadratic phenotypic optimization platform. Science Translation Medicine, 10(453), eaan0941.
25. Lin A. et al. (2019). Off-target toxicity is a common mechanism of action of cancer drugs undergoing clinical trials. Science Translation Medicine, 11(509), eaaw8412.
26. Wong C. H., Kein W. S., & Andrew W. L. (2019). Estimation of clinical trial success rates and related parameters. Biostatistics, 20(2), 273–286.
27. Harrer S., Shah P., Antony B., & Hu J. (2019). Artificial intelligence for clinical trial design. Trends in Pharmacological Sciences, 40(8), 577–591.
28. Rajkomar A. et al. (2018). Scalable and accurate deep learning with electronic health records. NPJ Digital Medicine, 1(18). https://doi.org/10.1038/s41746-018-0029-1
29. Hernández-Escobedo Q., Manzano-Agugliaro F., Gazquez-Parra J. A., & Zapata-Sierra A. (2011). Is the wind a periodical phenomenon? The case of Mexico. Renewable and Sustainable Energy Reviews, 15(1), 721–728.
30. Hernández-Escobedo Q., Manzano-Agugliaro F., & Zapata-Sierra A. (2010). The wind power of Mexico. Renewable and Sustainable Energy Reviews, 14(9), 2830–2840.
31. Henriksen L. C. (2010). Wind energy literature survey no 16. Wind Energy, 13(4), 524–526.
32. EWEA. (2006–2015). Wind in power European statistics. http://www.ewea.org/statistics
33. Wang L. X. & Mendel M. (1992). Generating fuzzy rules by learning from examples. IEEE Transactions on Systems, Man and Cybernetics, 22(6), 1414–1426.
34. Wang L. X. (1996). A course in fuzzy systems and control. Prentice-Hall Inc, Upper Saddle River, NJ.

35. Lapedes A. & Farber R. (1987). Nonlinear Signal Processing Using Neural Networks: Prediction and System Modeling. Technical report LA-UR-87-2662. Los Alamos National Laboratory, Los Alamos, NM.
36. Monfared M., Rastegar H., & Kojabadi H. M. (2009). A new strategy for wind speed forecasting using artificial intelligent methods. Renewable Energy, 34(3), 845–848.
37. Potter C. W. & Negnevitsky M. (2006). Very short-term wind forecasting for Tasmanian power generation. IEEE Transactions on Power Systems, 21(2), 965–972.
38. Saurenergy.com. https://www.saurenergy.com/solar-energy-articles/the-role-of-ai-and-ml-in-solar-energy
39. Trappey A. M. C., Chen P. P. J., Trappey C. V., & Ma L. (2019). A machine learning approach for solar power technology review and patent evolution analysis. Applied Sciences, 9(7), 1478.
40. AI-Alawi S. M. & AI-Hinai H. A. (1998). An ANN-based approach for predicting global radiation in locations with no direct measurement instrumentation. Renewable Energy, 14(1–4) 199–204.
41. Sharma N. K., Tiwari P. K., & Sood Y. R. (2012). Solar energy in India: Strategies, policies, perspectives and future potential. Renewable and Sustainable Energy Reviews, 16(1), 933–941.
42. Ramedani Z., Omid M., Keyhani A., Shamshirband S., & Khoshnevisan B. (2014). Potential of radial basis function based support vector regression for global solar radiation prediction. Renewable and Sustainable Energy Reviews, 39, 1005–1011.
43. Kumar A., Kumar K., Kaushik N., Sharma S., & Mishra S. (2010). Renewable energy in India: Current status and future potentials. Renewable and Sustainable Energy Reviews, 14(8), 2434–2442.
44. Dawn S., Tiwari P. K., Goswami A. K., & Mishra M. K. (2016). Recent developments of solar energy in India: Perspectives, strategies and future goals. Renewable and Sustainable Energy Reviews, 62, 215–235.
45. Philibert C. et al. (2014). Technology Roadmap: Solar Thermal Electricity. International Energy Agency, Paris, France.
46. Li C.-H., Zhu X.-J., Cao G.-Y., Sui S., & Hu M.-R. (2009). Dynamic modeling and sizing optimization of stand-alone photovoltaic power systems using hybrid energy storage technology. Renewable Energy, 34(3), 815–826.
47. Baños R., Manzano-Agugliaro F., Montoya F. G., Gil C., Alcayde A., & Gómez J. (2010). Optimization methods applied to renewable and sustainable energy: A review. Renewable and Sustainable Energy Reviews, 15(4), 1753–1766.
48. Bosch J. L., López G., & Batlles F. J. (2008). Daily solar irradiation estimation over a mountainous area using artificial neural networks. Renewable Energy, 33(7), 1622–1628.
49. Veerachary M. & Narri Y. (2000). ANN based peak power tracking for PV supplied DC motors. Solar Energy, 69(4), 343–350.
50. Rehman S. & Mohandes M. (2008). Artificial neural network estimation of global solar radiation using air temperature and relative humidity. Energy Policy, 36(2), 571–576.
51. Premalatha N. & Valan Arasu A. (2016). Prediction of solar radiation for solar systems by using ANN models with different back propagation algorithms. Journal of Applied Research and Technology, 14(3), 206–214.
52. Mellit A. & Kalogirou A. S. (2011). ANFIS-based modelling for photovoltaic power supply system: a case study. Renewable Energy, 36(1), 250–258.
53. IEA.org. https://www.iea.org/reports/global-energy-review-2020/renewables
54. Ren21.net. https://www.ren21.net/what-are-the-current-trends-in-renewable-energy/
55. CSiS.org. https://www.csis.org/analysis/optimizing-indias-electricity-grid-renewables-using-ai-and-machine-learning-applications

56. Wang H., Lei Z., Zhang X., Zhou B., & Peng J. (2019). A review of deep learning for renewable energy forecasting. Energy Conversion and Management, 198, 111799. DOI: 10.1016/j.enconman.2019.111799.
57. Jhaa S. Kr., Jasmin B., Anju J., Nilesh P., & Han Z. (2017). Renewable energy: Present research and future scope of artificial intelligence. Renewable and Sustainable Energy Reviews, 77(C), 297–317.
58. Zhao X., Wang C., Su J., & Wang J. (2019). Research and application based on the swarm intelligence algorithm and artificial intelligence for wind farm decision system. Renewable Energy, 134, 681–697.
59. RESET.org. https://en.reset.org/blog/germany-artificial-intelligence-making-wind-turbines-more-bird-friendly-02162020
60. Fu T. & Wang C. (2018). A hybrid wind speed forecasting method and wind energy resource analysis based on a swarm intelligence optimization algorithm and an artificial intelligence model. Sustainability, 10(3913), 1–24.
61. Mellita A. & Kalogirou A. S. (2008). Artificial intelligence techniques for photovoltaic applications: A review. Progress in Energy and Combustion Science, 34(5), 574–632.
62. Pillai I. R. & Banerjee R. (2009). Renewable energy in India: Status and potential. Energy, 34(8), 970–980.
63. Caputo D., Grimaccia F., Mussetta M., & Zich R. E. (2010). Photovoltaic plants predictive model by means of ANN trained by a hybrid evolutionary algorithm. In: Proceedings IEEE JCNN Barcelona, 1–6.
64. Ji W. & Chee K.-C. (2011). Prediction of hourly solar radiation using a novel hybrid model of ARMA and TDNN. Solar Energy, 85(5), 808–817.
65. Ogliari E., Grimaccia F., Leva S., & Mussetta M. (2013). Hybrid predictive models for accurate forecasting in PV systems. Energies, 6(4), 1918–1929.
66. Goswami D. Y. (2008). A review and future prospects of renewable energy in the global energy system. In: Goswami D. Y. & Zhao Y. (eds.), Proceedings of ISES World Congress 2007 (Vol. I–Vol. V), Springer, Berlin, Heidelberg.

2 Role of AI in Renewable Energy Management

Anupama Sharma[1], *Sanjeev Kumar Prasad*[2], *and Rashmi Chaudhary*[3]
[1]Department of Information Technology, AKGEC, Ghaziabad, Uttar Pradesh, India
[2]School of Computer Science and Engineering, Galgotias University, NCR Delhi, India
[3]Department of Computer Science and Engineering, Dr. Shyama Prasad Mukherjee International Institute of Information Technology, Naya Raipur, Chhattisgarh, India

CONTENTS

2.1 INTRODUCTION

We live in a world where every device is energy-dependent, whether it be at home, workplace, or any other place such as the place of entertainment, educational institutes, and hospitals. With the advancement of technology, energy dependency will rise in every field. Energy is the key to progress and development of our nation and the world. The environment provides us all the necessary support to live; energy is the most important of them. Every unit of living and nonliving things takes energy from the environment. Energy is necessary for living beings to live, grow, and survive. Nonliving things such as matters require energy to change their form. The energy produced from naturally replenished sources is clean and unlimited. Figure 2.1 represents the origins of natural power.

The inherent variability in weather conditions is a major challenge in the production of renewable energy. Weather cannot be predicted with a hundred percent accuracy, and this inaccuracy in prediction sow the seeds of imperfect disruption in a whole power supply chain generated from renewable sources. Examples of weather-dependent energies are energies produced from wind or solar. This uncertainty of weather conditions can be resolved better with the succor of machine learning in artificial intelligence (AI). Machine learning algorithms are able to analyze large web-based satellite or sensor data to produce better forecasts about weather conditions. Computation agents play a major role in uncertainty planning and make enhancements in predictability. Machine learning techniques can process historical data to analyze the past traditions of weather; this analysis can be used in the optimization of present processes and resources, and on the basis of all these observations, future directions, processes, demand, and supply all can be predicted. These intelligent technologies assist renewable energy at every step of its production and distribution, such as energy forecasting, energy efficiency, and accessibility.

Other major issues are efficient energy storage technologies, effective strategic approaches to optimum consumptions, automated assistance, guidelines for efficient energy management, etc. To deal with the these issues and challenges, the primary objective of AI-enabled setup is to accept and curtail the weather forecast defiance. The next step is to design strategic policies for intelligent technology-enabled renewable systems. The topmost important step is to develop self-learning-based automated assistants to distribute optimum controlled usage-based devices for general usages.

Intelligent storage systems are able to collect and store vast amounts of data from widely spread sensors. Energy consumption can also be optimized with AI-integrated systems. These intelligent systems deal with effective energy management during

FIGURE 2.1 Origins of natural power.

actual usages. All such issues related to renewable energy production setups, manufacturing processes, and storage procedures are discussed and analyzed in detail throughout this chapter.

Following are six major origins of natural power:

1. **Solar energy** is obtained from the sun. It may be directly consumed or, with technology, stored for customized usage.
2. **Wind energy** is a kind of native power. Wind turbines convert the kinetic energy of wind to other forms.
3. **Hydropower energy** is the most commonly used source of energy. The kinetic energy of flowing water is converted to electricity.
4. **Geothermal energy** can be captured from the whole globe's heat. It can mostly be used to heat or cool any building.
5. **Biomass energy** can be produced with the use of organic materials such as animal dung, sewage, agriculture and forestry residues, and industrial residues.
6. **Tidal energy** is produced through natural rise and fall of ocean water, known as tides, which are more commonly predictable in comparison to sun and wind.

2.1.1 Supremacy of Renewable Energy

Renewable energy is naturally available and it is sufficient to meet the requirements of our daily activities of life. There are many benefits of renewable energy. Figure 2.2 represents the main benefits of renewable energy.

A list of various benefits of renewable energy is as follows:

1. It is a **clean energy** because of the low emission of carbon in the process of production and use; hence, it spreads very less pollution in the environment compared to fossil fuels.
2. It is a **safe** form of energy and is less dangerous to produce, collect, and use.
3. There are **ample** sources available on the mother earth to produce and collect such energy. Energy requirements for one whole year can be produced from one-day sunlight.
4. It is a **green energy** because it supports to enhance a **healthy environment**.
5. **Several formats** and **several locations** simultaneously can be used to produce and collect renewable energy.

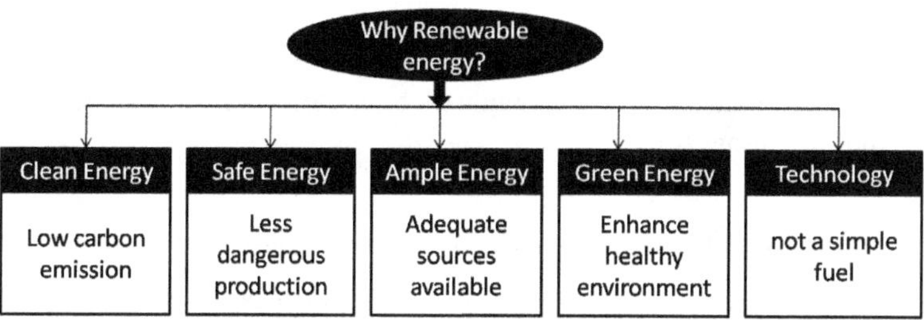

FIGURE 2.2 Benefits of renewable energy.

6. Waste products can be **recycled** in some kind of renewable production process; moreover, renewable energy collectors are very **easy to maintain**; hence, it is **profitable energy** for us.
7. Renewable energy production is a **technology**, not simply a fuel. It may help in building **energy-independent nations**.
8. **Rural development** can become a major gain for any of the nations due to renewable energy. Biomass energy sources are sufficiently present in these areas. Lots of open space suitable for renewable energy production are also available here.
9. It helps in **the economic growth of the nation** by creating lots of jobs.

The above-stated points are the primary benefits of renewable energy systems, but the secondary and the hidden power of these energies are endless.

2.1.2 Stumbling Block to Renewable Energy

Renewable energy has several benefits, which somehow depend on the environment. Hence, the benefits fluctuate with the weather conditions. Some hindrances are as follows:

1. The renewable power supply is **weather dependent**. Sunlight is not available 24/7 and the cloudy sky may also interrupt the power supply. Wind turbine blades need sufficient wind to turn.
2. **Geographic dependence** is a limitation of such energies; hence, commercially they are not viable. A strong distribution network must be created to transfer it as per the need.
3. **High setup cost** is another disadvantage in renewable energy production.
4. **Space requirement** is **very high**; for example, 100-hectare solar panels are required for generating approximately 50-megawatt energy. Three acres of wind turbines are required to generate 4-megawatt energy.
5. **Large storage capacity** is required to store such energy, which is expensive.
6. Renewable energy is **harmful to wildlife**. For example, wind turbine blades kill some birds, dams are harmful for fish life, and solar farm mirrors create high heat that is harmful to birds and other nearby animals.

Every coin has two sides, so does renewable energy. There are some obstacles to these energy systems but most of them can be addressed with proper use of technology. Technology is being used to produce renewable energy and people who have sufficient knowledge about such systems are using and encouraging others to use it. There is a need to educate every individual about the benefits of renewable energy systems and their usage.

2.1.3 Issues and Challenges Faced by the Renewable Power Sector

The renewable power sector has numerous benefits such as climate-friendliness and minimal pollution. The expenditure on infrastructure becomes a one-time investment, and with regular maintenance costs, the whole setup is capable of producing power for a long time. Enhanced weather forecasting and efficient scheduling of power resources are important for managing the demand and supply balance of renewable power. Supply chain management is a big challenge because it suffers from the unpredictability of the weather. Efficient dealing with issues and challenges with the support of emerging technologies such as AI, machine learning, and Internet of Things (IoT) can help renewable energy become the next-generation energy source. Issues and challenges associated with renewable energy affect maturation and resilience of this field. Some of the challenges associated with renewable energy are as follows:

1. **The unpredictability of the weather** is a major challenge faced when using natural resources. The situation becomes very challenging if there is a demand for power supply and renewable energy is unavailable due to cloudy weather or low wind speed. Especially when it comes to producing solar and wind energy, accurate weather prediction is of great help.
2. Implementation of **energy storage** for renewable energy is still in the nascent stage, which needs to be developed intelligently and, essentially, tested in depth. Renewables are not reliable as wind speed may drop or increase frequently and power ramp may not be able to store excess power generated by wind turbines due to suddenly increased wind speed.
3. The decision on **infrastructure** is a big challenge. Challenging issues in the initial stage during the setup process include decisions related to the size of the renewable park, storage capabilities required, etc.
4. Such fluctuating power parks are distributed across large geographical areas and **management of large distributed areas** is a challenging task.
5. **Electric grid management** is also a challenge and requires technical support to work in an optimized way.
6. Renewable energy development and production processes suffer from **technical barriers**, i.e., inadequate technology usages.
7. Many other challenges are related to **policies and social issues**.

Most of the challenges stated above are dealt with by the use of adequate technology. There is huge research going on to optimize renewable energy systems. The use of intelligent technology starts with a very initial stage, i.e., weather forecasting. Many AI-based applications are available to forecast the weather very accurately. Storage of such energy is again a big issue, for which AI has efficient solutions. Supply chain management is also effectively handled by smart technologies. Smart apparatuses are implemented and integrated with electric supply to assist customers with optimized uses of energy.

2.1.4 Classification of Artificial Intelligence Applications

The prerequisite to use AI in a renewable power system is the digitalization of this sector with large, valuable dataset. AI helps renewable energy systems, starting from the design to policy-making. Energy estimation, optimization, management, and distribution assisted by AI-based smart technologies are discussed in refs. [1–4].

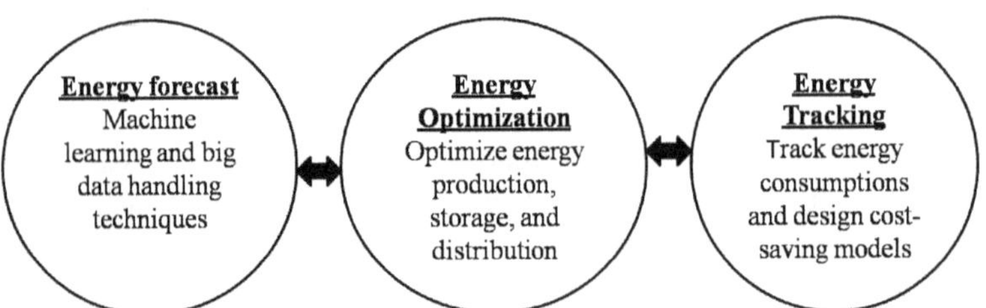

FIGURE 2.3 Artificial intelligence support to different stages of the renewable energy system.

AI greatly helps in resolving issues and challenges faced by the renewable energy systems and in optimizing this industry by fast investigation of a large amount of valuable data. Figure 2.3 represents the main categories of renewable energy systems where AI assists and provides technical support to enhance the whole production.

To address the whole technology support, we summarized it in the following broad categories, where AI helps in this sector:

1. **Energy forecast**: AI helps in speeding up and making accurate weather forecasts due to deep learning and big data handling capabilities.
2. **Energy optimization**: AI is used to optimize energy production, storage, and distribution.
3. **Energy tracking**: AI also helps in tracking energy consumptions and designing cost-saving models. Efficient recommendations for intelligent homes are possible with intelligent technologies.

2.2 ARTIFICIAL INTELLIGENCE APPLICATIONS FOR RENEWABLE ENERGY SYSTEM

AI has many applications for the above-listed categories of renewable energy production, distribution, and tracking. AI applications are based on four different smart techniques of AI, i.e., fuzzy logic, genetic algorithms (GAs), expert systems, and neural networks.

2.2.1 Applications for Energy Forecast

The main sources of renewable energy are wind and solar. Other sources are also important and used as and when needed. These sources are weather-dependent and hence fast and accurate weather forecast is the most important factor. The Institute of Urban Meteorology (IUM) and Sinovation Ventures arranged an AI-based weather forecasting contest in August 2018 [5]. Approximately 1000 teams from around the world joined this event, and 250 participants successfully presented their weather forecast models. The highest accurate results were presented by five teams, and these teams were declared winners in the final round for their accurate forecast experiments. This contest showed that AI-based weather forecast models are best-suited weather forecast models to enhance the forecast results. It also proved that large data can be processed very fast with intelligence techniques. Results show the importance of data processing techniques. The model composite structure is equally important to evaluate improved forecast results.

There are many AI applications that are present in the literature related to weather forecasts for renewable; some of them are given in refs. [6–8]. Some of the companies properly examined AI applications such as Nnergix, which is a weather forecasting startup launched in 2013. Xcel Energy is another company working in the area of renewable energy. Following are some smart weather forecast applications that are being used widely.

2.2.1.1 Nnergix

Renewable energy is weather-dependent; hence, power supply may suffer if weather predictions are not accurate. Accurate weather forecast is very helpful to make a balance between the demand and the supply of renewable power sectors. AI-based application Nnergix works well to support these sectors. Nnergix is a web-based energy forecast tool based on the data mining principle. It uses satellite and industry data to train the tool with the help of machine learning algorithms. A region-wise atmosphere can be predicted more accurately because machine learning algorithms analyze large data very correctly and build weather models. These weather models can work on a very small scale also and predict weather very fast.

2.2.1.2 Xcel

Xcel is a tool based on AI to predict weather forecasting accurately. It is a promising tool to catch weather fluctuations more accurately and anticipate when power sources will not be available so that the power storage facility can be improved and used as needed. It provides detailed and accurate weather report as AI works on a combination of local satellite data, different weather station data, and wind park data. Analysis of such a large amount of data in less time can be done with AI only. Xcel (in Colorado) generates detailed weather reports with high accuracy. Xcel works on multiple aims based on machine learning technologies. It collects data from thousands of sensors, which is trained with an artificial neural network (ANN) to reduce pollution at a plant. Xcel worked in collaboration with the National Oceanic and Atmospheric Association to forecast wind turbine power results. To inspect failures in the whole infrastructure, it uses drones and machine learning algorithms to automate failure prevention and detection.

Not only these two, but there are many AI-based applications proposed in the literature by many researchers to predict the weather with more accuracy.

Zhao et al. [9] implemented a wind energy decision system based on swarm intelligence optimization. Wind power estimation is the first module of the proposed system, whereas the forecast of wind speed is the second module. Gestalt intelligence is the main concept used in the first module to enhance the parameters of the Weibull distribution. This system provides enhanced forecast results with a great reduction in wind speed forecast errors.

An extensive wind energy prediction model is updated recently through NCAR (National Center for Atmospheric Research) [10] in collaboration with Xcel Energy to take up the customer's demands. This enhancement is achieved by intelligent integration of machine learning algorithms and numerical weather predictions. This improved system produces short-term forecast results with more accuracy. Improved probabilistic forecasting covers extreme events predictions like hail storms or snow, etc. Data quality control has been managed with the quantile approach in the empirical power conversion machine learning algorithms, and it significantly improves the accuracy of used algorithms. Forecast variability is assessed using a combination of an analog approach, i.e., an expert system and the variational Doppler radar analysis model.

A novel forecast model with three levels of accuracy filtering is proposed by Çevik et al. [11], which provides the prediction of next hour wind power. Historical data is used to predict the direction and speed of wind in the first stage. Empirical mode decomposition and stationary wavelet decomposition (SWD) methods are preprocessing methods, and artificial neuro-fuzzy inference system (ANFIS), ANN, and support vector regression are forecast technologies with past four years' data to train the system and past two years' data to test the system. To enhance the prediction, results obtained from the first level are again put into the same forecast methods, and resultant forecast values are upgraded. The third level is a correction level, which is used to upgrade the outcomes. In the first level, SWD-ANFIS provides best-estimated outcomes, and ANN works very well in terms of the performance at the second level. The last correction process provides the weighted average of the outcomes of all the previous three methods.

Jafarian-Namin et al. [12] proposed a new enhanced model to forecast wind power generation through the combination of the Box–Jenkins modeling and the neural network modeling approaches. The new model's results indicated the enhancements among the other tested ANN models and ANN-GA.

Much other work has been done to predict wind speed using AI techniques generally using ANNs, GAs, etc. Feed-forward back-propagation neural network is used in the estimation of wind power by Mabel and Fernandez [13]. Li and Shi [14] compared these approaches and found that performance depends on the location of wind farms. Hence, according to suitability, these approaches can be used as per the requirement.

Mohandes et al. [15] worked on wind speed till 100-m height using ANFIS, with training data of wind speed at heights 10, 20, 30, and 40 m. Yang et al. [16] used ANFIS method on data collected from 12 wind farms in China for interpolating the missing wind data.

Malik and Garg [17] used ANN for solar radiation prediction to enhance the forecast. The ANN used is feed-forward with back-propagation, with back-propagation being used as the learning algorithm. A three-layer network has been used with one hidden layer. The model worked for 19 inputs and the data used was from 67 cities in India.

Benali et al. [18] used ANN and smart persistence (SP) to enhance the forecasting horizon. They realized that forecast during winters and summers are more reliable than during autumn and spring seasons. Three components of solar luminosity, i.e., global horizontal, beam normal, and diffuse horizontal, are measured on the site of Odeillo, France, characterized by high meteorological variability, and hourly solar irradiations are predicted.

2.2.2 Artificial Intelligence Applications for Energy Tracking and Optimization of Energy Consumption

These applications intelligently connect clients to a smart electricity system to track electricity usage at their homes or workplaces, etc. Smart home services are networked to collect usage patterns of power-operated devices. These systems provide valuable advice regarding electricity saving and cost-cutting in electricity bills with usages of household devices.

2.2.2.1 Google DeepMind

Nowadays, machine learning is used in most of the devices such as smartphones and is required in day-to-day activities. It is also successfully used to provide efficient solutions to challenging issues such as efficient energy consumption. Smart technologies proved its importance in 2014 when DeepMind and Google together started the application Google DeepMind to enhance energy usages. They accomplished this task by using historical data collected from sensors of the huge data center. On the basis of this data, an ANN is trained on distinct operating frameworks and on the average future PUE (power usage effectiveness). Google DeepMind managed to reduce power consumption by 40% at Google's data servers. Figure 2.4 shows the graph of energy usages when the application is on and off.

DeepMind system is configured to train neural networks on sensors and satellite weather forecasts with the combination of historical turbine data. This system predicts the next 36 hours of wind energy output. Based on these predictions, a model is developed to recommend optimized hourly delivery to the power grid. This is important because energy sources that can be scheduled are often more valuable to the grid. Figure 2.5 represents the results of Google/DeepMind field study done in 2018 to show the benefit of this application.

2.2.2.2 Verdigris Technologies

A company named Verdigris Technologies presented an AI-based software platform to optimize power utilization. This application is basically focused on optimizing energy consumption in big enterprises. To set up a forecasting system, IoT hardware

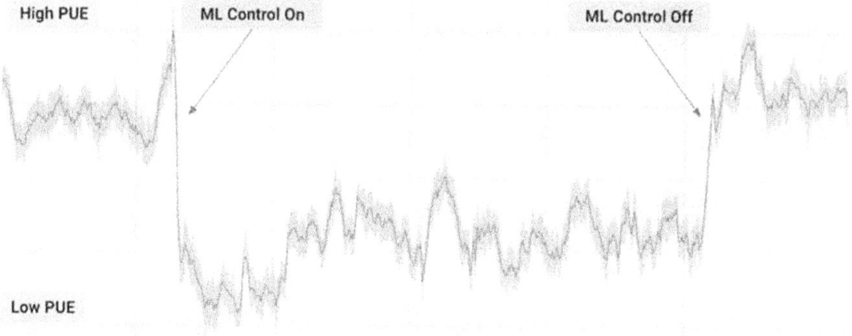

FIGURE 2.4 Energy usages graph of Google's data center. (From DeepMind.)

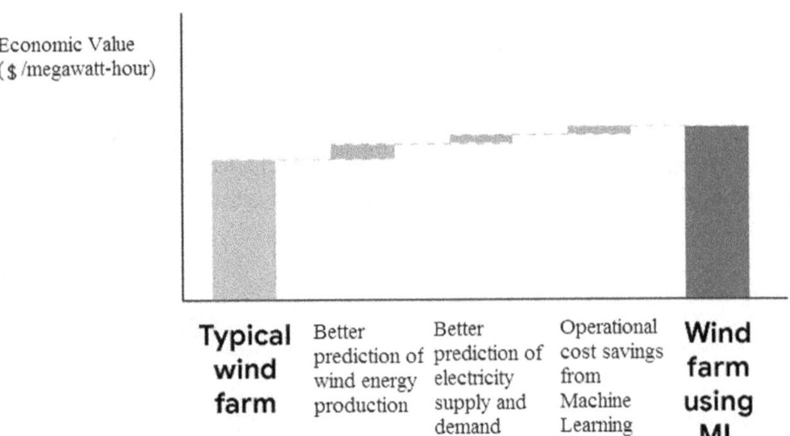

FIGURE 2.5 Results of Google's DeepMind application applied to wind power. (From DeepMind.)

should be installed all over the building. Smart sensors put on the client's site are able to track power usages at the client's site. Sensors collect power consumption data that is forwarded to the cloud, and after evaluation on the required dimension, it is again made available to the client so that the client can understand the inefficiency area in terms of energy consumption. Many companies utilized this application to optimize energy consumption, such as W Hotel San Francisco. The management of the hotel accepted the importance of application as they identified flaws in electricity consumption which cost over $13,000 of annual loss.

2.2.2.3 Verv

Verv is an AI-enabled application in the smart home category developed by London-based Green Running Ltd. This application assists clients in managing energy at their homes by supplying energy data for each and every home appliance. Users of this application are able to see the pattern of energy usages by their home appliance. Hence, they may be able to manage and regulate their energy usages before bills due. Verv also has several safety features, such as it gives indications if any apparatus has been left on unusually for more time. Moreover, it advises on decreasing the energy usage and hence reduces electricity cost. The app is available for tablets, laptops, and smartphones.

2.2.2.4 PowerScout

It is also an AI-based application to assist users regarding correct usages of electric devices and help in cost savings. It uses data analytics to improve the home assistance system and track energy usages in home appliances.

AI becomes a tool to advise and provide recommendations to assist customers in making decisions regarding buying renewable energy technology-oriented devices for their homes. Some of PowerScout partners are the US Department of Energy and Google.

2.2.3 Artificial Intelligence-Based Technologies to Assist Other Parts of the Renewable Life Cycle

The electric grid is a complex part of the energy generation system, and when we talk about the smart grid, it is more complex but provides improvements in renewable energy plants. To expand the process of intelligent integration of different renewable energies to the existing power grids, some expertise solutions are required. SYNERGYLABS is a company that delivers expertise AI-based machine learning tools to natural language processing, predictive and data analytics, etc., to fulfill the business demands of their clients. Its AI researchers and developers use the latest technologies, frameworks, and hi-tech methodologies to offer impeccable and quality services globally. There are many other companies such as Google DeepMind, PowerScout, and Xcel that provide efficient solutions in this field, as discussed in the previous section.

In the coming time, the research work of such companies will definitely accelerate the uses of renewable and then renewable energy will become the leader of this sector. This is how AI technologies will enhance the trustworthiness of renewable power and modernize the present grids to efficient usages and outcomes. Some of the other smart applications are as follows.

2.2.3.1 Intelligence Consolidated Control Centers

Devices and sensors of a grid are mostly interconnected to collect huge amounts of data. When AI is integrated into the system, data may be evaluated for excellent outcomes to the grid manager. AI technologies provide flexibility in the hand of the power suppliers; hence, they may intelligently manage or control the supply with demand chain for power. An intelligent load balance or adjustment processes with the required hardware equipment, such as automated industrial furnaces or self-controlled large AC units installed, as and when required which may establish balance during low power supply. Along with this, making weather and load predictions with the help of smart sensors and advanced sensors will enhance the blending effectiveness of this green power.

2.2.3.2 Intelligent Incorporation of Micro-grids

AI-based control enhancements may resolve energy supply issues and bottlenecks. Intelligent algorithms have the potential to provide quick command on variable behavior of the grid and help to control all the substations autonomously. Self-learning artificial intelligent technology-oriented algorithms help in improving system optimization.

2.2.3.3 Improved Safety and Reliability

AI is not only able to maintain the intermittency, but also compatible to make enhancements in safety and security. These technologies help companies to evaluate the energy usage patterns, identify the issues, and advise to maintain devise lifetime. If power leakage occurs, it may be identified by an automated process without visiting the place.

2.2.3.4 Make the Market for Renewable Energy

AI collaboration supports suppliers to find the market for renewable energy consumption by developing customized new service models. Data related to energy collection is analyzed with the help of smart technologies. Such technologies also guide the efficient consumption of power. Retailers benefit from such insights to target new consumer markets.

2.3 CONCLUDING THOUGHTS AND FUTURE OUTLOOK

The wonders of AI proficiency are ever-growing. AI is launching new opportunities for diversified sectors and industries to tap into unmapped data and connect it to the decentralized energy resources. This technology is efficient in acquiring data from huge structured, semi-structured, and unstructured sources. Extracted data is then parsed and identified for different interesting and usable patterns that exist in them. This information is then analyzed to make predictions and recommendations. Hence, AI provides better insights to develop accurate, automated, and real and better solutions for smart applications.

The energy sector is a big sector of the present era because all over the world advancement and development directly depend on energy. There are many options available to generate energy, i.e., oil, natural gases, and renewable sources. It is not possible for present energy system based on fossil fuels to meet the future energy requirements. Moreover, fossil fuels create carbon footprints that are responsible for the degradation of our natural environment. The present situation of development style is harmful to the earth and after some time it will not let us survive for more. We may visualize these changes in our atmosphere very easy and all living beings are facing health issues due to such development process. This is the time to step up for better solutions. Renewable resources are already available in plenty; we need to utilize them in an efficient and correct manner.

At the economic front, renewable energy will become a big option for future generations. It may become an effective strategy to enhance the survivability of the environment. AI is a great tool to integrate intelligence and self-learning in every field of work; it enhances the capacity of data analytics too. AI technologies can transfigure the renewable energy sector with improved efficiency and globalization of its usages. The adoption of renewable will remove the barrier of limitation of fuel in the development of every country, irrespective of the development area: whether it is going on in an urban or rural area. There is a need to educate people in this field of its advantages; this step will definitely accelerate its adoption worldwide. AI-enabled green energy systems are able to handle all the issues and challenges of renewable sources such as inconsistent weather conditions and integration with current power generation systems. Prediction and handling of dynamic conditions of nature will provide great strength to this sector and build the supplier's confidence in this field, instead of traditional energy sources, to fulfill consumer's requirements.

AI-driven renewable energy generation platforms are able to hold all the commitments regarding the energy sector. A better process of energy generation planning, automated setups, energy storage, distribution, and demand and supply chain can be maintained with the integration of AI technologies.

According to a recent paper published by DNV GL, AI will make its strong roots in the automation of green energy systems in the coming time, whether it be solar, wind, or another type of renewable source industry. AI amplifies renewable energy system's working and optimizes production and all the other processes. An increasing number of machine language-enabled sensors are expected to be installed at the places of renewable systems to automated monitoring of the setup, data processing, etc., which will enhance effectiveness and provide optimum outcomes.

Finally, it can be stated that with the collaboration of technology, which is making rapid advancements, the renewable energy sector has made significant progress in the last decade. However, there are a few challenges that still prevail that can be addressed with the help of AI and machine learning.

REFERENCES

1. Elsheikh, A. H., & Elaziz, M. A. (2019). Review on applications of particle swarm optimization in solar energy systems. International Journal of Environmental Science and Technology, 16(2), 1159–1170.
2. Chavan, S., & Chavan, M. (2020). Recent trends in ICT-enabled renewable energy systems. In: Tuba, M., Akashe, S., & Joshi A. (eds), Information and Communication Technology for Sustainable Development. Advances in Intelligent Systems and Computing, Vol. 933 (pp. 327–332). Springer, Singapore.
3. Andoni, M., Robu, V., Flynn, D., Abram, S., Geach, D., Jenkins, D., & Peacock, A. (2019). Blockchain technology in the energy sector: A systematic review of challenges and opportunities. Renewable and Sustainable Energy Reviews, 100, 143–174.
4. Cali, U., & Fifield, A. (2019). Towards the decentralized revolution in energy systems using blockchain technology. International Journal of Smart Grid and Clean Energy, 8(3), 245–256.
5. Ji, L., Wang, Z., Chen, M., Fan, S., Wang, Y., & Shen, Z. (2019). How much can AI techniques improve surface air temperature forecast?—A report from AI Challenger 2018 Global Weather Forecast Contest. Journal of Meteorological Research, 33(5), 989–992. doi: 10.1007/s13351-019-9601-0
6. Notton, G., Voyant, C., Fouilloy, A., Duchaud, J. L., & Nivet, M. L. (2019). Some applications of ANN to solar radiation estimation and forecasting for energy applications. Applied Sciences, 9(1), 209.
7. Mohanty, S., Patra, P. K., Mohanty, A., Viswavandya, M., & Ray, P. K. (2019). Artificial intelligence based forecasting & optimization of solar cell model. Optik, 181, 842–852.
8. Bermejo, J. F., Fernández, J. F. G., Polo, F. O., & Márquez, A. C. (2019). A review of the use of artificial neural networks models for energy and reliability prediction. A study for the solar PV, hydraulic and wind energy sources. Applied Sciences, 9, 1844.
9. Zhao, X., Wang, C., Su, J., & Wang, J. (2019). Research and application based on the swarm intelligence algorithm and artificial intelligence for wind farm decision systems. Renewable Energy, 134, 681–697.
10. Kosovic, B., Haupt, S. E., Adriaansen, D., Alessandrini, S., Wiener, G., Delle Monache, L., & Politovich, M. (2020). A comprehensive wind power forecasting system integrating artificial intelligence and numerical weather prediction. Energies, 13(6), 1372.
11. Çevik, H. H., Çunkaş, M., & Polat, K. (2019). A new multistage short-term wind power forecast model using decomposition and artificial intelligence methods. Physica A: Statistical Mechanics and Its Applications, 534, 122177.

12. Jafarian-Namin, S., Goli, A., Qolipour, M., Mostafaeipour, A., & Golmohammadi, A. M. (2019). Forecasting the wind power generation using Box–Jenkins and hybrid artificial intelligence: A case study. International Journal of Energy Sector Management, 13(4), 1038–1062.
13. Mabel, M. C., & Fernandez, E. (2008). Analysis of wind power generation and prediction using ANN: A case study. Renewable Energy, 33(5), 986–992.
14. Li, G., & Shi, J. (2010). On comparing three artificial neural networks for wind speed forecasting. Applied Energy, 87(7), 2313–2320.
15. Mohandes, M., Rehman, S., & Rahman, S. M. (2011). Estimation of wind speed profile using an adaptive neuro-fuzzy inference system (ANFIS). Applied Energy, 88(11), 4024–4032.
16. Yang, Z., Liu, Y., & Li, C. (2011). Interpolation of missing wind data based on ANFIS. Renewable Energy, 36(3), 993–998.
17. Malik, H., & Garg, S. (2019). Long-term solar irradiance forecast using artificial neural network: Application for performance prediction of Indian cities. In: Applications of Artificial Intelligence Techniques in Engineering (pp. 285–293). Springer, Singapore.
18. Benali, L., Notton, G., Fouilloy, A., Voyant, C., & Dizene, R. (2019). Solar radiation forecasting using artificial neural network and random forest methods: Application to normal beam, horizontal diffuse and global components. Renewable Energy, 132, 871–884.

3 AI-Based Renewable Energy with Emerging Applications

Issues and Challenges

Omkar Singh[1], Mano Yadav[2], Preeti Yadav[1], and Vinay Rishiwal[1]
[1]Department of CS & IT, M. J. P. Rohilkhand University, Bareilly, India
[2]Department of Computer Science, Bareilly College, Bareilly, India

CONTENTS

3.1 INTRODUCTION

Among the different kinds of energies, electrical energy plays a major role in today's era due to its demand for globalization in the world. Earlier, oil, coal, and natural gas were the only energies used to produce electrical energy [1]. Though electricity petition is being contented by these cradles of energies, its enormous usage has instigated enormous reduction in these relic energies and ecological glitches [2]. The unit of electricity from remaining energy borders has produced foremost effluence in terms of carbon dioxide (CO_2) release and conservatory gas or greenhouse gas (GHG) emanation, therefore variations occur in everywhere in the world [3]. Keeping in opinion of the proof, routine of temporary foundations of energy to meet electrical request has been discovered intensively [4]. Amid these substitute possessions, renewable energy has gained main attention universally. Energy generated from renewable energy sources (RESs) is ecologically approachable, with precise GHG production, long-lasting and fewer charges than conservative dynamism [5] sources. Among various kinds of RESs, photovoltaic (PV) plays an important role with distribution degree in energy arcades, to receive the solar energy at maximum possible level from the atmosphere [6]. As per the study, the amount of energy received from solar radiation is 1.8×10^{11} MW. It is large, probable, and maintainable energy generated from solar radiation, which has gained attention of investors, policy-makers, environment technologists, and governments [7]. Therefore, PV has large prospective in town and pastoral electrification [8]. Solar energy is being used in various forms such as cooling, heating, energy creation, mutual power, and unreceptive system [9]. PV system also includes minimum preservation cost, prolonging lifetime, and connection cost compensation within a specific time [10]. Key contributions of the chapter are shown in Figure 3.1.

3.2 RELATED WORK

Gautam et al. [11] proposed a framework to enhance solar energy power to grid and solar on related node using the same switching technology with the help of decision tree machine learning (ML)-based algorithm in the python framework. The hybrid ML technique with big data analysis proposed by Sharmila et al. [12] for optimization and distribution of the existing energy sources targeted to smart power management. Alkhnadari et al. [13] developed a ML-based hybrid model called machine learning and statistical hybrid model (MLSHM), which merges ML techniques using the statistical technique for further precise forecasting of future production of

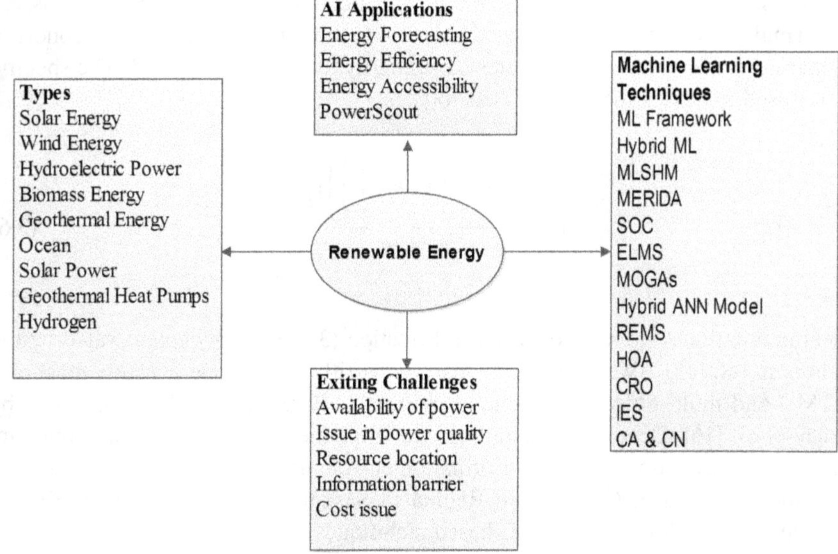

FIGURE 3.1 Key contributions chart.

solar energy using energy plants based on renewable energy. Traditional functions are used to develop the MLSHM model, which are given below:

$$r(t) = \sigma\left(w_r x(t) + u_r h(t-1) + b_r\right) \tag{3.1}$$

$$z(t) = \sigma\left(w_z x(t) + u_z h(t-1) + b_z\right) \tag{3.2}$$

$$\hat{h}(t) = \sigma(w_h x(t) + u_h h\left(r(t) * h(t-1) + b_h\right) \tag{3.3}$$

$$h(t) = \left(1 - z(t)\right) * h(t-1) + z(t) * \hat{h}(t) \tag{3.4}$$

The appropriate symbols used in Equations (3.1)–(3.4) are symbolized by authors in ref. [13].

An intelligent system model MERIDA is implemented by Marijana et al. [14], which assimilates the collection of big data and forecast model of power utilization for every power source in public construction. The computation function is used to develop this model as:

$$y_c = f\left(\sum_{i=1}^{n} w_i x_i\right) \tag{3.5}$$

where

y_c = calculated output

x_i and w_i = elements of input and weight vectors X and W

Som et al. [15] developed heuristic methods called state of charge (SOC) to assess the formal responsibility of lead–acid succession through optimum energy cohort in a separate amalgam wind-solar renewable energy distribution method. The specific vector used to design the method is as follows:

$$v_i^t = \frac{\left(m_{i+\frac{n}{2}} + \varepsilon\left(m_{i+\frac{n}{2}}\right)\right) v_{i+\frac{n}{2}}}{m_i + m_{i+\frac{n}{2}}} \tag{3.6}$$

Specific notations and symbols used in Equation (3.6) have been elaborated by the authors in ref. [15]. Two various ML-based algorithms, extreme learning machines (ELMs) and multi-objective genetic algorithms (MOGAs), have been proposed by Ronay et al. [16]. Developed techniques are implemented for immediate airstream rapidity forecast from an actual information customary of hourly airstream rapidity capacities for area of Canada and Regina in Saskatchewan. The basic prediction intervals used to develop both ML-based techniques are as follows:

$$P_r\left(L(x)\right) < y(x) < U(x) = 1 - \alpha \tag{3.7}$$

$$PICP = \frac{1}{n_p} \sum_{i=1}^{n_p} c_i \tag{3.8}$$

$$NMPIW = \frac{1}{n_p} \sum_{i=1}^{n_p} \frac{\left(U(x_i) - L(x_i)\right)}{y_{\max} - y_{\min}} \tag{3.9}$$

The appropriate symbols and notations used in Equations (3.7)–(3.9) have been described by the authors in ref. [16]. Musaylh et al. [17] developed an artificial neural network (ANN) model that applies forecast variables for 6-hour (h) and daily prediction power demand (G) conjecturing. The response parameters comprised six utmost pertinent weather parameters SILO (Scientific Information for Land Owners and 51 re-investigation parameters attained from ECMWF (European Centre for Medium-Range Weather Forecast models. To develop the ANN model, a particular technique is used as:

$$y(x) = F\left(\sum_{i=1}^{L} w_i(t).x_i(t) + b\right) \tag{3.10}$$

Symbols and notations mentioned in Equation (3.10) have been summarized by the authors in ref. [17]. A residential energy management system (REMS) technique has been proposed by Prakash et al. [18] that efficiently changes conceivable lots to renewable thrilled native power storing centered based on charge-discharge dealings and grid obtainability, thus plunging energy ingesting from grid. Perera et al. [19]

proposed a hybrid optimization algorithm (HOA) technique by uniting a replacement and actual engineering model (AEM) in order to haste up optimization procedure through upholding correctness. The efficiency of solar panel used in the proposed technique can be calculated as follows:

$$\eta_n^{SPV} = p^{SPV}\left[q^{SPV}\left(\frac{G_t^{\beta}}{G_0^{\beta}}\right)+\left(\frac{G_t^{\beta}}{G_0^{\beta}}\right)^{m^{SPV}}\right]$$

$$\left[\begin{array}{l} 1+r^{SPV}\left(\frac{\theta_t^{SPV}}{\theta_0^{SPV}}\right)+s^{SPV}\left(\frac{AM}{AM_0}\right)+ \\ \left(\frac{AM}{AM_0}\right)^{u^{SPV}} \end{array}\right], \forall t \in T \quad (3.11)$$

Specific notations and symbols used in Equation (3.11) are described by the authors in ref. [19]. A coral reefs optimization (CRO) approach has been developed by Sanz et al. [20], which syndicates dissimilar exploration appliances into a solitary technique, giving a worldwide exploration process of extraordinary excellence. A wrapper method is used as:

$$\sigma^o = \arg\min_{\sigma,\alpha}\left(\int v\left(y, f\left(x^*\sigma,\alpha\right)\right)dp(x,y)\right) \quad (3.12)$$

Notations and symbols used in Equation (3.12) have been elaborated by the authors in ref. [20]. Zhang et al. [21] proposed an integrated energy system (IES) technique for monitoring the power adaptation, and developed an operative technique for refining grid litheness and plummeting functioning charge of IESs. The specific function is used to design the IES technique.

$$\theta_{th}(t) = COP_{ave}\Delta P_{HP}(t) \quad (3.13)$$

Ahmad et al. [22] developed diverse regions, definite conservational and power ingesting statistics are acquired for effort specifications assortment and demonstrating examination. The specific formula used for developing diverse regions is as follows:

$$\begin{aligned} b^n = {} & g^n(X^n g^{n-1} g^{n-2}(X^2 g^1\left(X^1 q + c^1\right) \\ & + c^2 + c^{n-1}) + c^n) \end{aligned} \quad (3.14)$$

Precise symbolizations used in Equation (3.14) have been elaborated by the authors in ref. [12]. An assessment of various ML techniques used in renewable energy is given in Table 3.1.

TABLE 3.1
Assessment of Various ML Techniques in Renewable Energy

ML Technique	Advantages	Disadvantages
ML Framework [11]	Enhances solar power usages, collecting real-time data, data prediction strategy exported, reduces electricity cost.	Employees on large scale affect other environmental factors.
Hybrid ML [12]	Substantial gain ensures, smart energy management leads, relevant data collecting.	Electricity streamlining data demand is very high.
MLSHM [13]	Achieves higher accuracy, collects data very efficiently.	Training set parameters need to be tested.
MERIDA [14]	Improves energy efficiency, enables reconstruction management plan, and minimizes power utilization.	Potential intelligent power management using macro and micro ML techniques should be enabled.
SOC [15]	Utilizes renewable standalone power, optimizes renewable sources.	Enhances electricity and designed cost.
ELMs and MOGAs [16]	Employee's good precision prediction, predicts airstream.	Unable to find out methodical analysis impact on hidden neurons.
Hybrid ANN Model [17]	Calculates forest indecision, provides high accuracy, and covers multiple horizons.	Need to improve forecasting accuracy. Consumes more electricity.
REMS [18]	Minimizes power utilization, provides better accuracy.	On large scale, performance of REMS decreases.
HOA [19]	Provides better accuracy, reduces operational time, and saves energy.	Unable to provide energy necessity generation on regional scale.
CRO [20]	Improves wind speed direction, gives admirable prediction.	Need to explore new methods on deep learning in big feature selection problem (FSP).
IES [21]	Solves DM problems, adapts energy conversion ratio.	Unable to work with multi agent.
CA & CN [22]	Improves prediction accuracy, reduces operational cost.	Performance degrades with the selection of ultra-short and selection-term (ST) energy.

3.3 RENEWABLE ENERGY AND ITS TYPES

Renewable energy, frequently mentioned as spotless energy, originates from natural cradles or through procedures that are continually replaced. For example, wind speed or sunlight reserves and their availability is dependent on weather and time [23, 24].

3.3.1 Solar Energy

Hominids bind solar power for many years to produce crops, stay intense, and foods. As per NREL, new power from the sun sprays on the ground in a single hour and is used by everybody in the ecosphere throughout the year. In today's era, sun rays are being used through different types, including warm hospices and trades, warm marine or energy devices [25].

3.3.2 Wind Energy

We have resumed from windmills of old fashioned, now day's turbines are being used to produce wind energy and they work closely as extensive in wideness standpoint at civility everywhere in the ecosphere. For example, Breeze power turbine's knife-edges, which fodders a power-driven producer and harvests power. The Airstream, expands more than 6% of U.S. cohort, has developed the inexpensive power source in various portions for the nation [26].

3.3.3 Hydroelectric Power

Hydropower is known as a prevalent renewable power cradle for power in the United States; however, airstream power is rapidly predictable to obtain large revenue. Hydropower trusts on water as usual quick oceanic in a big stream or quickly down oceanic from a high theme tune and changes the strength of marine into power by rotating a generator's turbine knife-edges [27].

3.3.4 Biomass Energy

Biomass is a carbon-based substance which is derived from florae and faunae and comprises trees, waste wood, and crops. Once biomass is scorched, the biochemical energy is unconfined as warmth and can produce power with a vapor turbine. Biomass is frequently incorrectly understood as spotless, renewable firewood, and an olive green substitute to firewood, and further, relic coals for creating power [28].

3.3.5 Geothermal Energy

If you have always unperturbed in warm mainspring, you are suing geothermal energy. The ground's core is as warm as sun's superficial because of sluggish deterioration of harmful atoms in pillars at midpoint of the earth. Puncturing bottomless shafts carries very warm subversive water to superficial as a hydrothermal source, formerly impelled over a turbine to generate power [29].

3.3.6 Ocean

Tidal and wave power energy is quiet in a growing stage, but oceanic energy will lined repeatedly by moon's magnitude, which makes for beautiful selection. Certain tidal energy methods may damage environment, e.g., tidal bombardments that seem as barriers and are situated in a marine inlet or cove [30].

3.3.7 Solar Power

At a small scale, we can harness the sun's emissions to influence the entire community whether over PV cubicle plates or inert solar home-based enterprise. Inert solar households are intended to be comfortable in the sun over south-facing spaces, and then recollect the heat through concrete, tiles, bricks, and other resources that store heat [31].

3.3.8 Geothermal Heat Pumps

Geothermal system is new profits on a recognizable method and the circle at spinal of your fridge is a small heat induce, eliminating heat from the interior to save sustenance is reintroduced and composed. At home-based, geothermal impels endless temperature of the ground to calm households in summertime and intense households in wintertime to heat water [32].

3.3.9 Hydrogen

Hydrogen needs to join with extra essentials, such as oxygen to create water. While hydrogen is separated from an alternative component and handles petroleum and energy together [33]. Merits and demerits of different renewable energies are shown in Table 3.2.

TABLE 3.2
Renewable Energy Merits and Demerits

Type of Energy	Merit	Demerit
Solar	Infinite functionality of sun shine, unlimited supply of solar energy.	Solar energy has impractical cost for some households.
Wind	It is an unsoiled energy source, avoids air pollution; it does not produce CO_2 and other products which are harmful for environment.	It increases the cost in transition lines. Some cities oppose for raising noise pollution, certain birds are killed by striking in the turbine while flying.
Hydroelectric	It is used for the projects working on large scale such as hoover dam, and also covers small projects including small dams on small rivers.	It creates disturbance and negative distresses for animals and living lives. It also changes water status and ecosystems.
Geothermal	It signifies potential of energy; it avoids footprint on the earth.	It takes maximum cost to build infrastructures and susceptibility in earthquake.
Ocean	It is a predictable energy, more reliable and plentiful; it is a clean energy source.	It disturbs ocean environment, ocean habitats and sea life; in rough weather, it creates lower amount of energy.
Hydrogen	It is used as a clean scorching fuel and creates low amount of pollution.	It is incompetent when it originates to stop contamination.
Biomass	It used for personal use at home and business also in our daily lives.	Fresh plants require carbon dioxide to develop plants revenue time to improve.

3.4 AI APPLICATIONS IN RENEWABLE ENERGY

Researchers and organizations are exploring customs to affect artificial intelligence (AI) and recover the capability and suitability of expectable power technology. These tools work within three predictable energy compasses: energy forecasting, energy efficiency, and energy accessibility [34].

3.4.1 Energy Forecasting

3.4.1.1 Nnergix

It traces the energy source produced from the atmosphere-reliant power such as solar and wind and tends to challenge in renewable power. Nnergix is a predicting application based on web and data mining energy. Nnergix uses data from satellite and trains those data using ML for analyzing data of the companies to provide an accurate prediction [35].

3.4.1.2 Xcel

Xcel applies AI that goals at lecturing the contests allied with undependability of climate-reliant power sources such as wind and solar. Xcel can express energy source that will vary in métier. Xcel is applied in retrieving climate intelligences having sophisticated correctness and glowing comprehensiveness [36].

3.4.2 Energy Efficiency

3.4.2.1 Google DeepMind

This AI application is brought by Google in 2014 for improving energy usage. Its aim is to decrease power ingesting along with the subsequent releases when the power is recycled. This application is utilized for cooling of Google's information servers by minimizing power utilization and bills by 40% [37].

3.4.2.2 Verdigris Technologies

This scheme deals with a software technology that influences AI to enhance power ingesting. It is intended for big profitable constructions and executives of innovativeness amenities. The connection procedure instigates with IoT hardware connection. Shrewd sensors are straight devoted to customer's electrical board to discover power ingesting [38].

3.4.3 Energy Accessibility

3.4.3.1 Verv

It is power-based AI and is being used at homegrown subordinate in power organization. It provides data for household power utilization. Verv allows employer to understand records to utilize at household usage power. It also assists consumers to control their power expenditures [39].

3.4.3.2 PowerScout

It aims at refining punter tutoring and admittance to renewable power system. It utilizes AI-based model possible reserves on usefulness prices consuming manufacturing data. PowerScout influences analytics of data to recognize "shrewd homespun development" based on exclusive topographies and power custom at household customers [40].

3.5 CHALLENGES IN RENEWABLE ENERGY

Renewable energy is demarcated as a power composed of natural assets. Recently, there has been rising attention to renewable power and its conversion in energy unit. Still, there are numerous disputes and contests specified as follows [7].

3.5.1 Availability of Power

The major problem related to renewable power is its dependency on usual effects, which is uncontrollable by person. For example, solar power current is produced only when sunlight is available and goes off at nighttime; airstream energy is also influenced by the availability of airstream. Therefore, if wind speed is low, turbines will not seizure and resulting in zero energy movement to grid [9].

3.5.2 Issue in Power Quality

Dependably extraordinary energy excellence is desired to guarantee constancy and extraordinary competence of the system. The excellence of energy source permits the system for glowing with super dependability and inferior outlay [13].

3.5.3 Resource Location

Maximum renewable power florae that stake their power with the grid necessitate big space. Most of the renewable power generations are verbalized by position which can be repellent to consumers. Initially, certain renewable power generations are basically not obtainable in dissimilar areas [29].

3.5.4 Information Barrier

Improvement in this field is going on; still deficiency exist regarding information and consciousness almost the welfares and requirement of renewable power. Speculation and asset stipends have been completely available for performing of renewable energies. The vibrant is necessity to support and guidance for candidates and ensures that how extends renewable power [33].

3.5.5 Cost Issue

The unexpected initial fee of connection tends key sprints in development of renewable power. Though the enlargement of petroleum hub necessitates around 46 MW, it is clear that airstream and solar energy also require great speculation. Therefore, stowing methods of the produced power are affluent and signify an actual contest in terms of megawatt fabrication [38].

3.6 CONCLUSION

Current tendencies in energy production and dissemination technology demonstrate that smooth diffusion grid has enlarged significantly. End user utilizations are satisfying about energy excellence condition. This scenario offers methodical evaluation of sources of energy excellence difficulties connected with renewable dispersal produced system (airstream and solar power). Power reduces with airstream infiltration and intensification with solar infiltration. In this chapter, certain issues and challenges existing in the current renewable energy are discussed, which pave the way for future investigations for the researchers. The chapter also discusses various types of renewable energies with their merit and demerits. The most recent applications of renewable energy, which show the basic idea for researchers for further improvements in energy generation, are also discussed.

REFERENCES

1. M. N. Akhter, S. Mekhilef, H. Mokhlis and N.M. Shah. "Review on forecasting of photovoltaic power generation based on machine learning and metaheuristic techniques," *IET Renewable Power Generation*, vol. 13, pp. 1009–1023, 2019.
2. N. Phuangpornpitak and S. Tia, "Opportunities and challenges of integrating renewable energy in smart grid system," *Energy Procedia*, vol. 34, pp. 282–290, 2013.
3. J. Heinermann and O. Kramer, "Machine learning ensembles for wind power prediction," *Renewable Energy*, vol. 89, pp. 671–679, 2016.
4. M. A. F. B. Lima, P. C. M. Carvalho, L. M. Fernández-Ramírez and A. P. S. Braga. "Improving solar forecasting using deep learning and portfolio theory integration," *Energy*, vol. 195, pp. 1–25, 2020.
5. R. B. Ammar, M. B. Ammar and A. Oualha, "Photovoltaic power forecast using empirical models and artificial intelligence approaches for water pumping systems," *Renewable Energy*, vol. 153, pp. 1016–1028, 2020.
6. T. Ahmad, H. Zhang and B. Yan, "A review on renewable energy and electricity requirement forecasting models for smart grid and buildings," *Sustainable Cities and Society*, vol. 55, pp. 1–101, 2020.
7. S. M. Dawaoud, X. Lin and M. I. Okba, "Hybrid renewable microgrid optimization techniques: A review," *Renewable and Sustainable Energy Reviews*, vol. 82, pp. 2039–2052, 2018.
8. E. M. Sandhu and T. Thakur, "Issues, challenges, causes, impacts and utilization of renewable energy sources - grid integration," *International Journal of Engineering Research and Applications*, vol. 4, pp. 636–643, 2014.
9. A. Essl, A. Ortner and P. Hetteger, "Machine learning analysis for a flexibility energy approach towards renewable energy integration with dynamic forecasting of electricity balancing power", in *14th International Conference on the European Energy Market (EEM)*, Dresden, 2017, pp. 1–6.
10. A. Gligor, C. D. Dumitru and H. S. Grif, "Artificial intelligence solution for managing a photovoltaic energy production unit," *Procedia Manufacturing*, vol. 22, pp. 626–633, 2018.

11. M. Gautam, S. Raviteja and R. Mahalakshmi, "Household energy management model to maximize solar power utilization using machine learning," *Procedia Computer Science*, vol. 165, pp. 90–96, 2019.
12. P. Sharmila, J. Baskaran, C. Nayanatara and R. Maheswari. "A hybrid technique of machine learning and data analytics for optimized distribution of renewable energy resources targeting smart energy management," *Procedia Computer Science*, vol. 165, pp. 278–284, 2019.
13. M. Alkandari and I. Ahmad, "Solar power generation forecasting using ensemble approach based on deep learning and statistical methods," *Applied Computing and Informatics*, vol. 19, pp. 1–26, 2019.
14. M. Z. Sušac, S. Mitrović and A. Has, "Machine learning based system for managing energy efficiency of public sector as an approach towards smart cities," *International Journal of Information Management*, vol. 20, pp. 1–12, 2020.
15. T. Som, M. Dwivedi, C. Dubey and A. Sharma. "Parametric studies on artificial intelligence techniques for battery SOC management and optimization of renewable power," *Procedia Computer Science*, vol. 167, pp. 353–362, 2020.
16. R. Ak, O. Fink and E. Zio. "Two machine learning approaches for sort-term wind speed time-series prediction," *IEEE Transactions on Neural Networks and Learning Systems*, vol. 27, pp. 1734–1747, 2016.
17. M. S. Al-Musaylh, R. C. Deo, J. F. Adamowski and Y. Li. "Short-term electricity demand forecasting using machine learning methods enriched with ground-based climate and ECMWF reanalysis atmospheric predictors in southeast Queensland, Australia," *Renewable and Sustainable Energy Reviews*, vol. 113, pp. 3–22, 2019.
18. K. N, Prakash and P. D. Vadana. "Machine learning based residential energy management system," in *IEEE International Conference on Computational Intelligence and Computing Research (ICCIC)*, Coimbatore, pp. 1–4, 2017.
19. A. T. D. Perera, P. U. Wickramsinghe, V. M. Nik and J.-L. Scartezzini. "Machine learning methods to assist energy system optimization," *Applied Energy*, vol. 243, pp. 191–205, 2019.
20. S. S. Sanz, L. C. Bueno, L. Prieto et al. "Feature selection in machine learning prediction systems for renewable energy applications", *Renewable and Sustainable Energy Reviews*, vol. 90, pp. 728–741, 2018.
21. B. Zhang, W. Hu, D. Cao et al. "Deep reinforcement learning–based approach for optimizing energy conversion in integrated electrical and heating system with renewable energy", *Energy Conversion and Management*, vol. 202, pp. 1–13, 2019.
22. T. Ahmad, H. Chen, W. A. Shah "Effective bulk energy consumption control and management for power utilities using artificial intelligence techniques under conventional and renewable energy resources," *EPES*, vol. 109, pp. 242–258, 2019.
23. E. Hossain, I. Khan, F. U. Noor et al. "Application of big data and machine learning in smart grid, and associated security concerns: A review", 10.1109/ACCESS.2019.2894819, *IEEE Access*, pp. 1–40, 2017.
24. S. K. Jha, J. Bilalovic, A. Jha et al. "Renewable energy: Present research and future scope of Artificial Intelligence", *RSER*, vol. 77, pp. 297–317, 2017.
25. K. W. Kow, Y. W. Wong, R. K. Rajkumar et al. "A review on performance of artificial intelligence and conventional method in mitigating PV grid-tied related power quality events", *RSER*, vol. 56, pp. 334–346, 2016.
26. K. R. Kumar and M. S. Kalavathi "Artificial intelligence based forecast models for predicting solar power generation," *PMME*, vol. 5, pp. 796–802, 2018.
27. M. Borunda, O. A. Jaramillo, A. Reyes et al. "Bayesian network in renewable energy systems: A bibliographical survey", *RSER*, vol. 62, pp. 32–45, 2016.
28. M. A. M. Daut, M. Y. Hassan, H. Abdullah et al. "Building electrical energy consumption forecasting analysis using conventional and artificial intelligence methods: A review", *RSER*, vol. 16, pp. 1–11, 2016.

29. M. Ramezanizadeh, M. H. Ahmadi, M. A. Nazari et al. "A review on the utilized machine learning approaches for modeling the dynamic viscosity of nanofluids", *RSER*, vol. 114, pp. 1–15, 2019.
30. M. Sharifzadeh, A. S. Lock and N. Shah, "Machine-learning methods for integrated renewable power generation: A comparative study of artificial neural networks, support vector regression, and Gaussian process regression," *Renewable and Sustainable Energy Reviews*, vol. 108, pp. 513–538, 2018.
31. G. D. Sharma, A. Yadav and R. Chopra, "Artificial Intelligence and Effective Governance: A Review, Critique and Research Agenda," *AJTES*, vol. 20, pp. 1–16, 2020.
32. S. Sinha and S. S. Chandel, "Review of recent trends in optimization techniques for solar photovoltaic–wind-based hybrid energy systems," *RSER*, vol. 50, pp. 755–769, 2015.
33. A. Stetco, F. Dimohammadi, X. Zhao et al. "Machine learning methods for wind turbine condition monitoring: A review", *IJRE*, vol. 18, pp. 1–23, 2018.
34. C. Voyant, G. Notton, S. Kalogirou et al. "Machine learning methods for solar radiation forecasting: A review", *Renewable Energy*, vol. 105, pp. 569–582, 2017.
35. S. Walker, W. Khan, K. Katic et al. "Accuracy of different machine learning algorithms and added-value of predicting aggregated-level energy performance of commercial buildings", *Energy & Buildings*, vol. 209, pp. 1–14, 2020.
36. Z. Wang and R. S. Srinivasan, "A review of artificial intelligence-based building energy use prediction: Contrasting the capabilities of single and ensemble prediction models", *Renewable and Sustainable Energy Reviews*, vol. 16, pp. 1–13, 2016.
37. H. Wang, Z. Lei, X. Zhang et al. "A review of deep learning for renewable energy forecasting", *Energy Conversion and Management*, vol. 198, pp. 1–16, 2019.
38. H. Wang, Y. Liu, B. Zhou et al. "Taxonomy research of artificial intelligence for deterministic solar power forecasting", *ECM*, vol. 214, pp. 1–17, 2020.
39. J. H. Yousif, H. A. Kazem, N. N. Alattar et al. "A comparison study based on artificial neural network for assessing PV/T solar energy production", *Case Studies in Thermal Engineering*, vol. 13, pp. 1–13, 2019.
40. S. M. Zahraee, M. K. Assadi and R. Saidur, "Application of artificial intelligence methods for hybrid energy system optimization", *RSER*, vol. 66, pp. 617–630, 2016.

4 Foundations of Machine Learning

Neeta Nathani and Abhishek Singh
Gyan Ganga Institute of Technology and Sciences, Jabalpur, Madhya Pradesh, India

CONTENTS

4.1 INTRODUCTION TO ARTIFICIAL INTELLIGENCE

Artificial intelligence(AI) is an academic discipline founded in 1955 and was started as a research field in 1956 at a workshop in Dartmouth College, where John McCarthy coined the term "artificial intelligence" [1]. AI is a subpart of computer science concerned with how to give computers the sophistication to act intelligently and in increasingly wider realms. It participates thoroughly in computer science's passion for abstraction, programming and logical formalisms, and detail – for algorithms over behavioral data, synthesis over analysis, and engineering (how to do) over science (what to know) [2].

AI is the simulation of human intelligence in machines when the machine is assigned to perform some tasks, and it focuses on the development of certain human traits in machines, such as planning, learning, reasoning, natural language processing (NLP), problem identification, analysis, solving skills, knowledge representation, situation analysis, perception, motion, manipulation, social intelligence, and creativity.

Based on these aspects, AI is broadly categorized into two types:

1. Weak AI
2. Strong AI

Weak AI or *narrow AI* or *artificial narrow intelligence (ANI)* focuses solely on performing a single task and has limited capabilities or narrow scope, for example, a line follower robot or an obstacle detector robot. Another example is Alexa. Today, everyone is familiar with Amazon's Alexa. Just speak some song name and Alexa will play it for you and it is user friendly.

ANI is the only form of AI that is successfully synthesized till date. Apple's Siri and Google Assistant are other examples of ANI. But can Alexa or Siri or Cortana do everything you ask them to do? No. They can only perform those tasks which they are trained upon. For example, if you will ask Alexa to tell the traffic from your home to work, maybe she will not be able to tell because she was not trained for this type of work.

Facial recognition system, speech-to-text conversion, fingerprint detection, internet searching, driving an autonomous vehicle, weather or stock prediction, disease diagnosis in the medical field, surveillance, and social media monitoring tools are some of the successful and most widely used applications of ANI. ANI works on a certain set of algorithms with limitations and they pretend to be very intelligent in front of the common man. But in reality, ANI does not mimic human intelligence; it merely simulates human behavior based on a narrow range of parameters and contexts.

Strong AI, as the name suggests, can perform multiple tasks, and is further classified into two types:

1. Artificial general intelligence (AGI)
2. Artificial super intelligence (ASI)

AGI technology is almost of the human level. AGI can replicate what a human being can do, with its ability to learn and apply. One fine real-world example is the supercomputer Fujitsu-built K. However, to simulate one neural activity, it takes nearly 40 minutes. With tremendous advancements in image processing techniques, a practical AGI may be achieved soon in reality.

ASI technology is beyond the human mind and even far superior, but fortunately or unfortunately, they are only in fiction yet. One fine example is the robot "Ultron" from the movie *Age of Ultron*, which is a self-aware robot that can make relationships and can exhibit emotions as well. Greater memory power, faster processing ability, decision-making, and problem-solving capabilities surpassing human imagination are the characteristic features of ASI technology.

Many experts refer to AGI as strong AI and ASI as super AI. But since both of these technologies have not been achieved yet and are theoretical, many feel that they both lie in the category of strong AI. That is why, in many books, AI is categorized into two types, weak and strong, while some experts believe that there are three categories, weak, strong, and super.

At this point, a question comes in mind: Why is it so tough to achieve strong AI? And the answer is in the question itself. Yes, mind.

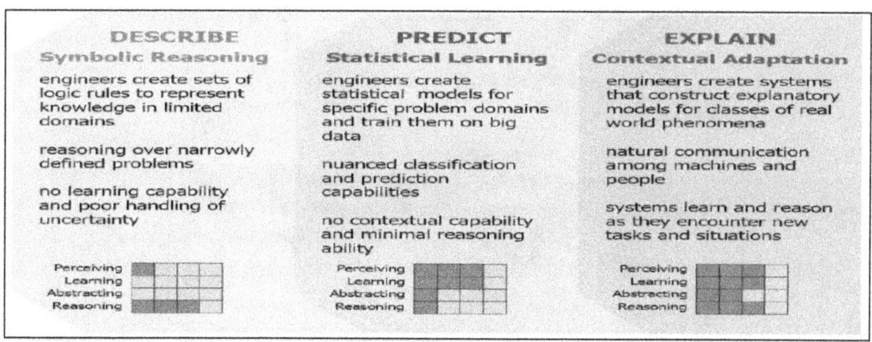

DESCRIBE
Symbolic Reasoning
engineers create sets of logic rules to represent knowledge in limited domains
reasoning over narrowly defined problems
no learning capability and poor handling of uncertainty
Perceiving
Learning
Abstracting
Reasoning
PREDICT
Statistical Learning
engineers create statistical models for specific problem domains and train them on big data
nuanced classification and prediction capabilities
no contextual capability and minimal reasoning ability
Perceiving
Learning
Abstracting
Reasoning
EXPLAIN
Contextual Adaptation
engineers create systems that construct explanatory models for classes of real world phenomena
natural communication among machines and people
systems learn and reason as they encounter new tasks and situations
Perceiving
Learning
Abstracting
Reasoning

FIGURE 4.1 Steps in AI research process.

The human brain is the study model for creating general intelligence in AI, and the lack of comprehensive knowledge on the functionality of the human brain has left researchers struggling to mimic the basic cognitive functions of sight and movement. David Gunning from the Defense Advanced Research Projects Agency (DARPA) characterizes the history of AI research in three steps, as shown in Figure 4.1. The first step was started in the 1950s and it was concerned with symbolic reasoning, like generating proofs of mathematical theorems. The second step was all about statistical learning, which deals with developing models. The third step was to adapt, in which communication between machine and human will take place.

Now, we have seen that our area of interest is ANI since itis the only AI category which is present around us. Let us explore this thing a little more.

ANI's machine intelligence comes from the use of NLP to perform tasks such as chatbots and similar AI technologies. By comprehending the speech and text patterns in natural language, ANI is computed to interact with humans in a natural, personalized manner. Let us see the example code of *speech-to-text* converter as shown in Figure 4.2. First, we need to install some modules such as *speech recognition* and *PyAudio*, as shown in Figure 4.3. Here, *Python Programming Language* using

```
import speech_recognition as sr
import webbrowser
r = sr.Recognizer()
with sr.Microphone() as source:
    print('Speak Something:')
    r.adjust_for_ambient_noise(source)
    audio = r.listen(source)
    try:
        text = r.recognize_google(audio)
        print('You said:{}'.format(text))
        webbrowser.open(text)
    except:
        print('Sorry!Voice not clear.')
```

FIGURE 4.2 Pythoncode of speech-to-text conversion.

```
!pip3 install --upgrade speechrecognition
import speech_recognition as sr
sr.__version__

Collecting speechrecognition
  Using cached https://files.pythonhosted.org/package
4c8cfaed973412f88ae8adf7893a50/SpeechRecognition-3.8.
Installing collected packages: speechrecognition
Successfully installed speechrecognition-3.8.1

'3.8.1'
```

```
!pip3 install PyAudio

Collecting PyAudio
Installing collected packages: PyAudio
Successfully installed PyAudio-0.2.11
```

FIGURE 4.3 Installation of modules: speech recognition and PyAudio.

Anaconda Package is employed. Python is one of the most influential programming languages all over the universe [3] and is widely used in AI development. The integrated development environment is *Jupyter Notebook*.

Once the required libraries are installed, the code can be written. We will not go into explaining the code here but just show how an AI code looks like.

This is how a basic *speech-to-text converter* code looks like. You speak something and it will get displayed as a text. It depends on the quality of the microphone as well.

ANI can be categorized as reactive, limited memory, self-aware, or theory of mind.

1. *Reactive ANI* is very basic and has no memory or data storage capabilities within it. It imitates the human mind's aptness to react to different kinds of impulses without using any experience to incur current decisions. "Deep Blue," the supercomputer of IBM which defeated grandmaster Gary Kasparov in the late 1990s, is an example of reactive ANI.
2. *Limited-memory ANI* is Class-II AI with more advanced features and is equipped with data storage and learning capabilities that enable machines to use historical data to inform decisions.
3. *Self-aware* and *theory of mind* development of ANI are the most advanced classes of AI where the machines will become conscious and become strong AI.

Mostly, ANI is limited-memory ANI, where machines use large volumes of data, and with this feature of memory comes the concept of machine learning (ML).

4.2 FOUNDATIONS OF MACHINE LEARNING

ML is a special type of ANI that learns by itself, and as it gets more and more data, it gets better at learning. All ML is AI, but not all AI is ML. If an AI technology does not learn by itself, then it is not ML. ML is a subset of AI, as shown in Figure 4.4.

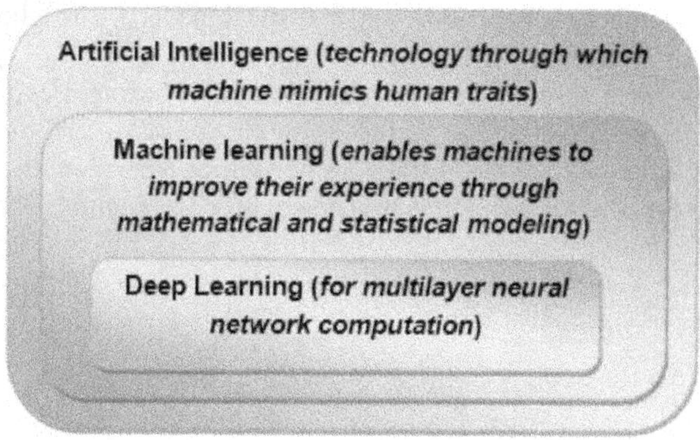

FIGURE 4.4 Machine learning: a subset of artificial intelligence.

The most fundamental difference between traditional and ML programming (as mentioned in Figure 4.5) is that in traditional programming data is applied as an input and is then programmed by the developer to generate the desired output, whereas in ML programming data is applied as an input and the output is also known. It is the responsibility of the machine or computer to program them.

The machine learns by analyzing data, improvising it, and then interacting with the outside world. This further makes the machine adapt itself correctly to the new environment, without being explicitly programmed.

4.2.1 Basic Concepts of Machine Learning

ML is like a statistical model, and the final aim is to analyze the structure of the data, i.e., to fit theoretical distributions to the data that are well understood. With statistically defined models, there is a theory that is mathematically proven, but this also requires that the data must fulfill some strong assumptions too. To create an efficient ML system, data analyzing capabilities, understanding of both basic and advanced algorithms, knowledge of automation and iterative processes, scalability, and ensemble modeling are some common requirements.

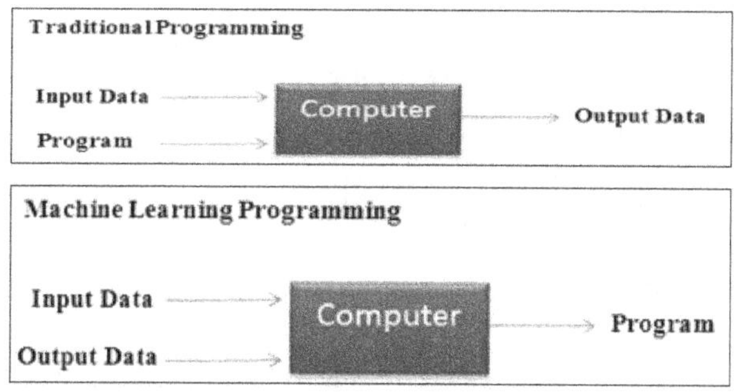

FIGURE 4.5 Traditional vs. machine learning programming.

The basic differences between ML and statistical model are as follows:

1. What we called a *dependent variable* in statistics becomes a *label* in ML. And they both are termed as *target.*
2. What we called a *variable* in statistics becomes a *feature* in ML.
3. What we called a *transformation* in statistics becomes *feature creation* in ML.

Because ML uses a continual approach to learn from data, the learning can be easily automated.

4.2.2 Applications of Machine Learning

ML applications go far beyond computer science. Examples of ML from companies across a wide spectrum of industries, all applying ML to the creation of innovative products and services, are listed as follows:

1. Recommendation engines. Example: Netflix viewing suggestions. Application area: media + entertainment + shopping
2. Sorted, tagged, and categorized photos. Example: reviewer-uploaded photos on Yelp. Application area: search + mobile + social
3. Self-driving cars. Example: Waymo cars use ML to understand surroundings. Application area: automotive + transportation
4. Gamified learning and education. Example: Duolingo's language lessons. Application area: education
5. Calculating customer lifetime value (CLTV) metrics. Example: ASOS uses CLTV to drive profit. Application area: fashion
6. Predicting when patients get sick. Example: KenSci assisting caregivers. Application area: healthcare
7. Determining credit worthiness: Example: Deserve's model for lending to students. Application area: finance
8. Targeted emails. Example: Optimail. Application area: marketing
9. Ranking posts on social media. Example: Twitter's new timeline. Application area: social media
10. Computer vision farming: Example: Blue River Technology's "See & Spray." Application area: agriculture

4.2.3 Classification of Machine Learning

There are different types of learning that anyone can face as a practitioner in the field of ML: from complete fields of study to certain specific techniques. Basically, ML algorithms are classified into four types, which are described in detail in the following sections and are shown in Figure 4.7. Supervised and unsupervised are mostly used by a lot of ML engineers and data geeks. Reinforcement learning is really powerful and complex to apply for problems. The semi-supervised kind of learning is used and applied to the same kind of scenarios where supervised learning is applicable. However, one must note that this technique uses both unlabeled and labeled data for training. The various subtypes of all ML types are shown in Figure 4.8.

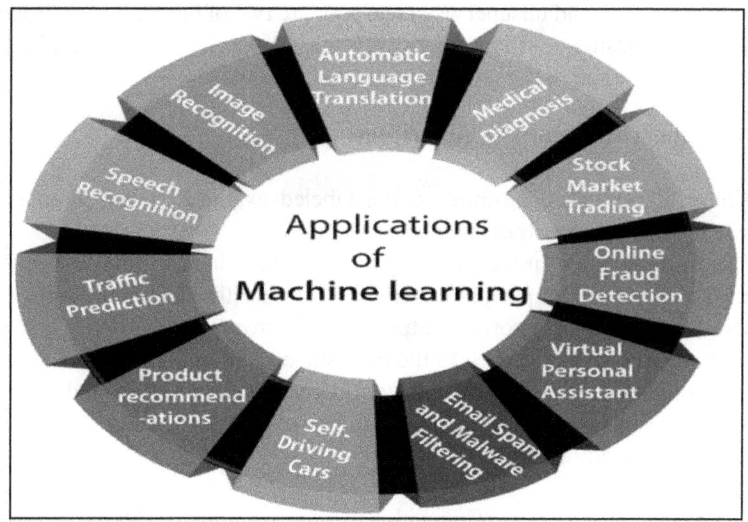

FIGURE 4.6 Disciplinary areas of machine learning [15].

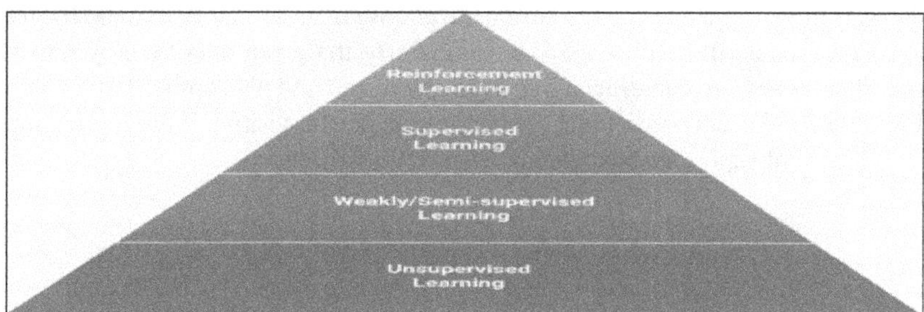

FIGURE 4.7 Different types of ML.

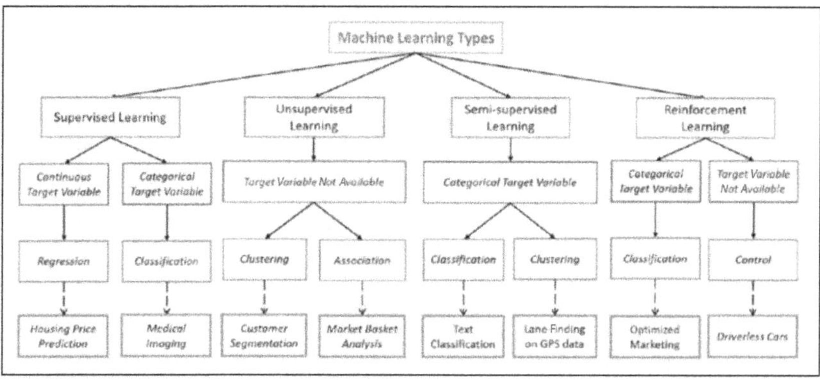

FIGURE 4.8 Subtypes of different types of ML.

Although supervised and unsupervised learning are two of the most widely accepted ML methods by businesses today, there are various other ML techniques. Following is an overview of some of the most accepted ML methods.

4.2.3.1 Supervised Learning Algorithms

1. These algorithms are trained using labeled examples, such as an input where the desired output is known.
2. The training algorithm receives a group of inputs alongside the corresponding set of correct outputs, and therefore the algorithm learns by comparing its actual output with correct outputs to find errors.
3. The model is then modified to the final stage accordingly.
4. Using methods such as classification, regression, prediction, and gradient boosting, supervised learning uses patterns to predict the values of the label on additional unlabeled data.

Supervised learning is like learning with a mentor: training dataset will be the mentor and the test dataset will train the model. For example, a person has visited the restaurants marked by red circle, as shown in Figure 4.9. The restaurants which the person has not visited are marked by blue circle.

Now, if this person has two restaurants to choose from, A and B, marked by green color, which one will he choose? We can classify the given data linearly into two parts. That means, we can draw a line segregating red and blue circle. This is shown in Figure 4.10. We can predict that chances of the person visiting B are more than A. This is a case of supervised learning.

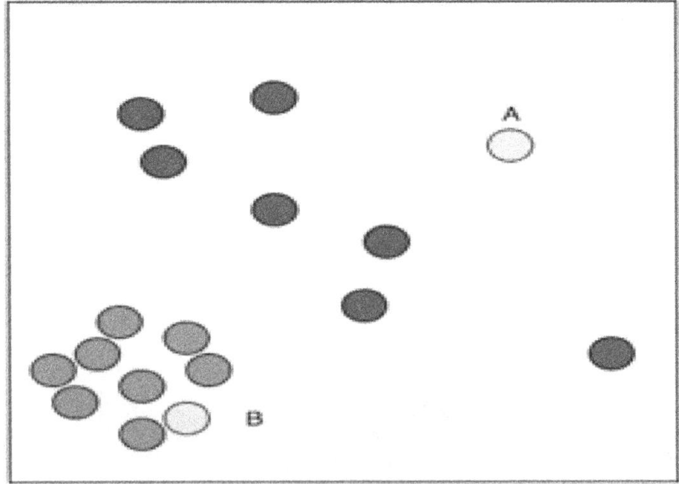

FIGURE 4.9 Circles showing restaurants visited by a person. Red color is visited and blue color is not visited. Green color is the restaurant he wishes to visit.

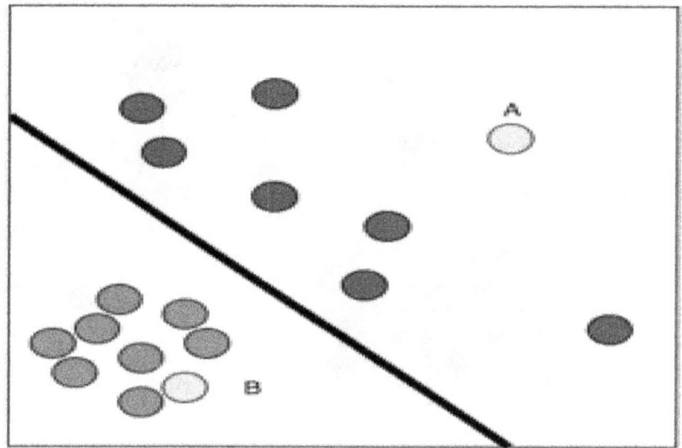

FIGURE 4.10 The person will choose green color B which is near red color circles because of supervised learning concept.

Types of supervised learning:

1. Classification: It is a supervised learning in which the model is trained to classify using different models – for example, classifying whether a person is suffering from a disease or not or an email is a spam or not.
2. Regression: It is a supervised learning task where the model is used to predict values –for example, predicting the price of house or stocks, etc.

Example of supervised learning algorithms:

i. Linear regression
ii. Nearest neighbor
iii. Gaussian naive Bayes
iv. Decision trees
v. Support vector machine (SVM)
vi. Random forest

4.2.3.2 Unsupervised Learning Algorithms

1. The algorithm is used against data that has no historical labels. The system is not told the "correct answer." The algorithm must find out what is being shown.
2. The goal is to explore the data and find some structure within.
3. Popular techniques that are widely used include self-organizing maps, nearest-neighbor mapping, K-means clustering, and singular value decomposition.

For example, a taxi driver has an option of accepting or rejecting the bookings. Figure 4.11 shows his accepted booking location on map with blue circles.

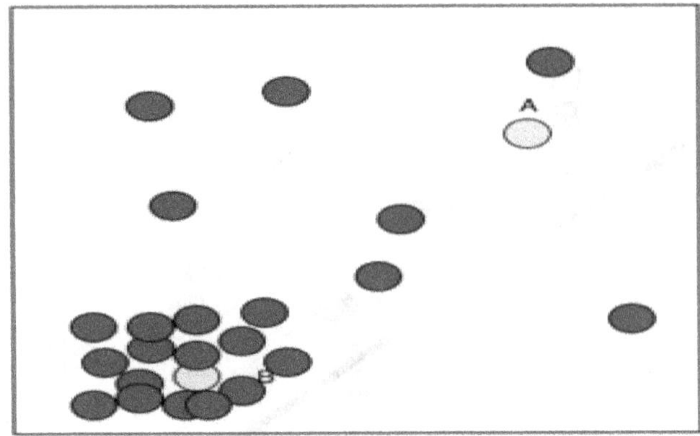

FIGURE 4.11 Blue circles indicate accepted booking locations of a taxi driver. Green colors indicate the next two booking he has to accept.

Now, the taxi driver has got two bookings A and B. Which one he will accept? If we observe the plot in Figure 4.12, we can see that his accepted booking shows a cluster at lower left corner.

Types of unsupervised learning:

1. Clustering: A clustering problem is where you want to discover the inherent groupings in the data such as grouping customers by purchasing behavior.
2. Association: An association rule learning problem is where you want to discover rules that describe large portions of your data such as people that buy X also tend to buy Y.

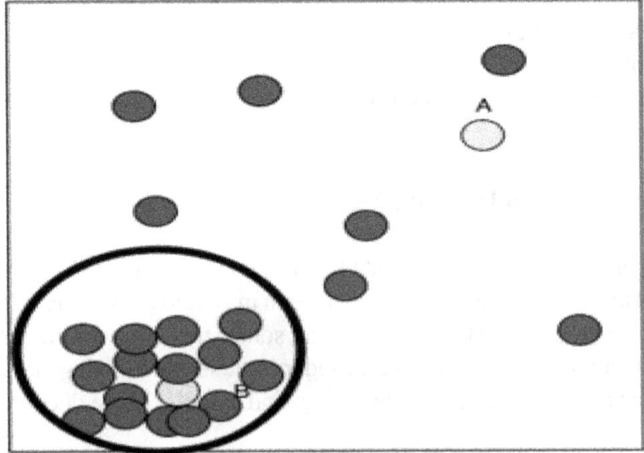

FIGURE 4.12 The big circle indicates that the taxi driver has accepted the booking for the place which he knows very well, that is, around dense blue circled.

4.2.3.3 Semi-supervised Learning Algorithm

1. The algorithm is used for the same applications as that in supervised learning. The heuristic approach of *self-training* (also known as *self-learning* or *self-labeling*) is historically the oldest approach to semi-supervised learning, with examples of applications starting in the 1960s [4, 5].
2. It uses both labeled and unlabeled data for training – typically, a small amount of labeled data with a large amount of unlabeled data. This is because unlabeled data is less expensive and takes less effort to acquire.
3. It can be used with methods such as classification, regression, and prediction.
4. Semi-supervised learning is useful when the cost associated with labeling is too high to allow for a fully labeled training process.
5. Identifying a person's face on a webcam is an example of this algorithm. Semi-supervised learning has recently become more popular and practically relevant due to the variety of problems for which vast quantities of unlabeled data are available – e.g., text on websites, protein sequences, or images [6].

Types of semi-supervised learning:

1. Classification: Points that are close to each other are more likely to share a label. This is also generally assumed in supervised learning and yields a preference for geometrically simple decision boundaries [6].
2. Clustering: The data tends to form discrete clusters, and points in the same cluster are more likely to share a label. This is a special case of the smoothness assumption and gives rise to feature learning with clustering algorithms [6].

4.2.3.4 Reinforcement Learning Algorithm

1. This algorithm is widely used for robotics, gaming, and navigation.
2. Reinforcement learning provides an intuitively appealing framework for addressing a wide variety of planning and control problems [7].
3. The algorithm uses trial and error to determine which actions yield the greatest rewards.
4. This type of learning has three primary components: the agent or the learner or decision-maker, the environment or the resources, which means everything the agent interacts with, and actions that describe everything the agent can do. This is shown in Figure 4.13.

The final objective is for the agent to choose actions that maximize the expected reward over a given amount of time.

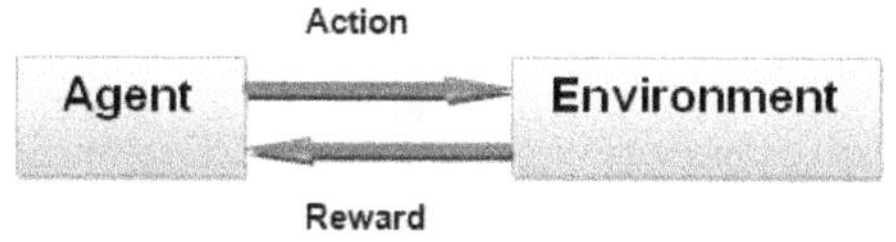

FIGURE 4.13 Interaction between an agent and an environment.

The agent will reach the goal much faster by following a good policy.

Therefore, the goal in reinforcement learning is to learn the best policy. Reinforcement learning models are much more popular in neurobiology [8].

Types of reinforcement learning:

1. Classification: Reinforcement learning [9, 10] algorithms enable an agent to learn an optimal behavior when letting it interact with some unknown environment and learn from its obtained rewards. Reinforcement learning is quite different from supervised learning where an input is mapped to a desired output by using a dataset of labeled training instances.
2. Control: The network is required to produce an appropriate control action in response to the current world state.

Whichever ML algorithm one chooses to implement, every ML algorithm possesses the following three components:

- *Representation*: It is the way to represent knowledge. The most commonly used examples include decision trees, sets of rules, instances, graphical models, neural networks, SVMs, model ensembles, etc.
- *Evaluation*: It is used to evaluate candidate hypotheses. The most commonly used examples include accuracy, prediction and recall, squared error, likelihood, posterior probability, cost, margin, entropy K-L divergence, etc.
- *Optimization*: It is used to generate the search process of the hypothesis or programs. Examples include combinatorial optimization, convex optimization, and constrained optimization.

It must be kept in mind that all ML algorithms are combinations of these three components.

4.2.4 Machine Learning Process Architecture

Linear regression, logistic regression, and decision trees are just names of the algorithms. These are just theoretical concepts that describe what to do in order to achieve the specific effect. A model is a mathematical formula which is a result of ML algorithm implementation. It has measurable parameters that can be used for prediction. Models can be trained by modifying their parameters in order to achieve better results. It is possible to say that models are representations of what an ML system has learned from the training data, as mentioned in Figure 4.14.

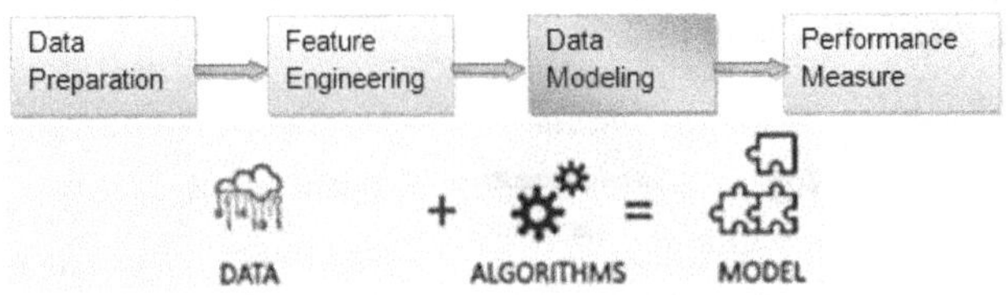

FIGURE 4.14 Diagram depicting the relationship between ML model and ML algorithm.

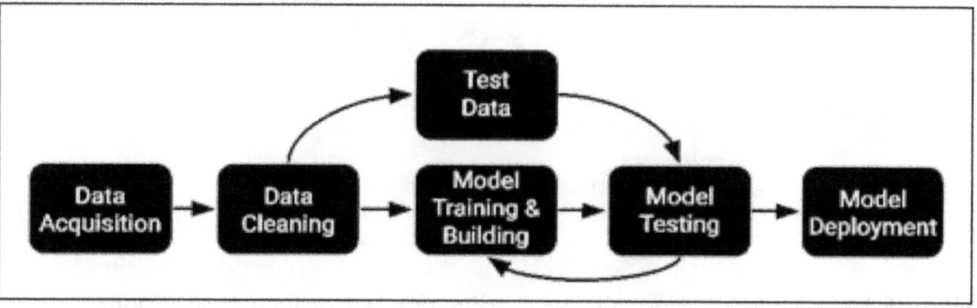

FIGURE 4.15 Block diagram of machine learning process architecture.

The most important process in ML is processing data. Before being able to become a predicting model, the data has to go through various steps. The general process architecture is shown in Figure 4.15 and described in the following:

1. *Data acquisition* is nothing but getting data. This stage is also termed as *data preprocessing stage* and it involves several steps such as collecting data and then segregating the data based on specific features and then sending this segregated data to the processing unit where further categorization is performed.
2. In the *data cleaning* process, corrupted, incomplete, duplicated, and improperly formatted data is removed or modified for further analysis. This data provides inaccurate results and therefore cleaning this data is very important. Data normalization, transformation, encoding, and memory processing requirements are also performed in this step.
3. The next step is *splitting* the data. The data is divided into two sets, *training set*, which is used for training the data, and *test set*, which is used to test the performance of the model. During the training process, the model is continuously evaluated, and thus, there is another set known as *validation set*, which validates the data through error rate calculations at regular time instants and thus its accuracy is determined. The split can be in the ratio of 70/30, 80/20, 75/25, or anything depending on the need. This ratio means that from 100% data, 70% data will be treated as training data and the remaining 30% data will be the testing data. Note that the validation set is not independent. It is just the training set evaluated at some specific time. It is used to test the machine periodically until the desired output is achieved.
4. Once the datasets are ready, *data modeling* is started using different modeling algorithms. An optimized algorithm is needed to generate maximum possible output from the model and to achieve maximum performance from the machine. The output then becomes capable of providing the required data for the machine to make decisions.
5. Finally, the machine is deployed.

Let us understand ML with a simple example. The first step is to have some data. There are numerous sources from where data can be made available, such as Kaggle,

```
import pandas as pd
import seaborn as sns
import numpy as np
import matplotlib.pyplot as plt

data = pd.read_csv("People Charm case.csv")
data.head()
```

FIGURE 4.16 Python code to call modules, libraries, and file for data acquisition.

GitHub, and Reddit. One can also have his/her dataset. For this example, a dataset known as *People Charm Case.csv* is used. It is a CSV file available on Kaggle. It can be easily downloaded from the internet. Again, to access the file for data processing, there can be many programming languages. Here, *Python* language and *Pandas* library have been used. "Pandas" is a short name for *panel distribution*, a term widely used in the field of statistics. We will not go deep into how *Pandas* works, but will show you how the above-explained process of ML works. Let us begin with accessing the data, i.e., *data acquisition*, as shown in Figure 4.16.

It can be seen that there are many libraries included in the code. These libraries are required at later stages to perform certain data manipulations. *Numpy* is used for performing numerical operations. *Seaborn* and *Matplotlib* are used for displaying results in the form of graphs and figures. At this point, library is not needed other than the *Pandas*, written in the first line. Let us see the output which is generated at this point. It is a very large file.

Using the *.head ()* function as shown in Figure 4.17, only the first 5 rows can be seen. A copy of the data can be made using *.copy ()* function so that our original file is safe. Using *.info ()* function, one can check a summary of the file content, as shown in Figure 4.18.

It can be clearly seen that earlier the file was called by the variable *data* but now it has been copied to another variable *data1.* The *info ()* function is giving the information that there are 14,999 entries in 10 columns each. Other information that can be retrieved is that there are no null values, data types involved, column names, and memory usage.

After this, the next step is *data cleaning.* This process is necessary to check and find any form of unwanted data. The term *unwanted* is a vague term and it depends on the requirement of the user. Data that is *unwanted* for someone can be *wanted*

	satisfactoryLevel	lastEvaluation	numberOfProjects	avgMonthlyHours	timeSpent.company	workAccident	left	promotionInLast5years	dept	salary
0	0.38	0.53	2	157	3	0	1	0	sales	low
1	0.80	0.86	5	262	6	0	1	0	sales	medium
2	0.11	0.88	7	272	4	0	1	0	sales	medium
3	0.37	0.52	2	159	3	0	1	0	sales	low
4	0.41	0.50	2	153	3	0	1	0	sales	low

FIGURE 4.17 Output snippet of using *head ()* function displaying first five rows and columns of the data file.

```
data1 = data.copy()

data1.info()

<class 'pandas.core.frame.DataFrame'>
RangeIndex: 14999 entries, 0 to 14998
Data columns (total 10 columns):
satisfactoryLevel          14999 non-null float64
lastEvaluation             14999 non-null float64
numberOfProjects           14999 non-null int64
avgMonthlyHours            14999 non-null int64
timeSpent.company          14999 non-null int64
workAccident               14999 non-null int64
left                       14999 non-null int64
promotionInLast5years      14999 non-null int64
dept                       14999 non-null object
salary                     14999 non-null object
dtypes: float64(2), int64(6), object(2)
memory usage: 1.0+ MB
```

FIGURE 4.18 Output snippet of using *info()* function.

data for another person. Regardless, let us see how to do *data cleaning.* Formatting data, checking for duplicate values, checking for non-numeric values, etc., are some of the important steps in *data cleaning.*

It can be observed from the code in Figure 4.19 that here duplicate values are checked and the first value is kept and all other duplicate values are removed. The number of entries after removing duplicates has reduced to 11,991 from 14,999.

```
data1.drop_duplicates(keep = 'first', inplace = True)
data1.info()

<class 'pandas.core.frame.DataFrame'>
Int64Index: 11991 entries, 0 to 14998
Data columns (total 10 columns):
satisfactoryLevel        11991 non-null float64
lastEvaluation           11991 non-null float64
numberOfProjects         11991 non-null int64
avgMonthlyHours          11991 non-null int64
timeSpent.company        11991 non-null int64
workAccident             11991 non-null int64
left                     11991 non-null int64
promotionInLast5years    11991 non-null int64
dept                     11991 non-null object
salary                   11991 non-null object
dtypes: float64(2), int64(6), object(2)
memory usage: 936.8+ KB
```

FIGURE 4.19 Output snippet of using *info ()* function after removing all duplicate values from the data.

```
from sklearn.model_selection import train_test_split
from sklearn.neighbors import KNeighborsClassifier
from sklearn import metrics
from sklearn.linear_model import LinearRegression
from sklearn.metrics import mean_squared_error

data1_omit = data1.dropna(axis = 0)
data1_omit = pd.get_dummies(data1_omit, drop_first = True)
x1 = data1_omit.drop(['left'], axis = 'columns', inplace = False)
y1 = data1_omit['left']
X_train, X_test, y_train, y_test = train_test_split(x1,y1,test_size = 0.25, random_state = 0)
print(X_train.shape, X_test.shape, y_train.shape, y_test.shape)

knn = KNeighborsClassifier(n_neighbors = 2)
knn.fit(X_train, y_train)
y_pred = knn.predict(X_test)
print(metrics.accuracy_score(y_test, y_pred))

(8993, 18) (2998, 18) (8993,) (2998,)
0.9439626417611742
```

FIGURE 4.20 Output snippet of using various function of *Scikit-learn* machine learning library.

Thus, it can be seen why data cleaning is important. Such large duplicate values would have given an inaccurate result in model prediction.

Now moving on to the third step, i.e., *model training and testing*, here, first of all, certain libraries and learning algorithms are needed. Here, *Scikit-learn* library is used which is a very popular, powerful, and open-source software tool used in ML. As per the theory, we need to split our data into training set and test set. Here, 75% of data is used as a training set and 25% of the data as a test set. Let us see how it is coded down, as shown in Figure 4.20.

Now, there is extensive use of libraries because many operations are being performed here. In the tenth line of the code, it can be observed how data is split into training and testing sets. Here, k-nearest neighbor algorithm, which is a supervised learning algorithm, is used to implement the model. It is observed that the predicted model has 94% accuracy in the output. There are numerous approaches and algorithms which can help to yield a better result.

4.3 DEEP LEARNING

With the availability of more and more data called Big Data and more memory, another thought referred to as *deep learning* came into existence. Instead of depending on hard-coded rules to solve problems, an ML algorithm is trained by feeding it with the real-world data. ML then develops a model that searches for patterns between the data given and the data that is predicted. That model can make predictions for new things which it has never seen before. As the model is exposed to more and more training data, its accuracy gets better and better. On the other hand, deep learning is a subset of ML where algorithms are created and function similarly to ML, but there are many levels of these algorithms, and each level provides a different

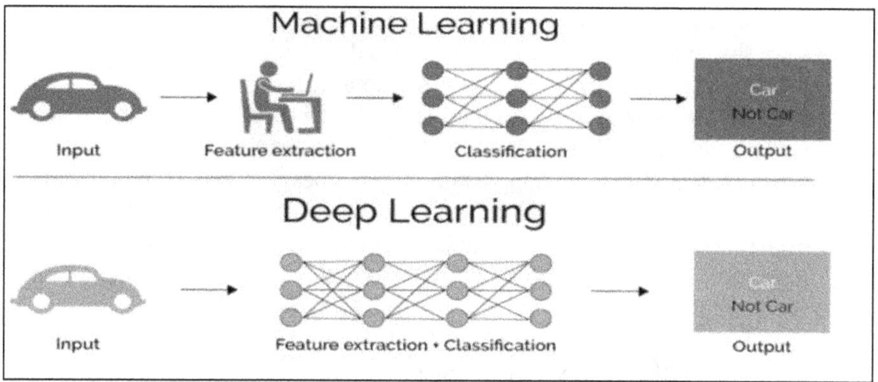

FIGURE 4.21 Machine learning vs. deep learning.

interpretation of the data, as shown in Figure 4.21. This network of algorithms is called artificial neural networks (ANNs). In simple words, it resembles the neuronal connections in the human brain.

4.3.1 Differences between Deep Learning and Machine Learning

1. The most important distinction between deep learning and ML is the way data is presented in the system. ML algorithms almost always require structured data, whereas deep learning networks rely on layers of ANN.
2. ML algorithms are developed to learn *acting* by understanding the labeled data and then use it to produce new results with more datasets. However, when the result is incorrect, there is a need to *teach* them.
3. Deep learning networks do not need human mediation, as structured layers in neural networks place the data in pecking order of various ideas that finally learn from their own mistakes. However, even they can be wrong if the data quality is not good enough.
4. Data decides everything. It is the standard of the data that ultimately determines the standard of the result.

Example: Let us say we have got a set of pictures of dogs and cats and we have to recognize the images of dogs and cats separately using ML algorithms and deep learning neural networks.

1. *Solution using ML*: How does the algorithm know which one is dog and which is the image of cat? The answer is the availability of structured data, as we discussed in the definition of ML above. Simply mark the images of dogs and cats in order to determine the characteristics of both animals. These data will be sufficient for training an ML algorithm, and then the model will continue to work on the basis that it understands about the markings and classifications of millions of other images of animals that it had studied earlier.

2. *Solution using deep learning*: Deep learning uses a different approach to solve this problem. The main advantage is that it does not need structured/tagged image data to classify the two animals. In this case, the input data (image data) is sent through different levels of neural networks, and each network determines the specific features of the images in specific order. This is exactly the same approach our brain does to solve any problem by going through step-by-step solution using available concepts and building new concepts if necessary to find the answer. After processing the data through different levels of neural networks, the system finds appropriate attributes to classify both the animals by their images.

Thus, the ML algorithm needs labeled/structured data to know the variations between images of cats and dogs, study the classification, and then draw a conclusion. Deep learning, on the other hand, is able to classify the pictures of both animals from data processed within the layers of the network. This did not require any labeled/structured data, because it relies on different outputs processed by each layer, which is then combined to make a single way of classifying images. Big Data is also used in ML; it is the use of neural networks which distinguishes ML from deep learning. It gives the best results among all ML types. Thus, we can say that deep learning is a subset of ML which is a subset of AI, as given in Figure 4.22. *Tensor Flow* and *Keras* are very popular deep learning libraries. Other than using *neural networks*, a very important point that distinguishes the three technologies of AI, ML, and deep learning is "interpretation ability" or "interpretability."

"Interpretability" is the ability to explain what prediction was made by the machine and how much this prediction is explainable. While AI and ML make predictions that are sometimes interpretable, the predictions made by deep learning are not interpretable, although they are more accurate as compared to predictions made

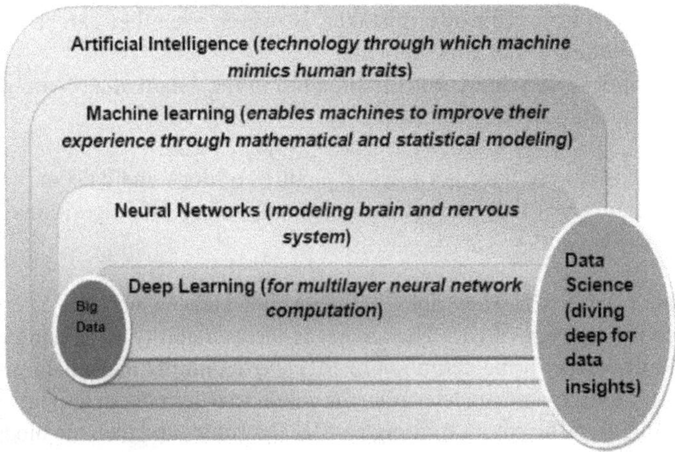

FIGURE 4.22 Relationship between AI, ML, neural networks, deep learning, Big Data, and data science.

by AI and ML. Therefore, in cases where responsibility for a decision is needed, interpretability matters, and in such cases, it is always advisable to use other types of ML over deep learning. One such field is disease diagnosis. There will be a probability of getting a less accurate result when not using deep learning but the benefit will be that you will be able to get the answer of why it gave such a diagnosed result.

4.4 HOW MACHINE LEARNING IS REVOLUTIONIZING THE RENEWABLE ENERGY SECTOR?

With the availability of more and more data to explore and with recent advancements in the field of AI, ML, and deep learning, the models of power generation, pricing, and consumption are structurally redefined, causing significant changes in the green energy sector. Innovative, smarter ways of monitoring, modeling, analyzing, and predicting energy generation and usage are being employed to achieve sustainable energy goals as the global population is facing an unprecedented environmental challenge.

4.4.1 Use Case

Better computational power and hybrid transaction/analytical processing systems (HTAPS) are enabling ML algorithms to optimize the energy and also the power sector on an outsized scale. Big Data is functioning in synchronization with AI, ML, and deep learning and is improving the energy sector generation-consumption quantitative relation by activity bound actions on datasets such as: capturing data, storing data, updating, querying and visualizing the data. Power consumption has become one of the critical concerns in design of electronic computing systems. High power consumption degrades system reliability, increases the cooling cost for high performance systems, and reduces the service time of batteries in portable devices [11]. Dynamic power management (DPM), defined as the selective shut-off or slow-down of system components that are idle or underutilized, has proven to be an effective technique for reducing power dissipation at system level [12]. An effective DPM policy should minimize power consumption while maintaining performance degradation to an acceptable level. Design of such DPM policies has been an active research area. The following technologies are widely being explored to optimize the energy and power sector.

1. AI and grid management:
   a. One among the foremost fascinating uses of AI in energy is grid management.
   b. Electricity is delivered to customers through a power grid. The difficult issue regarding the power grid is that power generation and power demand must match every time. Otherwise, problems like blackouts and system failures will arise.
   c. Although there are numerous ways to store energy, the foremost common way is the ancient but still an efficient method of pumped hydroelectric storage. It works by pumping water to a selected elevation and then harnessing it again by permitting it to fall onto turbines.

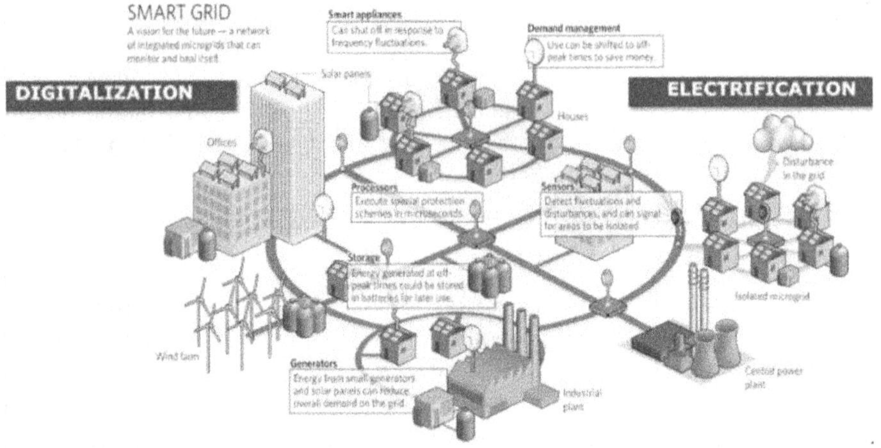

FIGURE 4.23 The future of electricity: new technologies transforming the grid edge.

d. When dealing with renewable energy, it is difficult to predict the grid's electricity production capacity. After all, it depends on several factors such as sunlight and wind.

e. This is where smart grids step in. Smart grids are power grids that combine the strengths of Internet of Things (IoT), AI, and Big Data to create a digital power grid that enables two-way communication between customers and utility companies. Smart grids are equipped with smart meters, sensors, and alerting devices that incessantly gather and display data to consumers in order that they will improve their energy consumption behaviors, as shown in Figure 4.23.

It can even be fed to ML algorithms to predict demand, improve performance, scale back prices, and forestall system failures. Although smart grids are being adopted in many developed countries, we still have a long way to go before switching to 100% renewable energy sources, AI-controlled power distribution, and grid management.

2. Demand response:
   a. When large swings in demand occur, it is often very expensive for countries that produce most of their energy through renewable energy sources.
   b. With most countries shifting towards green energy, responding effectively to swings in demand is becoming even harder.
   c. Germany, for example, plans to cover 80% of its electricity consumption using renewable energy by 2050.
   d. There are two main issues that countries like Germany will face. The first is swings in demand. It is common for electricity demand to skyrocket on a particular day or period of the year (say, on Christmas or on New Year eve). The second issue is climate volatility. If there is no wind, or the sky is covered, it can be challenging to fulfill the electricity demand.

   e. In both cases, supplemental stations or fossil fuel-powered facilities need to make up for the excess demand.
   f. To solve these issues, countries are collaborating with companies to analyze and predict weather data, climate conditions, electricity demand, and so on.
   g. Germany initiated a project with *EWeLiNE*, which aims at forecasting what proportion of wind and solar power to expect at a given time. This allows the country to make up for excess electricity demand by using nonrenewable energy whenever necessary.
   h. In order to accurately match supply and demand, they use large historical datasets to train their ML algorithms – as well as data collected from the wind turbines or solar panels – to effectively forecast weather and power changes.
3. Energy source exploration:
   a. Besides meteorology and renewable energy source optimization, AI and data science are being used in fuel energy source exploration and drilling.
   b. A few years ago, ExxonMobil teamed up with MIT to produce self-learning submersible robots to explore the ocean surface [13].
   c. These robots will be equipped with ML algorithms not only to help them learn from their mistakes while conducting explorations but also to carry out the same work that a scientist would do, without the risk.
   d. The robot will explore, record data about the ocean floor, and make an analysis based on the data. This will allow for exploring new locations to drill oil and natural gas along the ocean floor.
4. Predictive maintenance:
   a. Other than helping to achieve energy production with energy consumption, AI is becoming a serious driver in assuring the reliability and robustness of power grids.
   b. In 2003, a massive blackout in Ohio was caused by a low-hanging high-voltage power line brushing against an overgrown tree. The power system alarm failed and there was no indication that the incident had occurred.
   c. The electric company did not discover anything until three more power lines started failing for similar reasons. Ultimately, this oversight caused a cascade effect, leading the complete grid to shut down.
   d. The blackout lasted for 2 days and affected 50 million people. Additionally, 11 people died and there have been about $6 billion in losses incurred.
   e. Unfortunate events like the Ohio blackout can now be completely avoided. ML algorithms are efficiently predicting machine failures and helping the energy companies to smoothly transit from a reactive maintenance stand to a predictive maintenance stand.

Opting renewable energy resources can be good for the governments and electric companies to focus on. Companies such as Google and Microsoft are trying to make an impact on the environment by lowering their overall energy consumption. Google has massive data centers all around the world and they produce a great amount of heat, and require a massive amount of electricity to cool down. To address this issue, DeepMind AI used ML algorithms to reduce energy cooling on its Google data centers by 40% [14].

4.5 CONCLUSION

In the presence of modeling, data-driven ML tools can speed up the design cycle, reduce the complexity and cost of implementation, and improve the performance of known algorithms. To this end, ML can use efficiently the available data and computing resources in many engineering domains, including modern communication systems. Supervised, unsupervised, and reinforcement learning paradigms lend themselves to different tasks depending on the availability of examples of desired behavior or of feedback. The applicability of learning methods hinges on specific features of the problem under study, including its time variability and its tolerance to errors. As such, a data-driven approach should not be considered as a universal solution, but rather as a useful tool whose suitability should be assessed on a case-by-case basis. Furthermore, ML tools allow for the amalgamation of traditional model-based engineering techniques with the existing domain knowledge in order to use the synergy of the two solutions to the fullest.

REFERENCES

1. Kaplan, A., and Haenlein, M. (2019). "Siri, Siri, in my hand: Who's the fairest in the land? On the interpretations, illustrations, and implications of artificial intelligence." Business Horizons. 62:15–25.
2. Nilsson, N. (1980). Principles of Artificial Intelligence. Palo Alto, CA: Tioga Press.
3. Singh, A. (2020). Python Programming Universe v 3.8: With Walrus Operator, Positional Argument Operator and Many More… (Vol. 1). Independently Published.
4. Scudder, H. (1965). "Probability of error of some adaptive pattern-recognition machines." IEEE Transactions on Information Theory. 11(3):363–371.
5. https://en.wikipedia.org/wiki/Semi-supervised_learning#cite_note-survey-7.
6. Zhu, X. (2008). "Semi-supervised learning literature survey." Technical Report 1530. University of Wisconsin-Madison.
7. Lagoudakis, M. G., and Parr, R. (2003)."Reinforcement learning as classification: Leveraging modern classifiers." In: Proceedings of the Twentieth International Conference on Machine Learning (ICML-2003), Washington, DC.
8. Roelfsema, P. R., and van Ooyen, A. (2005)."Attention-gated reinforcement learning of internal representations for classification", Neural Computation. 17:2176–2214.
9. Kaelbling, L. P., Littman, M. L., and Moore, A. W. (1996)."Reinforcement learning: A survey." Journal of Artificial Intelligence Research. 4:237–285.
10. Sutton, R. S., and Barto, A. G. (1998) Reinforcement Learning: An Introduction. Cambridge MA: The MIT Press.
11. Wang, Y., Xie, Q., Ammari, A., and Pedram, M. (2011). "Deriving a near-optimal power management policy using model-free reinforcement learning and Bayesian classification." In: Proceedings of the 48th Design Automation Conference (DAC 2011), San Diego, CA.
12. Benini, L., Bogliolo, A., and De Micheli, G. (2000)."A survey of design techniques for system level dynamic power management." IEEE Transactions on VLSI Systems. 8(3):299–316.
13. https://energyfactor.exxonmobil.com/news/mit-collaboration.
14. https://deepmind.com/blog/article/deepmind-ai-reduces-google-data-centre-cooling-bill-40.
15. https://javatpoint.com/applications-of-machine-learning.

5 Introduction of AI Techniques and Approaches

Namrata Dhanda and Rajat Verma
Amity University Uttar Pradesh, Lucknow, India

CONTENTS

5.1 INTRODUCTION

Artificial intelligence (AI) is a term that depicts the intelligence exhibited by the machines [1]. Systems that think like humans can be considered as a form of AI. It works in a contrasting fashion to the intelligence that is depicted by all living objects. The broad field of AI can be trifurcated into subfields, namely, neural network, machine learning (ML), and deep learning. Currently, AI is being implemented in the areas of security and surveillance [2–4], sports analytics [5], shopping and fashion [6, 7], retail [8, 9], etc. AI makes complete use of diverse algorithmic rules that work in an amalgamated manner to improve the efficiency of the corresponding proposed objective. AI examines the environment and performs the tasks accordingly to improve the efficacy of the technique involved [10]. A few popular examples that have revolutionized the technological domain are ALVINN [11], Deep Blue [12], DART [13], PROVERB [14], etc.

The full form of ALVINN is autonomous land vehicle in a neural network. It was trained to steer a car and keep it in a particular lane. The total distance that was to be traveled was 2850 miles in which 98%, i.e., 2793 miles, were traveled by the machine itself and only 2%, i.e., 57 miles, were taken by humans [15, 16]. Similarly, the Deep Blue program created by IBM won a chess match in 1997 in an exhibition match by defeating the chess grandmaster Garry Kasparov [17, 18]. The DART program of AI, i.e., Dynamic Analysis and Replanning Tool, had the purpose of automated logistic scheduling and planning for transportation [19, 20]. The PROVERB program from the domain of AI is useful for solving the crossword puzzles [21].

5.1.1 Weak, Strong, and Superintelligence Aspects of AI

The three subgroups—weak, strong, and superintelligence aspects of AI lead to the foundation of automated systems.

5.1.1.1 Weak AI

The term "weak" in weak AI depicts the limited domain. Weak AI deals with creating some sort of AI that can solve some kinds of programs but not all, and works in some limited sections. An example of this weak AI is natural language processing (NLP). NLP acts as an interface between the human beings and the computing machines [22, 23].

5.1.1.2 Strong AI

In the year 1980, John Searle forwarded the concept of strong AI in his article "Minds, Brains, and Programs" [24]. As the limited word was mentioned in the weak perspective of AI, here in strong AI the word limited is removed, which means that in strong AI one can truly rely on solving problems, and the scope is not limited. It is also known as self-aware AI [25]. Strong AI can be bifurcated into two halves:

- **Human-like artificial intelligence**: In this, the program will work like a human mind (thinking and reasoning).
- **Nonhuman-like artificial intelligence**: In this, the program develops complete nonhuman sentience.

5.1.1.3 Superintelligence

In AI, the superintelligence aspect is an intelligent agent that has an outstanding capability to solve diverse problems. This mechanism increases its intelligence and passes the cognitive behavior of an average human being in a quick tenure [26, 27]. It is somehow infeasible also, since superintelligence may damage and harm humanity as AI can manipulate human beings and may not allow them to escape from a particular kind of a network.

A particular example of the superintelligence concept could be the emulation of a human mind that runs and performs on faster hardware in comparison to the actual human brain.

5.1.2 Approaches of Artificial Intelligence

In AI, there are four types of approaches:

- **Acting humanly**: For acting humanly, Alan Turing proposed the Turing test in 1950 [28–30]. For this, some parameters are required, which are as follows:
  - NLP [31]
  - Knowledge representation [32]
  - Automated reasoning [33]
  - ML [34]
- **Thinking humanly:** For thinking humanly, the theory of the human mind needs to be examined [35].
- **Thinking rationally:** The concept of "right thinking" was proposed by Aristotle [36].
- **Acting rationally:** Acting rationally is another approach of AI [37].

5.1.3 Goals of Artificial Intelligence

There are primarily two goals of AI, which are as follows:

- **Creating expert systems**: The primary goal of AI is to create a system that can show intelligent behavior, demonstrating and explaining advice to people/users.
- **Implementation of human intelligence in machines**: This goal depicts the creation of systems that can act, behave, and think as human beings do.

5.1.4 Applications of AI

There are different areas where AI has made its mark and some of them are as follows:

- **Healthcare**: There is no doubt about the fact that ML can play a vital role in research. Research is becoming efficient in the field of healthcare through AI as well [38]. The contribution of AI to healthcare includes:
  - The amalgamation of cognitive sciences, software, and medical fields.
  - The invention of tools and mechanisms that enhance decision-making processes as well as research perspectives.
  - The examination, representation, and cataloging of medical information.
- **Business**: The diverse ML algorithms are integrated with the business and analytics industry to serve customers. Chatbots are an example of it as it automates the processes when a user visits a website and performs multiple tasks simultaneously [39].

- **Education**: Automatic grading according to some specific standards and assessing students have enhanced the field of education through AI [40].
- **Autonomous vehicles**: The concept of self-driving cars is now possible because of AI. ALVINN has been a great example of this [41].
- **Energy industry**: For enhancing the energy sector through AI, digitizing the energy sector is a requirement. Large datasets for evaluation are also required for enhancing this sector through AI.

5.1.5 Contribution of AI in Energy

In the field of energy, AI plays a vital role in enhancing the efficiency of energy systems. Some applications of AI in the field of energy are power grids [42], power consumption units [43], virtual power plants [44], electricity trading [45], etc. Power grids include sector coupling, smart grids, and monitoring and coordination of grids. Power consumption includes smart home and smart meters. Virtual power plants include the forecasting processes and the monitoring and co-ordination of decentralized plants. A brief illustration of applications of AI in the energy sector is shown in Figure 5.1.

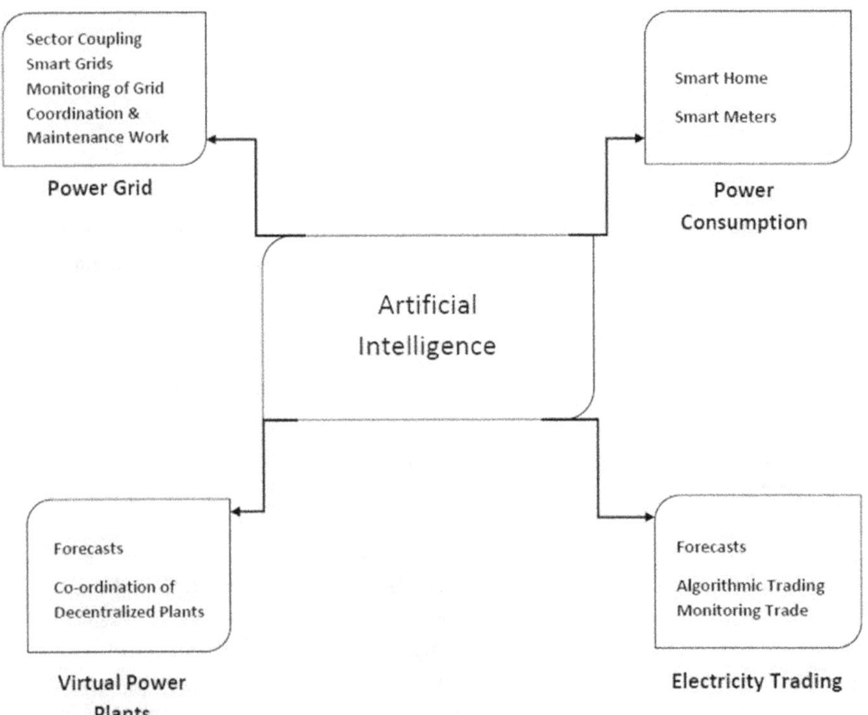

FIGURE 5.1 Applications of AI in energy sector.

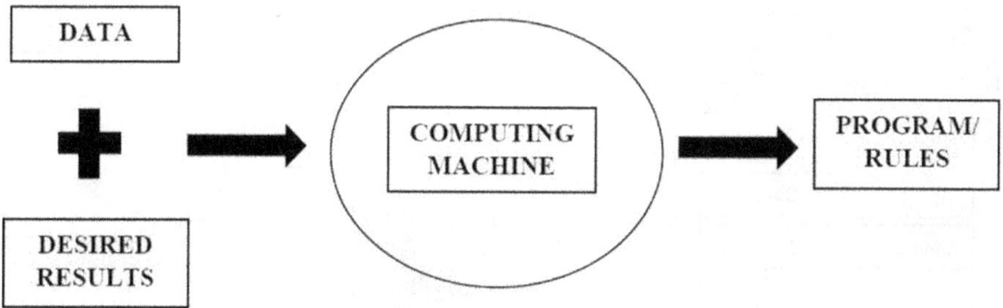

FIGURE 5.2 Machine learning.

5.1.6 Machine Learning

The most popular subgroup of AI is ML. ML signifies the learning depicted by machines, and it enhances the performance of the machine through experiences [46]. In ML, the attributes of the input are the data and the desired results, and the produced program/rules are considered as the output attributes. The block diagram of ML is shown in Figure 5.2.

It is different from the traditional learning perspective, as traditional learning takes the data and program/rules as an input attribute and produces the results as an output. The block diagram of traditional learning is shown in Figure 5.3.

5.1.6.1 Supervised Learning

The most common subgroup of ML is supervised learning. It can transform a variety of techniques. These techniques include classification (for discrete data), regression (for continuous data), artificial neural networks (ANNs), support vector machines (SVM), etc., and are discussed in the chapter. In supervised learning, a teacher is required to supervise the model.

Generally, the new professionals of the ML industry will start their research work as well as their implementations using the approach of supervised learning only. Supervised learning is trained using the labeled data [47]. It contains an input-output pair that is used as an input attribute and is passed to the learning algorithm module with which a new model is built. Then a new input is given to that model and the

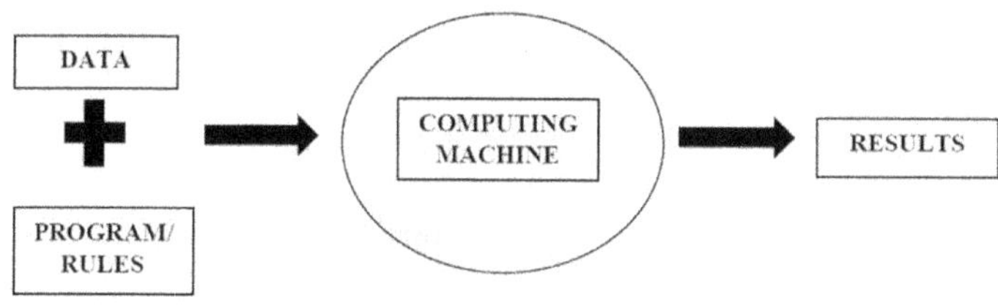

FIGURE 5.3 Traditional learning.

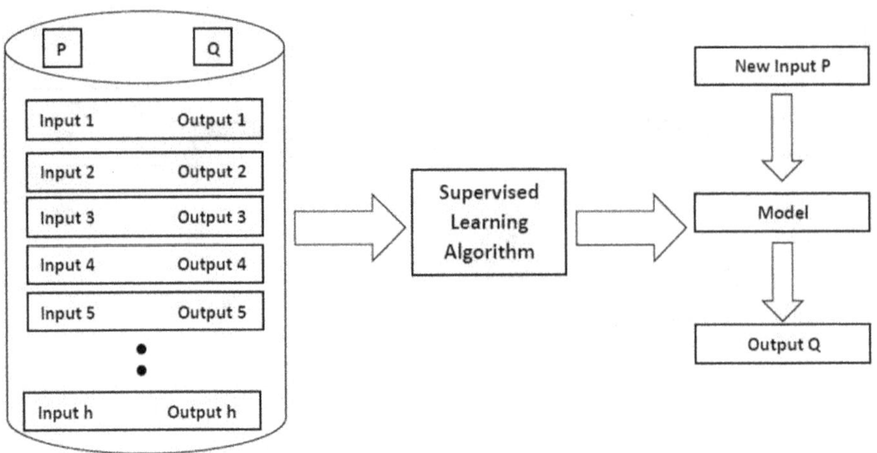

FIGURE 5.4 Block diagram of supervised learning.

output is calculated. If the output belongs to a predetermined pattern, then the model is said to be accepted; otherwise, the model is rejected. Its block diagram is depicted in Figure 5.4.

The figure shows that in supervised learning, a given set of input attributes (i.e., $P_1, P_2, P_3, P_4, \ldots, P_h$) along with their output attributes (i.e., $Q_1, Q_2, Q_3, Q_4, \ldots, Q_h$) are kept in a knowledge dataset. The learning algorithm takes an input P_i and execute with its model and produce the result Q_i as the desired output.

An example of supervised learning could be some input parameters of renewable resources, such as the capability of regeneration or maintaining sustainability, and its corresponding output, i.e., renewable or nonrenewable resources. This entire data is passed on to the learning algorithm module so that the machine could be trained. Now, a new parameter is passed as an input attribute to the model. If the machine can successfully categorize the data into renewable or nonrenewable resource, then the model is correct and successfully built; otherwise, the model is rejected.

5.1.6.2 Unsupervised Learning

In unsupervised learning, there is no requirement to supervise the model, as, in this model, the system is allowed to work on its own. It is different from the supervised form of learning. The block diagram of unsupervised learning is shown in Figure 5.5. In the unsupervised mode of learning, the inputs are provided that are passed on to the learning algorithm and the cluster or groups are formed. In this form, the teacher is not required.

The figure shows that in unsupervised learning the inputs are collected as a set of features that are described as $P_1, P_2, P_3, P_4, \ldots, P_h$. However, the output features are not available. The input parameters are passed to a learning algorithm module and diverse groups, called clusters, are formed [48–52].

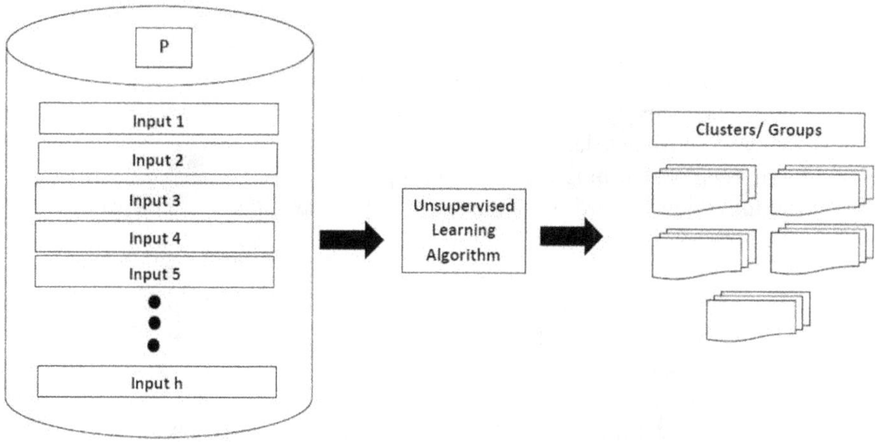

FIGURE 5.5 Block diagram of unsupervised learning.

5.1.6.3 Reinforcement Learning

Reinforcement learning was invented by John Andreae in 1963 when he made the system known as STELLA [53, 54]. The third form of learning, i.e., reinforcement learning, follows a dynamic approach and works on the concept of feedbacks [55–57]. In reinforcement learning, the utility of the agent is predicted using the reward function [58]. The main aim of this form of learning is to maximize the rewards. The block diagram of reinforcement learning is shown in Figure 5.6.

Figure 5.6 illustrates the block diagram of reinforcement learning, which shows that there is an agent or machine that performs certain actions in the environment. The interpreter interprets the actions performed by the agent or machine in the environment and returns reward to the agent for its action. There is a state that is fed again as an input to the agent, and this process goes on forever.

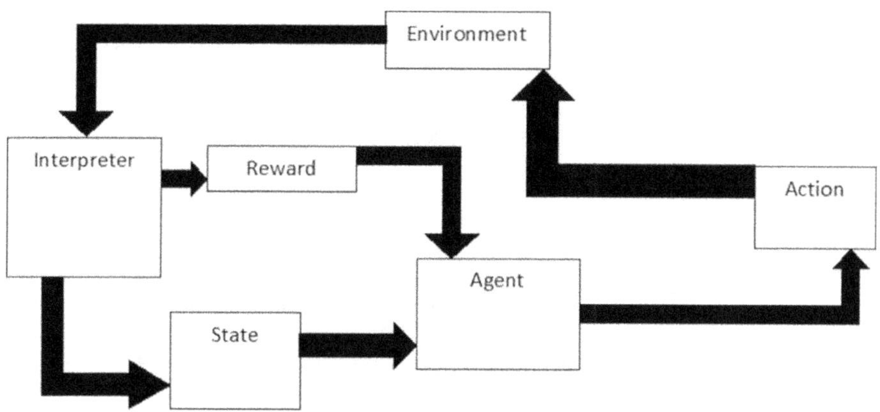

FIGURE 5.6 Block diagram of reinforcement learning.

The main points of reinforcement learning are as follows:

- The initial state of the learning process is the input attribute.
- The output of this process could be various, depending on the feedback received.
- The input attributes deal with the training processes.
- Reinforcement learning is a continuous process.
- The maximum reward in reinforcement learning is the best positive feedback for the agent.

As an example, a scientist is predicting whether a resource is renewable or nonrenewable. He performs some tests to predict the same. In a few tests, the results were positive, as it satisfied the minimum requirements to be a renewable or a nonrenewable resource. This gave him positive feedback. He gets motivated and performed more tests to discover new things. Now, in most tests, he did not get anything. Here, the feedbacks are negative, which illustrate that the scientist must have done something wrong. Now, he will work to remove the errors and improve the diverse parameters.

5.2 CLASSIFICATION

Classification is a task in ML that deals with the organized process of assigning a class label to an observation from the problem domain. The traditional classification algorithm was invented by a Swedish botanist Carl Von Linnaeus [59]. It is a subgroup of the supervised form of ML [60, 61]. As an example of the classification approach, a resource has to be classified as a renewable resource or a nonrenewable resource. It is shown in Figure 5.7.

In Figure 5.7, there are two types of symbols: one is a triangle and the other one is a rectangle. The triangles represent the renewable resources and the rectangles represent the nonrenewable resources. There is a hyperplane (partition line) that bifurcates the two entities depicting classification.

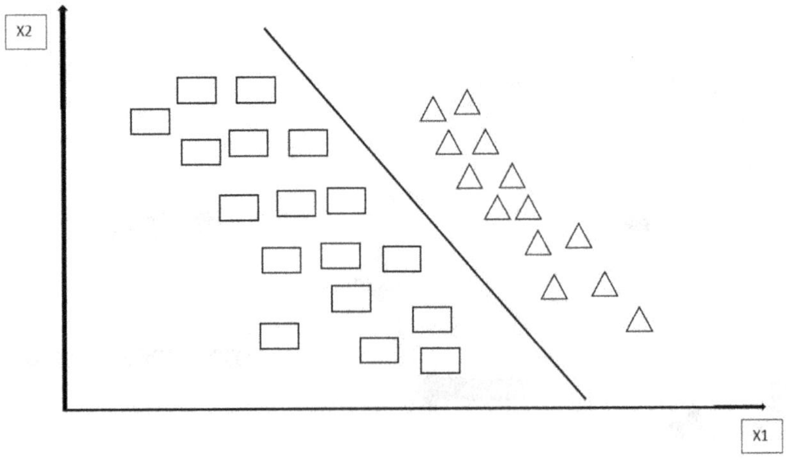

FIGURE 5.7 Classification.

Generally, classification is of four types, which are as follows:

- **Binary classification**: This type of classification includes two class labels (dichotomous), out of which one class is in the normal state and the other class is in the abnormal state. The above example of a resource that could be classified into renewable or nonrenewable resource category is an example of binary classification [62, 63]. Many algorithmic rules follow only dichotomous class labels and not more than that.
- **Multilabel classification**: This type of classification has more than two class labels. Here, for each example, either one or more class labels need to be identified [64–66].
- **Multiclass classification**: This type of classification has more than two class labels [67–69].
- **Imbalanced classification**: This type of classification deals with the unequal number of examples in each class [70–72].

5.3 REGRESSION

Regression is one of the simplest tools to find the relationship between the continuous independent attributes and the dependent attribute [73, 74]. It is a subgroup of the supervised form of ML. It is of the following types:

- **Linear regression**: Used for predictive analytics [75–77]
- **Logistic regression**: Used when the dependent variable is dichotomous [78, 79]
- **Polynomial regression**: It is used for curvilinear data [80, 81]
- **Stepwise regression**: It works with predictive models [82, 83]
- **Ridge regression**: Used for multiple regression data [84, 85]
- **Lasso regression**: Used for the purpose of variable selection and regularization [86, 87]
- **Elastic net regression**: Used when the penalties of lasso and ridge method are combined [88]

In statistics, there are equations for simple linear regression as well as multiple linear regression [89, 90]:

- **Linear regression (simple):** $Q = n + cP + i$
- **Linear regression (multiple):** $Q = n + c_1P_1 + c_2P_2 + c_3P_3 + \ldots + c_hP_h + i$

where

- Q is known as dependent variable
- P or $P_{o\in h}$ are independent variable
- n is an intercept
- c or $c_{o\in h}$ are slope variables
- i is regression residual
- h and o are any natural number

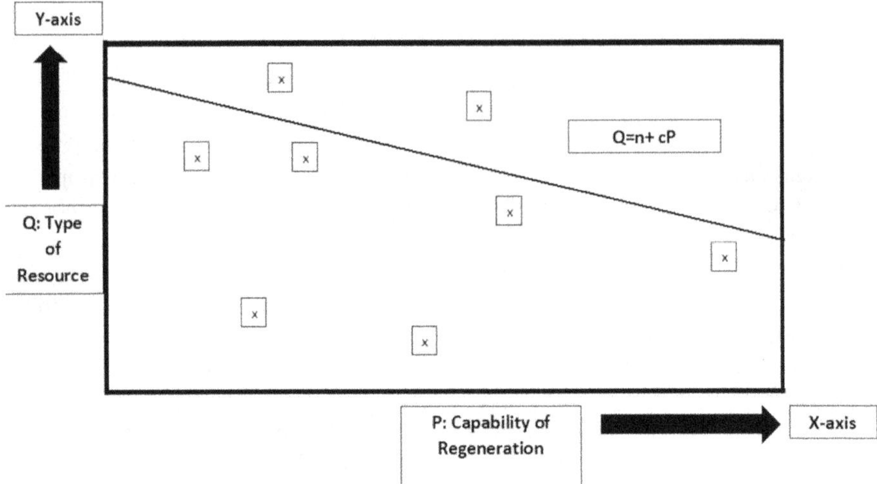

FIGURE 5.8 Linear regression.

An example of the regression approach could be of predicting the resource as a renewable resource or a nonrenewable resource, having attributes such as the capability of regeneration (independent attribute, P), and Q can be the categorized version of a renewable or nonrenewable resource (dependent attribute). The entire scenario could be depicted in Figure 5.8.

In Figure 5.8, a y-axis represents the type of resource (renewable or nonrenewable) depending on the input provided by the input attribute (capability of regeneration). Multiple input parameters could also be used here.

5.4 SUPPORT VECTOR MACHINE

The original SVM algorithm was invented by Vladimir N. Vapnik and Alexey Ya. Chervonenkis in 1963 [91]. It is a subgroup of the supervised form of ML. SVM is used for analyzing data that can be used for the process of regression as well as classification [92, 93]. The main aim of SVM is to find a hyperplane that divides and classifies the data points distinctly. Taking a similar example of the bifurcation of renewable and nonrenewable resources but giving it a more detailed view, SVM is depicted in Figure 5.9.

Figure 5.9 depicts SVM that involves the hyperplane, support vectors, maximum margins, and data points that belong to either renewable resource or a nonrenewable resource.

Support vectors are the data points that are very near to the hyperplane and that affect its position also. If the support vectors are deleted, then the position as well as the orientation of the hyperplane will be changed and the maximum margin will also be affected [94–96]. The maximum margin is the distance between the closest points in both classes.

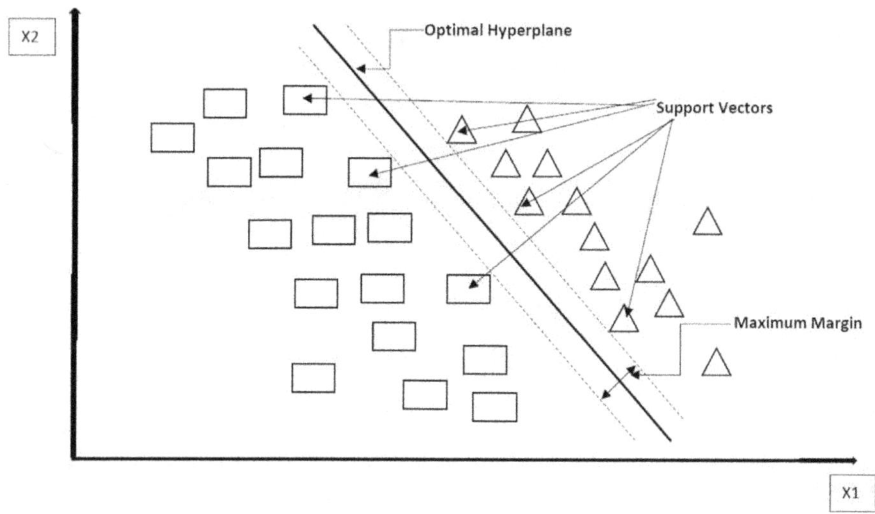

FIGURE 5.9 Support vector machine.

5.5 ARTIFICIAL NEURAL NETWORKS

ANN is a very important application of AI and was invented by psychologist Frank Rosenblatt in 1958 [97]. It imitates the functioning of a biological human brain. It has an input layer, numerous hidden or processing layers, and an output layer. The outputs that are provided by the neural networks are not restricted to the input attributes provided to them. They also do not need a database to store the input values; rather, they store everything in their networks. The main aim of ANN is to solve computations and problems as the human brain does.

The structure of an ANN is shown in Figure 5.10.

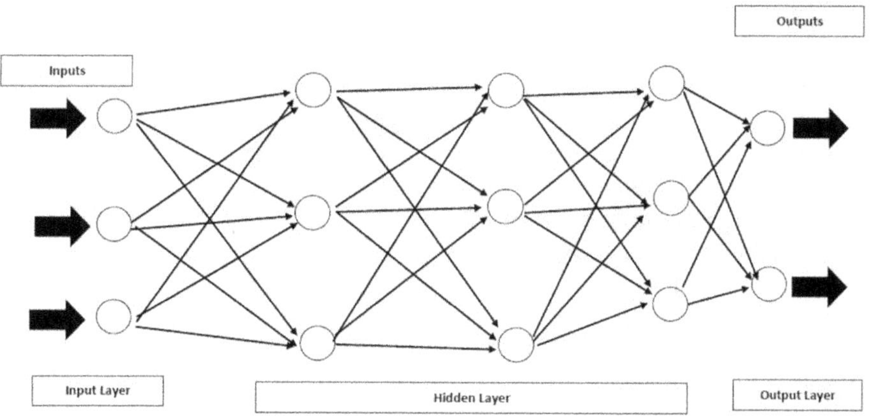

FIGURE 5.10 Artificial neural networks (general).

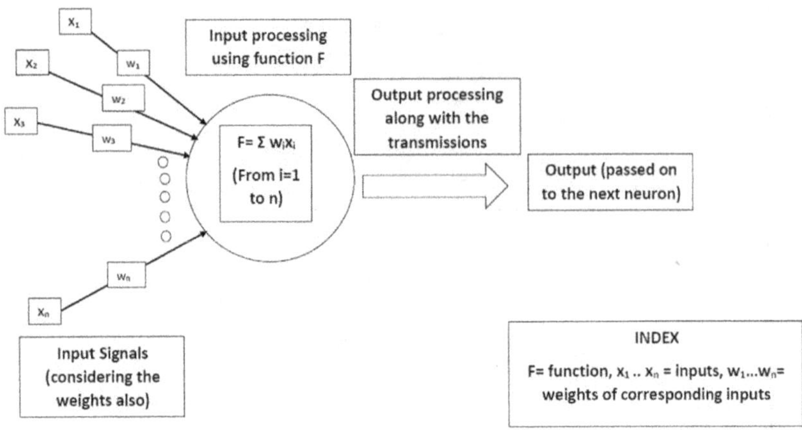

FIGURE 5.11 Artificial neural networks (detailed look).

Figure 5.10 illustrates input attributes in the input layer, numerous processing or hidden layers, and the output attributes present in the output layer. In this figure, only three processing layers are there; however, processing layers can be many depending on the requirement by the machine and situation. A detailed scenario of ANN is depicted in Figure 5.11.

In Figure 5.11, the inputs are given in the form of x_1, x_2, x_3, …, x_n. They also include weights in the form of w_1, w_2, w_3, …, w_n. The processing is done using the function *F*. The function *F* can also be represented in the form of an equation:

$$F = w_1x_1 + w_2x_2 + w_3x_3 + w_4x_4 + \ldots + w_nx_n.$$

Then, the output is processed along with the transmissions. The output that is obtained after processing acts as an input for the next neuron. This process goes on till the completion.

There are seven types of ANN:

- Multilayer perceptron [98]
- Convolutional neural network [99]
- Recursive neural network [100]
- Recurrent neural network [101]
- Long short-term memory [102]
- Sequence-to-sequence model [103]
- Shallow neural network [104]

5.6 NATURAL LANGUAGE PROCESSING

NLP began with the article "Machine and intelligence" that was published in 1950 [105]. Natural language is how we humans communicate with others. It includes speech as well as text. Human beings see text each day in one or the other form. It includes e-mails, webpages, menus, signs, etc.

NLP [106, 107] is a field of AI that deals with the interaction between humans and computers. It is becoming the driving force of every industry nowadays.

It deals with the analysis, i.e., understanding of data obtained from humans in a useful manner. Some applications of NLP are as follows:

- Summarizer [108]
- Sentiment analysis [109]
- Creation of ChatBot [110]
- Automatic generation of keywords [111]
- Reduce words to their roots [112]
- Identification of type of entity extracted [113]
- Social media monitoring tool [114, 115]

5.7 CONCLUSION AND FUTURE SCOPE

The era of AI has brought many changes to the world and had a deep impact on the lives of the people. This will continue to grow in the upcoming as well as far future. The techniques and approaches of AI are well depicted in this chapter. The subsets of AI, i.e., ML and its subgroups, are also highlighted in this chapter. The different types of supervised learning techniques, namely, classification, regression, neural networks, NLP, and SVMs, are also discussed in this chapter.

Subsequently, the research will be continued in the field of AI to find out its growing contribution to the research sector.

REFERENCES

1. Lei, Y., He, Z., & Zi, Y. (2008). A new approach to intelligent fault diagnosis of rotating machinery. *Expert Systems with Applications, 35*(4), 1593–1600.
2. Dilek, S., Çakır, H., & Aydın, M. (2015). Applications of artificial intelligence techniques to combating cyber crimes: A review. Preprint arXiv:1502.03552.
3. Yampolskiy, R. V. (Ed.). (2018). *Artificial intelligence safety and security*. Boca Raton, FL: CRC Press.
4. Demertzis, K., & Iliadis, L. (2015). A bio-inspired hybrid artificial intelligence framework for cyber security. In *Computation, Cryptography, and Network Security* (pp. 161–193). Cham: Springer.
5. Brefeld, U., & Zimmermann, A. (2017). Guest editorial: Special issue on sports analytics. *Data Mining and Knowledge Discovery, 31*(6), 1577–1579.
6. Yu, Y., Choi, T. M., & Hui, C. L. (2011). An intelligent fast sales forecasting model for fashion products. *Expert Systems with Applications, 38*(6), 7373–7379.
7. Kim, H. S., & Cho, S. B. (2000). Application of interactive genetic algorithm to fashion design. *Engineering Applications of Artificial Intelligence, 13*(6), 635–644.
8. Semenov, V. P., Chernokulsky, V. V., & Razmochaeva, N. V. (2017, October). Research of artificial intelligence in the retail management problems. In *2017 IEEE II International Conference on Control in Technical Systems (CTS)* (pp. 333–336).
9. Wong, C., Guo, Z. X., & Leung, S. Y. S. (2013). *Optimizing decision making in the apparel supply chain using artificial intelligence (AI): From production to retail*. Cambridge: Elsevier.

10. Fethi, M. D., & Pasiouras, F. (2010). Assessing bank efficiency and performance with operational research and artificial intelligence techniques: A survey. *European Journal of Operational Research, 204*(2), 189–198.
11. Pomerleau, D. A. (1989). ALVINN: An autonomous land vehicle in a neural network. In *Advances in neural information processing systems 1* (pp. 305–313). San Francisco, CA: Morgan Kaufmann.
12. Campbell, M., Hoane Jr, A. J., & Hsu, F. H. (2002). Deep blue. *Artificial Intelligence, 134*(1–2), 57–83.
13. Bennett, J. S., & Hollander, C. R. (1981, August). DART: An Expert System for Computer Fault Diagnosis. In *Proceedings of the VIIth IJCAI*, Vancouver (pp. 843–845).
14. Littman, M. L., Keim, G. A., & Shazeer, N. M. (1999, July). Solving crosswords with PROVERB. In *Proceedings of the Sixteenth National Conference on Artificial Intelligence (AAAI-99)*, Orlando, FL (pp. 914–915).
15. Pomerleau, D. A. (1993). Knowledge-based training of artificial neural networks for autonomous robot driving. In *Robot learning* (pp. 19–43). Boston, MA: Springer.
16. Li, Z., Wang, J., Li, B., Gao, J., & Tan, X. (2014). GPS/INS/Odometer integrated system using fuzzy neural network for land vehicle navigation applications. *The Journal of Navigation, 67*(6), 967–983.
17. Newborn, M. (2012). *Kasparov versus Deep Blue: Computer chess comes of age.* Berlin: Springer Science & Business Media.
18. Seirawan, Y., Simon, H. A., & Munakata, T. (1997). The implications of Kasparov vs. Deep Blue. *Communications of the ACM, 40*(8), 21–25.
19. Rashmi, K. V., & Gilad-Bachrach, R. (2015, May). DART: Dropouts meet multiple additive regression trees. In *Proceedings of the 18th International Conference on Artificial Intelligence and Statistics (AISTATS)*, San Diego, CA (pp. 489–497).
20. Hedberg, S. R. (2002). DART: revolutionizing logistics planning. *IEEE Intelligent Systems, 17*(3), 81–83.
21. Borrajo, D., Ríos, J., Pérez, M. A., & Pazos, J. (1990). Dominoes as a domain where to use proverbs as heuristics. *Data & Knowledge Engineering, 5*(2), 129–137.
22. Aggarwal, M. (2011). Information retrieval and question answering NLP approach: An artificial intelligence application. *International Journal of Soft Computing and Engineering (IJSCE), 1*(NCAI2011).
23. Jacobs, P. S. (1992, March). Joining statistics with NLP for text categorization. In *Proceedings of the Third Conference on Applied Natural Language Processing* (pp. 178–185).
24. Searle, J. R. (1980). Minds, brains, and programs. *Behavioral and Brain Sciences, 3*(3), 417–424.
25. Omohundro, S. M. (2008, February). The basic AI drives. In *Proceedings of the 2008 Conference on Artificial General Intelligence* (Vol. 171, pp. 483–492).
26. Bostrom, N., Dafoe, A., & Flynn, C. (2016). *Policy desiderata in the development of machine superintelligence.* Future of Humanity Institute, University of Oxford. Retrieved June 8, 2018.
27. Gill, K. S. (2016). Artificial super intelligence: beyond rhetoric. *AI and Society, 31*, 137–143.
28. Moor, J. (Ed.). (2003). *The Turing test: The elusive standard of artificial intelligence* (Vol. 30). Berlin: Springer Science & Business Media.
29. Saygin, A. P., Cicekli, I., & Akman, V. (2000). Turing test: 50 years later. *Minds and Machines, 10*(4), 463–518.
30. Hodges, A. (2012). *Alan Turing: The Enigma.* London: Random House.
31. Friedman, C., & Hripcsak, G. (1999). Natural language processing and its future in medicine. *Academic Medicine, 74*(8), 890–895.

32. Brachman, R. J., & Levesque, H. J. (1985). *Readings in knowledge representation.* Los Altos, CA: Morgan Kaufmann Publishers Inc.
33. Wos, L., Overbeck, R., Lusk, E., & Boyle, J. (1984). *Automated reasoning: introduction and applications.* Englewood Cliffs, NJ: Prentice-Hall.
34. Bratko, I. (1993). Machine learning in artificial intelligence. *Artificial Intelligence in Engineering, 8*(3), 159–164.
35. Tan, U. (2007). The psychomotor theory of human mind. *International Journal of Neuroscience, 117*(8), 1109–1148.
36. Macaulay, M., & Arjoon, S. (2013). An Aristotelian-Thomistic approach to professional ethics. *Journal of Markets & Morality, 16*(2), 507–527.
37. Miiller, Y. (1990). Decentralized artificial intelligence. In: *Decentralised A.I.*(pp. 3–13). Amsterdam: North Holland.
38. Yu, K. H., Beam, A. L., & Kohane, I. S. (2018). Artificial intelligence in healthcare. *Nature Biomedical Engineering, 2*(10), 719–731.
39. Hamscher, W. (1994). AI in business-process reengineering. *AI Magazine, 15*(4), 71.
40. Beck, J., Stern, M., & Haugsjaa, E. (1996). Applications of AI in education. *XRDS: Crossroads, The ACM Magazine for Students, 3*(1), 11–15.
41. Rychtyckyj, N. (1999, July). DLMS: Ten Years of AI for Vehicle Assembly Process Planning. In *Proceedings of the* AAAI/IAAI (pp. 821–828).
42. Li, W., Logenthiran, T., Phan, V. T., & Woo, W. L. (2016, November). Intelligent multi-agent system for power grid communication. In *2016 IEEE Region 10 Conference (TENCON)* (pp. 3386–3389).
43. Sozontov, A., Ivanova, M., & Gibadullin, A. (2019). Implementation of artificial intelligence in the electric power industry. In E3S Web of Conferences (Vol. 114, p. 01009).
44. Hernández, L., et al. (2013). A multi-agent system architecture for smart grid management and forecasting of energy demand in virtual power plants. *IEEE Communications Magazine, 51*(1), 106–113.
45. Szkuta, B. R., Sanabria, L. A., & Dillon, T. S. (1999). Electricity price short-term forecasting using artificial neural networks. *IEEE Transactions on Power Systems, 14*(3), 851–857.
46. Alpaydin, E. (2020). *Introduction to machine learning.* Cambridge, MA: MIT Press.
47. Miyato, T., Maeda, S. I., Koyama, M., & Ishii, S. (2018). Virtual adversarial training: a regularization method for supervised and semi-supervised learning. *IEEE Transactions on Pattern Analysis and Machine Intelligence, 41*(8), 1979–1993.
48. Baldi, P. (2012, June). Autoencoders, unsupervised learning, and deep architectures. In *Proceedings of ICML Workshop on Unsupervised and Transfer Learning* (pp. 37–49).
49. Srivastava, N., Mansimov, E., & Salakhudinov, R. (2015, June). Unsupervised learning of video representations using LSTMs. In *Proceedings of the 32nd International Conference on Machine Learning* (pp. 843–852).
50. Niebles, J. C., Wang, H., & Fei-Fei, L. (2008). Unsupervised learning of human action categories using spatial-temporal words. *International Journal of Computer Vision, 79*(3), 299–318.
51. Lee, H., Grosse, R., Ranganath, R., & Ng, A. Y. (2011). Unsupervised learning of hierarchical representations with convolutional deep belief networks. *Communications of the ACM, 54*(10), 95–103.
52. Memisevic, R., & Hinton, G. (2007, June). Unsupervised learning of image transformations. In *2007 IEEE Conference on Computer Vision and Pattern Recognition* (pp. 1–8).
53. Andreae, J. H. (1995, November). The future of associative learning. In *Proceedings of the 1995 Second New Zealand International Two-Stream Conference on Artificial Neural Networks and Expert Systems* (pp. 194–197).

54. Hugh, A. J., & Lawrence, J. P. (1967). *U.S. Patent No. 3,355,713*. Washington, DC: U.S. Patent and Trademark Office.
55. Abbeel, P., & Ng, A. Y. (2004, July). Apprenticeship learning via inverse reinforcement learning. In *Proceedings of the Twenty-First International Conference on Machine Learning* (p. 1).
56. Wiering, M., & Van Otterlo, M. (2012). Reinforcement learning. *Adaptation, Learning, and Optimization, 12*, 3.
57. Ziebart, B. D., Maas, A. L., Bagnell, J. A., & Dey, A. K. (2008, July). Maximum entropy inverse reinforcement learning. In *Proceedings of the Twenty-third AAAI Conference on Artificial Intelligence* (Vol. 8, pp. 1433–1438).
58. Rothkopf, C. A., & Dimitrakakis, C. (2011, September). Preference elicitation and inverse reinforcement learning. In *Joint European Conference on Machine Learning and Knowledge Discovery in Databases* (pp. 34–48). Berlin, Heidelberg: Springer.
59. Anderson, M. J. (2009). *Carl Linnaeus: Father of classification*. Berkeley Heights, NJ: Enslow Publishing, LLC.
60. Ye, Q., Zhang, Z., & Law, R. (2009). Sentiment classification of online reviews to travel destinations by supervised machine learning approaches. *Expert Systems with Applications, 36*(3), 6527–6535.
61. Jain, P., Garibaldi, J. M., & Hirst, J. D. (2009). Supervised machine learning algorithms for protein structure classification. *Computational Biology and Chemistry, 33*(3), 216–223.
62. Unler, A., & Murat, A. (2010). A discrete particle swarm optimization method for feature selection in binary classification problems. *European Journal of Operational Research, 206*(3), 528–539.
63. Ball, C. A., & Tschoegl, A. E. (1982). The decision to establish a foreign bank branch or subsidiary: An application of binary classification procedures. *Journal of Financial and Quantitative Analysis, 17*(3), 411–424.
64. Ghamrawi, N., & McCallum, A. (2005, October). Collective multi-label classification. In *Proceedings of the 14th ACM International Conference on Information and Knowledge Management* (pp. 195–200).
65. Trohidis, K., Tsoumakas, G., Kalliris, G., & Vlahavas, I. P. (2008, September). Multi-label classification of music into emotions. In *ISMIR 2008* (Vol. 8, pp. 325–330).
66. Tsoumakas, G., & Vlahavas, I. (2007, September). Random k-labelsets: An ensemble method for multilabel classification. In *European Conference on Machine Learning* (pp. 406–417). Berlin, Heidelberg: Springer.
67. Hastie, T., Rosset, S., Zhu, J., & Zou, H. (2009). Multi-class AdaBoost. *Statistics and Its Interface, 2*(3), 349–360.
68. Ou, G., & Murphey, Y. L. (2007). Multi-class pattern classification using neural networks. *Pattern Recognition, 40*(1), 4–18.
69. Amit, Y., Fink, M., Srebro, N., & Ullman, S. (2007, June). Uncovering shared structures in multiclass classification. In *Proceedings of the 24th International Conference on Machine Learning* (pp. 17–24).
70. Huang, C., Li, Y., Change Loy, C., & Tang, X. (2016). Learning deep representation for imbalanced classification. In *Proceedings of the IEEE Conference on Computer Vision and Pattern Recognition* (pp. 5375–5384).
71. Garcı, S., Triguero, I., Carmona, C. J., & Herrera, F. (2012). Evolutionary-based selection of generalized instances for imbalanced classification. *Knowledge-Based Systems, 25*(1), 3–12.
72. Sáez, J. A., Luengo, J., Stefanowski, J., & Herrera, F. (2015). SMOTE–IPF: Addressing the noisy and borderline examples problem in imbalanced classification by a re-sampling method with filtering. *Information Sciences, 291*, 184–203.

73. Criminisi, A., Shotton, J., & Konukoglu, E. (2011). *Decision forests for classification, regression, density estimation, manifold learning and semi-supervised learning. Microsoft Research Technical Report TR-2011-114*, *5*(6), 12.
74. Amini, M. R., & Gallinari, P. (2002, July). Semi-supervised logistic regression. In *Proceedings of the 15th European Conference on Artificial Intelligence* (pp. 390–394).
75. Ritter, M. A., Harty, L. D., Davis, K. E., Meding, J. B., & Berend, M. E. (2003). Predicting range of motion after total knee arthroplasty: clustering, log-linear regression, and regression tree analysis. *The Journal of Bone and Joint Surgery*, *85*(7), 1278–1285.
76. Breiman, L., & Friedman, J. H. (1997). Predicting multivariate responses in multiple linear regression. *Journal of the Royal Statistical Society: Series B (Statistical Methodology)*, *59*(1), 3–54.
77. Zou, K. H., Tuncali, K., & Silverman, S. G. (2003). Correlation and simple linear regression. *Radiology*, *227*(3), 617–628.
78. Kleinbaum, D. G., Dietz, K., Gail, M., Klein, M., & Klein, M. (2002). *Logistic regression*. New York, NY: Springer-Verlag.
79. Menard, S. (2002). *Applied logistic regression analysis* (Vol. 106). Thousand Oaks, CA: Sage.
80. Demartines, P., & Hérault, J. (1997). Curvilinear component analysis: A self-organizing neural network for nonlinear mapping of data sets. *IEEE Transactions on Neural Networks*, *8*(1), 148–154.
81. Max, T. A., & Burkhart, H. E. (1976). Segmented polynomial regression applied to taper equations. *Forest Science*, *22*(3), 283–289.
82. Bendel, R. B., & Afifi, A. A. (1977). Comparison of stopping rules in forward "stepwise" regression. *Journal of the American Statistical Association*, *72*(357), 46–53.
83. Mahmood, Z., & Khan, S. (2009). On the use of k-fold cross-validation to choose cutoff values and assess the performance of predictive models in stepwise regression. *The International Journal of Biostatistics*, *5*(1).
84. Hoerl, A. E., Kannard, R. W., & Baldwin, K. F. (1975). Ridge regression: some simulations. *Communications in Statistics-Theory and Methods*, *4*(2), 105–123.
85. Fearn, T. (1983). A misuse of ridge regression in the calibration of a near infrared reflectance instrument. *Journal of the Royal Statistical Society: Series C (Applied Statistics)*, *32*(1), 73–79.
86. Hans, C. (2009). Bayesian lasso regression. *Biometrika*, *96*(4), 835–845.
87. Zou, H., & Hastie, T. (2005). Regularization and variable selection via the elastic net. *Journal of the Royal Statistical Society: Series B (Statistical Methodology)*, *67*(2), 301–320.
88. Ogutu, J. O., Schulz-Streeck, T., & Piepho, H. P. (2012, December). Genomic selection using regularized linear regression models: Ridge regression, lasso, elastic net and their extensions. *BMC Proceedings*, *6*(S2), S10).
89. Preacher, K. J., Curran, P. J., & Bauer, D. J. (2006). Computational tools for probing interactions in multiple linear regression, multilevel modeling, and latent curve analysis. *Journal of Educational and Behavioral Statistics*, *31*(4), 437–448.
90. Preacher, K. J., & Rucker, D. (2003). A primer on interaction effects in multiple linear regression. Retrieved November 10, 2003, http://www.quantpsy.org/interact/interactions.htm.
91. Zahir, N., & Mahdi, H. (2015). Snow depth estimation using time series passive microwave imagery via genetically support vector regression (case study Urmia Lake Basin). *The International Archives of Photogrammetry, Remote Sensing and Spatial Information Sciences*, *40*(1), 555.

92. Maroco, J., Silva, D., Rodrigues, A., Guerreiro, M., Santana, I., & de Mendonça, A. (2011). Data mining methods in the prediction of dementia: A real-data comparison of the accuracy, sensitivity and specificity of linear discriminant analysis, logistic regression, neural networks, support vector machines, classification trees and random forests. *BMC Research Notes*, *4*(1), 299.
93. Rossi, F., & Villa, N. (2006). Support vector machine for functional data classification. *Neurocomputing*, *69*(7–9), 730–742.
94. Chang, C. C., & Lin, C. J. (2011). LIBSVM: A library for support vector machines. *ACM Transactions on Intelligent Systems and Technology (TIST)*, *2*(3), 1–27.
95. Hsu, C. W., & Lin, C. J. (2002). A comparison of methods for multiclass support vector machines. *IEEE Transactions on Neural Networks*, *13*(2), 415–425.
96. Shawe-Taylor, J., & Cristianini, N. (2000). *Support vector machines* (Vol. 2). Cambridge: Cambridge University Press.
97. Huang, G. B. (2015). What are extreme learning machines? Filling the gap between Frank Rosenblatt's dream and John von Neumann's puzzle. *Cognitive Computation*, *7*(3), 263–278.
98. Dimla Sr, D. E., & Lister, P. M. (2000). On-line metal cutting tool condition monitoring.: II: tool-state classification using multi-layer perceptron neural networks. *International Journal of Machine Tools and Manufacture*, *40*(5), 769–781.
99. Lawrence, S., Giles, C. L., Tsoi, A. C., & Back, A. D. (1997). Face recognition: A convolutional neural-network approach. *IEEE Transactions on Neural Networks*, *8*(1), 98–113.
100. Socher, R., Lin, C. C., Manning, C., & Ng, A. Y. (2011). Parsing natural scenes and natural language with recursive neural networks. In *Proceedings of the 28th International Conference on Machine Learning (ICML-11)* (pp. 129–136).
101. Mikolov, T., Karafiát, M., Burget, L., Černocký, J., & Khudanpur, S. (2010). Recurrent neural network based language model. In *Eleventh Annual Conference of the International Speech Communication Association*.
102. Ma, X., Tao, Z., Wang, Y., Yu, H., & Wang, Y. (2015). Long short-term memory neural network for traffic speed prediction using remote microwave sensor data. *Transportation Research Part C: Emerging Technologies*, *54*, 187–197.
103. Sak, H., Shannon, M., Rao, K., & Beaufays, F. (2017, August). Recurrent neural aligner: An encoder-decoder neural network model for sequence to sequence mapping. In *Interspeech 2017* (Vol. 8, pp. 1298–1302).
104. Soltanolkotabi, M., Javanmard, A., & Lee, J. D. (2018). Theoretical insights into the optimization landscape of over-parameterized shallow neural networks. *IEEE Transactions on Information Theory*, *65*(2), 742–769.
105. Chopra, A., Prashar, A., & Sain, C. (2013). Natural language processing. *International Journal of Technology Enhancements and Emerging Engineering Research*, *1*(4), 131–134.
106. Manning, C. D., Manning, C. D., & Schütze, H. (1999). *Foundations of statistical natural language processing*. Cambridge, MA: MIT press.
107. Collobert, R., Weston, J., Bottou, L., Karlen, M., Kavukcuoglu, K., & Kuksa, P. (2011). Natural language processing (almost) from scratch. *Journal of Machine Learning Research*, *12*, 2493–2537.
108. Aone, C., Okurowski, M. E., Gorlinsky, J., & Larsen, B. (1999). A trainable summarizer with knowledge acquired from robust NLP techniques. In *Advances in automatic text summarization* (71–80). Cambridge, MA: MIT Press.
109. Cambria, E., Poria, S., Gelbukh, A., & Thelwall, M. (2017). Sentiment analysis is a big suitcase. *IEEE Intelligent Systems*, *32*(6), 74–80.
110. Khanna, A., Pandey, B., Vashishta, K., Kalia, K., Pradeepkumar, B., & Das, T. (2015). A study of today's AI through chatbots and rediscovery of machine intelligence. *International Journal of u- and e-Service, Science and Technology*, *8*(7), 277–284.

111. Hulth, A. (2004). *Combining machine learning and natural language processing for automatic keyword extraction*. Doctoral dissertation, Institutionen för data-och systemvetenskap (tills m KTH).
112. Pustejovsky, J., & Boguraev, B. (1993). Lexical knowledge representation and natural language processing. *Artificial Intelligence*, *63*(1–2), 193–223.
113. Gotti, F., & Langlais, P. (2016, May). Harnessing open information extraction for entity classification in a French corpus. In *Canadian Conference on Artificial Intelligence* (pp. 150–161). Cham: Springer.
114. Johansson, F., Brynielsson, J., & Quijano, M. N. (2012, August). Estimating citizen alertness in crises using social media monitoring and analysis. In *2012 European Intelligence and Security Informatics Conference* (pp. 189–196).
115. Alam, F., Ofli, F., & Imran, M. (2020). Descriptive and visual summaries of disaster events using artificial intelligence techniques: case studies of Hurricanes Harvey, Irma, and Maria. *Behaviour & Information Technology*, *39*(3), 288–318.

6 A Comprehensive Overview of Hybrid Renewable Energy Systems

Amit Kumer Podder, Muhammed Zubair Rahman, Sujon Mia, and S M Fuad Hossain Fahim
Department of Electrical and Electronic Engineering, Khulna University of Engineering & Technology, Khulna, Bangladesh

CONTENTS

6.1 INTRODUCTION

One of the coherent indications of the development of any country is the higher proportion of the population that gets access to electricity. Approximately 0.836 billion people, which represents nearly about 11% of the world population in 2018, are deprived of access to electricity [1]. Unfortunately, the population occupying this percentage are living mainly in rustic or remote areas. Access to electrical energy for these rural communities or remote areas is brought about by utilizing fossil fuel, which is detrimental to the environment [2]. Compared to fossil fuel, renewable energy resources (RESs) are free of cost, available, and environment friendly. In a broad sense, when we are talking about green sources, we can mention wind, solar, biomass, hydro, geothermal, and ocean energy. However, to apply these sources instead of conventional sources, there originate some flaws. In comparison to coal, gas, or other nonrenewable energy sources, the high initial cost and dependency on weather make the renewable energy sources uncertain and intermittent. To solve this problem and to yield a cost-efficient, reliable, and sustained distribution of electrical energy, two strategies can be thought: one is to gear up the research to ameliorate the technology, modules, and raw materials to minimize the setup cost [3] and another one is to assemble two or more sources in proper combination in place of one so that deficiency of one source can be replenished by the strengths of the rest and vice versa [4, 5]. Thus, the hybrid renewable energy systems (HRESs) come to the front (as shown in Figure 6.1), which is composed of two or more RESs, along with or without conventional energy sources, and work in standalone or on-grid mode [6].

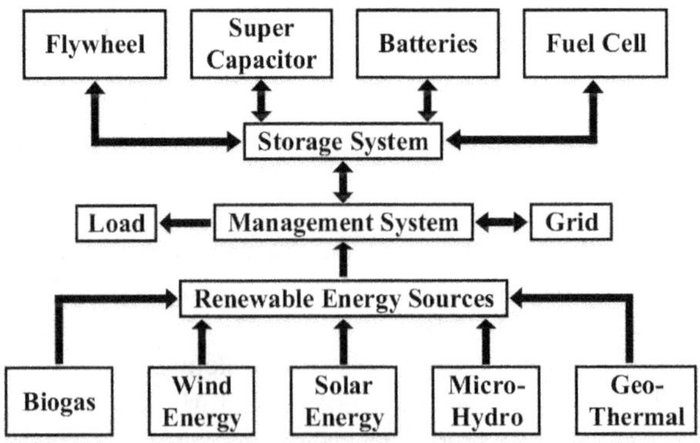

FIGURE 6.1 Basic block diagram of an HRES comprising renewable energy sources, storage systems, and management systems.

Since the HRESs are composed of multiple sources, storage devices, and consumers, the operations and optimization among components become quite complex. The configuration of the system also varies numerously, and it becomes perplexing to find out the appropriate one for a specified task. Furthermore, the stability of the system does not remain the same as in the case of the conventional interconnected power system. The maintenance of the HRESs also plays a crucial role in the sustainable operation of the system. Several optimization techniques based on classical and artificial intelligence (AI) are required for efficient operation. The performance of the system should also be verified before implementation. For this reason, detailed knowledge about the available performance-predicting simulation software is necessary. Considering all the above-mentioned facts, this chapter is designed to provide a comprehensive overview, challenges, and prospects of HRESs within a signal unit.

The chapter is organized into eight sections. The motivation behind the approach to HRESs rather than a single generation unit system is presented in Section 6.1. The definition, issues regarding HRESs, and pros and cons are detailed in Section 6.2. The configurations and stability issues and maintenance of the components of HRESs are described in Sections 6.3 and 6.4, respectively. The comparison among available optimization techniques and performance-predicting simulating software are presented in Sections 6.5 and 6.6, respectively. The challenges and prospects of HRESs are presented in Section 6.7. Finally, a conclusion is drawn in Section 6.8.

6.2 HRESs

HRESs consist of more than one RESs to raise the efficiency of the system and to provide better control in energy supply to meet the energy demand. However, in some worst periods of the year, it is vital to introduce some conventional sources (diesel or petrol generator) to make up the deficiency in the load demand. To make sure the system is stable during the time of overflow, the configuration must be integrated with a storage option and this can work as a backup. In broad, the systems with the efficient and alternative layout to the conventional energy production design, which consists of more than one nonrenewable energy systems and/or RESs along with storage support, and works in on-grid or off-grid mode can be termed as HRES [4, 7].

6.2.1 Issues Regarding HRESs

Since the availability of the renewable sources depends on the geographical position and weather, it is required to choose suitable sources certain in that locality. This can be done by properly analyzing a year-round database. Hence, the issues regarding HRESs are described in the following.

6.2.1.1 Information Assembly

The focus on designing the layout must take into consideration the rudimentary conditions of each locality such as present and future electricity demands, an abundance of natural resources, the supply of backup conventional fuels, existing infrastructure, and socioeconomic order of the people living there. Therefore, the database must include these.

6.2.1.2 Business-Friendly Analysis

Either government or any private agent installs such type of plants. Hence, specific economic features of the locality and adequate allocation of tariffs and subsidies must be kept in mind to assure a profit to the company or government after overcoming operation and management (O&M) cost while making it affordable to the customers [1].

6.2.1.3 Technological Viewpoint

This includes all types of infrastructure regarding generation and distribution, location of the modules, control equipment, and storage segment. However, properly trained manpower is also a key point to substantiate the plant from a technical point of view.

6.2.1.4 Operation, Maintenance, and Management

To ensure the longevity of HRES, it is important to keep a consistent eye by scrutinizing the components regularly; hybridizing more resources results in more complications. An optimal management plan must be proposed and uphold by the companies engaged with the plant.

6.2.1.5 Monitoring and Evaluation for Future

Since HRES consists of more than one source, there is more complexity and nonlinearity. Proper monitoring of the performance graph will provide enough information to evaluate the total system, rectify the delinquencies, and predict propositions for the future.

6.2.2 Pros and Cons of HRESs

HRESs are lucrative models used for several practices, particularly in off-grid power generation systems. An HRES can be developed to boost the utilization of renewable resources, ensuring a low-carbon-emission system [8]. It can be modeled to accomplish expectations at a negotiable cost since many researchers have already minimized the production cost of the modules and are certain that there will be a promising reduction shortly. Due to the independence of HRES from one energy source, the hybrid system draws more emphasis. Its complementary nature helps in replenishing deficiency of one source by the strength of the rest and vice versa. For example, during the monsoon, the lack of sunlight can be supplemented by the sound wind flow, microturbines, and fuel cells integrating heat and power operation, thereby increasing the overall plant efficiency [4].

However, a hybrid energy system introduces some hindrances also. Since it is developed by merging several energy sources and storage units, it is quite sophisticated in comparison to a single-source system. Again, its efficiency leans on the condition of weather. Sometimes, it requires storage units to support the system when there is overflow. The storage units, basically battery, need continuous monitoring and are of a short lifetime (at least present condition of batteries says so) resulting in increment of O&M cost. Disposition of batteries brings some negative impacts on the environment also. The system's stability can be disrupted by the sudden fluctuation of output power since the power is generated from different sources of HRES. Load sharing is often not associated with the scope or range of ratings of the sources.

6.3 CONFIGURATION OF HRES

The HRES configuration can be classified into several categories depending on the choice of common bus type, choice of converters, and integration scheme. They are briefly described in the following.

6.3.1 Choice of Common Bus Type

Various sources of energy can be coupled to a DC bus or an AC bus, or DC/AC bus, and can be categorized into the following types.

6.3.1.1 The Architectonics of DC Bus

In this hybrid energy system, power supplied from all the power sources are concentrated on a DC bus, which is shown in Figure 6.2(a). With the help of AC/DC inverter and DC/AC inverter, AC sources are connected with DC bus and DC bus supplies to AC loads, respectively. AC loads are covering from DC bus through an inverter with appropriate control action. To meet the peak-time loads, batteries are used as the reserve energy source. The foremost merit is the straightforward configuration of this system assortment [9].

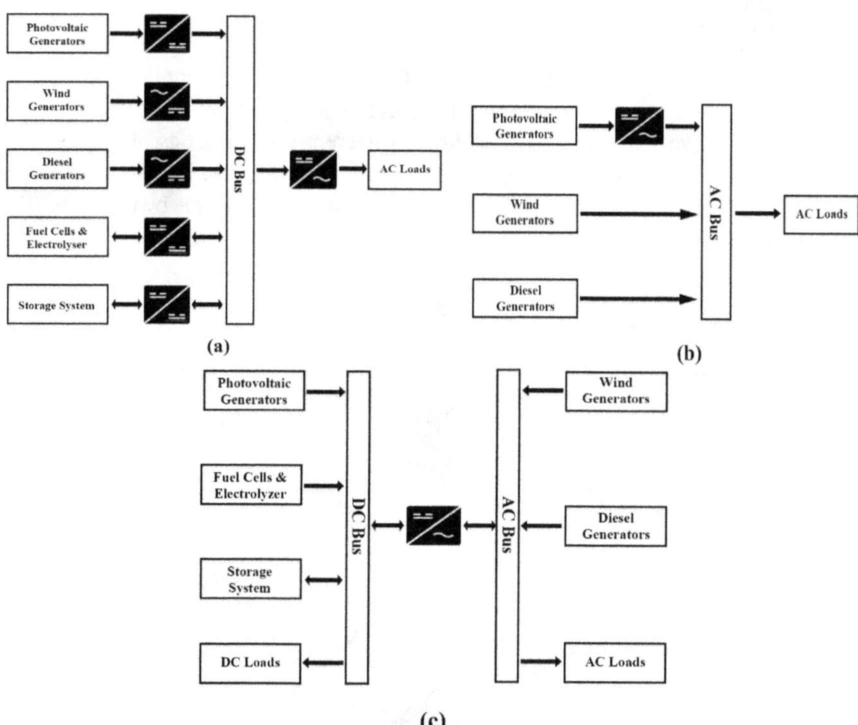

FIGURE 6.2 Configuration of hybrid system with different choices of common bus type. (a) The architectonics of DC bus, (b) the architectonics of AC bus, (c) the architectonics of DC/AC bus.

6.3.1.2 The Architectonics of AC Bus

This hybrid configuration in Figure 6.2(b) provides greater accomplishment compared with the former arrangement because every converter of this configuration can be synchronized with the generator. In the case of peak load hours or heavy load demand, other generators and storage devices perform in parallel. In the instance of synchronization between the charge voltages and the production voltages, this system is relatively complicated due to its parallel operation. Compared with DC-coupled topology, this system has some advantages also. For example, because of shorter dimensions, greater comprehensive efficiency, and reduction of conservative fetch and working time of the diesel generator, the best functioning of the diesel generator occurs [9].

6.3.1.3 The Architectonics of DC/AC Bus

Figure 6.2(c) shows the arrangement of DC and AC buses, which has greater functioning compared to the last two arrangements. In this topology, wind generators and diesel generators can supply as the crow flies to a bit of the AC loads. Bidirectional converter located between two buses and diesel generator can run in collateral or separately [9].

6.3.2 Choice of Converters

Power from the electrical energy sources to the receiver loads is supplied through the necessary power converters that convert required energy features. Figure 6.3 shows the different converters that are selected for different combinations of sources and loads. DC source covers DC and AC loads by using DC/DC and DC/AC converters, respectively. Similarly, the AC source also covers DC and AC loads by using AC/DC and AC/AC converters, respectively.

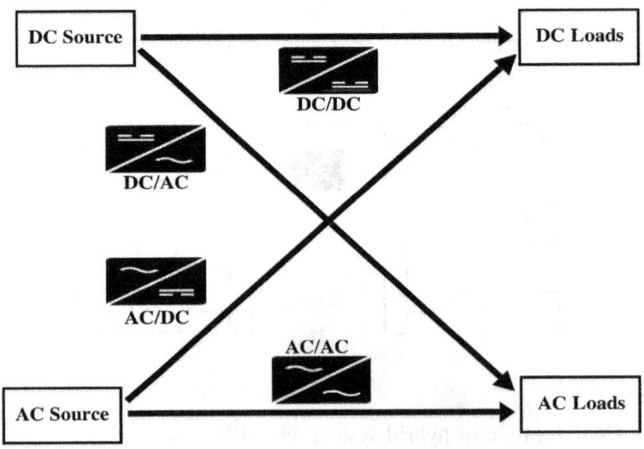

FIGURE 6.3 Sources and loads contributed by different converters.

6.3.3 Integration Scheme

Different possible configurations according to the integration of the energy sources are described in the following. There are mainly two parts: stand-alone hybrid systems and grid-connected systems [4].

6.3.3.1 Stand-Alone Hybrid Systems

In this hybrid system, different energy sources are integrated into a bus where the national grid is isolated from this hybrid system [10]. Therefore, this system is an individual hybrid system. All possible combinations are shown in Figure 6.4 and are described in the following three subsections, i.e., series hybrid system, parallel hybrid system, and switched hybrid system.

6.3.3.1.1 Series Hybrid System

Centralized DC bus and centralized AC bus are the two configurations of this hybrid system. Entire sources of energy, storage devices, and loads are coupled to a DC bus through proper electronic appliances such as rectifier, inverter, and DC–DC converter. This configuration is shown in Figure 6.4(a) and has different advantages, such as it removes the exigency for frequency and voltage regulations of the distinct wellspring of energies. When both the source and load are AC, this configuration performs in low efficiency because of the two-stage conversions of power in the way of passing [11].

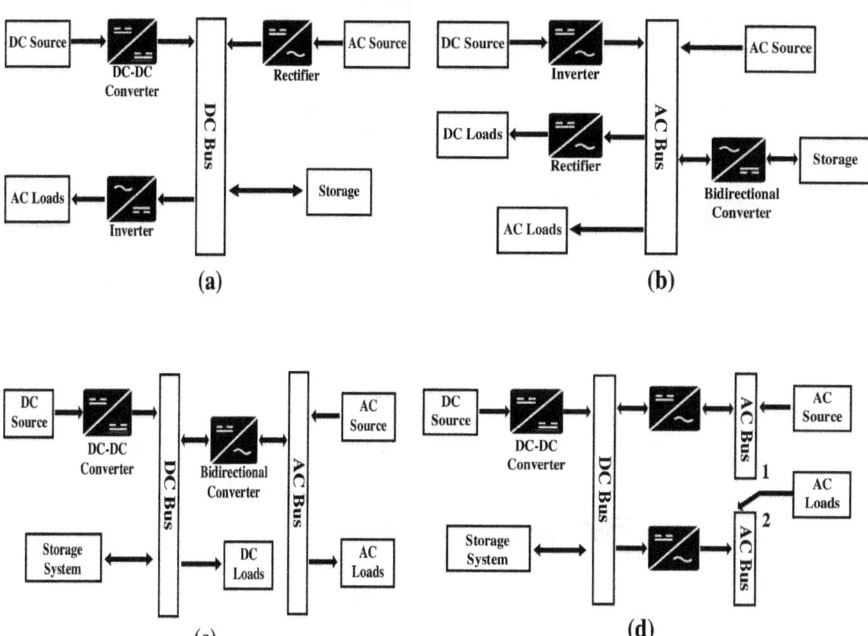

FIGURE 6.4 Different stand-alone hybrid system configurations. (a) DC series, (b) AC series, (c) parallel, (d) switched.

In Figure 6.4(b), the centralized AC bus shows that the entire sources of energy, loads, and storage devices are coupled to an AC bus through proper electronic devices. To run expanding energy stipulation and to maintain the voltage and frequency, the system needs synchronization between AC sources and inverters [12].

6.3.3.1.2 Parallel Hybrid System

Figure 6.4(c) shows the arrangement where the AC source and loads are attached to the AC bus directly and the DC source and DC loads are also attached to the DC bus directly. Between two buses, a two-way converter allows drift of energy in the middle of them [13].

6.3.3.1.3 Switched Hybrid System

The configuration structure of switched hybrid system is shown in Figure 6.4(d). It has some demerits such as only one source operates at a given instance and power supply gets interrupted during switching between sources.

6.3.3.2 Grid-Connected Systems

Three grid-connected hybrid arrangements are shown in Figure 6.5 wherein every system has a national grid connection. They are described in the following.

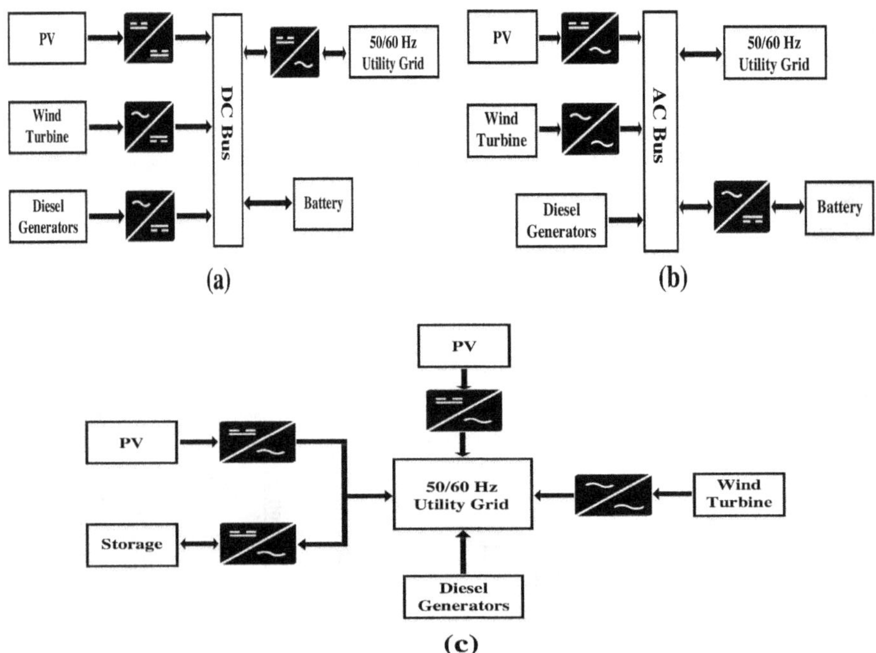

FIGURE 6.5 Different grid-connected configurations. (a) Centralized DC bus, (b) centralized AC bus, (c) distributed AC bus.

6.3.3.2.1 *Centralized DC Bus Architecture*

The centralized DC bus architecture is shown in Figure 6.5(a), where the entire sources of AC energy, such as diesel and wind generators, deliver power to the DC bus through rectifiers. Energy flows from DC bus to AC grid with the help of an inverter.

6.3.3.2.2 *Centralized AC Bus Architecture*

Figure 6.5(b) shows this system where all the energy sources and storage batteries are settled in one place and, before being connected to the grid, they are connected with AC bus through proper electronic devices.

6.3.3.2.3 *Distributed AC Bus Architecture*

Figure 6.5(c) shows the distributed AC bus architecture, where it is not necessary to settle the power sources near to each other; therefore, it does not have one main bus. Sources are connected to the grid supply from different distributed geographical positions and every source supplies power to the grid separately.

6.4 STABILITY ISSUES AND MAINTENANCE

The stability issues and maintenance of HRES are described in the following.

6.4.1 Stability Issues

The stability issue in a small distributed system is not the same as that in large interconnected systems. The individual components of a large system are small for the entire system. Therefore, sudden change of a generating unit or any change in the load consumption schedule does not cause an undesirable condition or change in system frequency because of the high inertia of the system [1]. In contrast, small HRES has low inertia and is more vulnerable to the individual elements of the system. Any change, decrease or increase, in the load demand and unavailability of any generation unit cause a significant change in system frequency, which makes the system unstable. This issue causes a challenging condition for penetrating a high amount of energy in HRES, as rainy days or bright sunlight in the solar photovoltaic (PV) system can quickly drop or increase energy production from the system [14]. Hence, to maintain the system's stability, the HRES should be capable of maintaining stability under any unbalanced conditions. The spinning reserve of the generation unit and the storage system or the battery, therefore, plays a key role in maintaining the system's stability based on the renewable energy penetration and architecture of the system. Moreover, the multilevel control strategy of the HRES helps in maintaining the system's stability and high power quality under the high and medium renewable energy penetration to the system [15]. The control strategy also helps in maintaining the reliability as well as the financial and environmental aspects of the system. An overall energy management system can ensure the optimum energy flow from the individual energy sources and maintain stability in the system.

6.4.2 Maintenance of HRES

For keeping the HRES in healthy condition and to determine any unwanted electrical or mechanical problems, regular maintenance is necessary. Sometimes, a successful project failed in a short duration due to lack of funds for maintenance. Therefore, an adequate maintenance fund should be considered in any successful project for conducting periodic maintenance over the long term. The maintenance of the HRES involves the maintenance of the different system components such as solar PV, wind turbine, diesel generator, battery, inverters, and so on. Some of these are discussed in the following.

6.4.2.1 Maintenance of Solar PV System

The solar PV system requires less maintenance than the other system components of HRES. The typical maintenance of the solar PV system is as follows [16–18]:

i. Washing the solar panel with cold water or specialized detergents to remove the dust and dirt from it.
ii. Visually inspecting for detecting any defects, lamination damage, or breakdown tempered glass due to any stress because the damage may increase the current leakage and eventually decrease the output.
iii. Checking the lamination of the module busbar connection to prevent degradation due to corrosion or moisture.
iv. Checking the condition of the module mounting frame to prevent degradation due to rusting of the bolt or insects chewing.
v. Checking and protecting the insulating joint of DC connectors from the salt, moisture, or dust and replacing the damaged connection with a new one.

6.4.2.2 Maintenance of Small Wind Turbine

Small wind turbines need more sophisticated maintenance than solar PV. The maintenance of it is presented in the following [19]:

i. Checking the screws or bolts and guy-wire once or twice a year.
ii. Greasing and oiling the moving parts at least twice a year.
iii. Inspecting the blades of the turbine to protect them from crack and stress caused by the moisture.

6.4.2.3 Maintenance of Diesel Generator

Diesel generators utilized in HRES usually have the capacity of 5 kW to several hundred kW. They require the following maintenance [1].

i. Checking the level of the fuel tank and often filling it.
ii. Changing the utilized oil in the engine every 4 months.
iii. Changing the element of fuel filter every 1000 hours of operation.

iv. Checking the condition of starting battery and its voltage at least every month.
v. Changing the fuel injectors and fuel injector pump every 1000 hours and 2000 hours of operation, respectively.
vi. Checking the loose connection of any connectors at least every month of operation.

6.4.2.4 Maintenance of Storage Systems

Storage systems such as lithium ion and lead acid battery require the following maintenance.

i. Battery life is highly influenced by its proper maintenance. Therefore, the state of charge and exterior temperature of the battery should be regularly checked to prevent lowering its life expectancy.
ii. The battery management system should be checked regularly to detect hidden danger.
iii. The reliability of connections of different connectors and the insulation of the cables should be inspected.

6.4.2.5 Maintenance of Power Electronic Components

The power electronic components include mainly the inverter DC/DC converter in HRES. These components require the following maintenance.

i. Removing the dirt or dust from the converter circuit, especially heatsink, by a brush or dry cloth.
ii. Checking the inverter functioning indicated by an LED at least once a month.
iii. Checking the fuses, grounding component, and mechanical connections regularly.
iv. Checking the exhaust fan's operation and the filter.

6.5 OPTIMIZATION TECHNIQUES FOR HRES

For maximizing the efficiency and decreasing the system tariff, different optimization techniques are needed. Complex problems of the system can be solved by using optimization methods [20, 21]. The traditional and AI-based optimization techniques are described in the following sections.

6.5.1 Traditional Optimization Techniques

The traditional optimization techniques include graphical construction, probabilistic and deterministic approach, and several classical techniques such as hill-climbing, linear programming, nonlinear programming, and dynamic programming. Their characteristics

TABLE 6.1
Different Traditional Optimization Techniques, Their Characteristics, and Applicable Fields

Optimization Techniques	Characteristics	Applicable Fields	Ref.
Graphical construction	Design variables are solved graphically	PV array and battery	[22–24]
Probabilistic approach	Provides an insight of the system utilizing assembled data	System of the hybrid combination	[25–27]
Deterministic approach	For resolving definite values, equations and constant parameters are used	Stand-alone PV with battery bank	[28, 29]
Classical techniques			
a. Linear programming model (LPM)	LPM studies in the linear objective function	Reliability and economic analysis of solar-wind system	[30]
b. Dynamic programming (DP)	A problem is solved as many subproblems and the solutions of subproblems are combined to find the problem solution		
c. Nonlinear programming (NLP)	Nonlinear parts restrain in the impartial function of the task		
d. Hill-climbing	Better solution found by incremental iteration from an arbitrary solution		[31]

along with their applicable fields are presented in Table 6.1, from which the researchers may identify the apposite one for the specified application.

6.5.2 AI-Based Optimization Techniques

AI-based optimization techniques generally include genetic algorithm (GA), particle swarm optimization (PSO), fuzzy logic, and artificial neural network (ANN) methods. Several metaheuristic methods such as simulated annealing (SA) and ant colony algorithm are also AI-based techniques. Hybrid techniques, i.e., combining different AI and metaheuristics methods, are utilized to enhance the optimization efficiency. Nowadays, the AI-based techniques play an important role in harnessing better output from the components of HRESs than do the traditional techniques. Table 6.2 presents the different AI, hybrid, and software-based optimization techniques, their characteristics, and applicable fields for easier understanding among methods.

TABLE 6.2
Different AI, Hybrid, and Software-Based Optimization Techniques, Their Characteristics, and Applicable Fields

Optimization Techniques	Characteristics	Applicable Fields	Ref.
Artificial intelligence			
a. GA	Variables of the function served as genetic ingredients	Hybrid renewable energy system (solar and wind) with a backup storage system	[32, 33]
b. PSO	This algorithm is like an appropriation algorithm for determining the finest and finest of the finest		[34, 35]
c. Fuzzy logic	An advanced quantum evolutionary process		[36]
d. ANN	This occupies the computational or mathematical model		[37]
Metaheuristic techniques			
a. SA	This works just like the metal-toughening operation	Size optimization for hybrid systems	[38]
b. Ant colony algorithm	Similar to the attitude of ants to maintain a certain fragrance for marking the shortest pavement toward food for other ants		[39]
Hybrid techniques			
a. SA–Tabu Search b. Monte Carlo simulation (MCS)-PSO hybrid iterative/GA c. Adaptive neuro-fuzzy inference system (ANFIS) d. ANN/GA/MCS e. PSO/DE (differential evolution)	For reducing the restraints of the single method from the above, some methods are integrated	Perform on the authentic and constructive remedy for hybrid renewable energy systems (HRESs)	[40–43]
Software-based			
HOMER	Designing the system in this simulation software with the requirement criteria	HRESs	[44, 45]

For designing optimal HRESs, different criteria have to be considered, which mainly are of two types: economic and technical. For minimizing the costs of HRESs, economic criteria are needed, and for ensuring reliability, efficiency, and environmental objectives, technical criteria are required. The optimization objectives for HRESs are presented in Table 6.3.

TABLE 6.3
Optimization Objectives for HRESs

Optimization Objectives	Optimization Parameters	Summary of Related Research Works: Optimization Techniques	Findings	Ref.
Cost optimization				
a. Energy cost minimization	Levelized cost of energy (LCE)	PSO, GA	Optimal models are designed for determining the LCE and reduce CO_2 emission with a combination of PV, battery storage, wind, and diesel generator	[46, 47]
b. Net present cost (NPC) minimization	NPC	PSO, ANN/GA	To meet the load demand, an optimized system is formed in India with a combination of diesel generator, solar, biomass, hydro, and biogas energy. Another system designed in Nigeria with India is PV/diesel/battery combination	[48, 49]
c. Other cost-related optimization	Levelized unit electricity cost (LUEC), life cycle cost (LCC), annualized cost of the system (ACS), average generation cost (AGC), total cost of the system (TCS), and capital cost (CC).	GA, PSO, multiobjective programing (MOP)	Different optimizations are designed to incorporate reliability and cost models, and calculate the size of battery storage sweep for different hybrid systems	[50, 51]
Technical optimization	Loss of power supply probability (LPSP) or loss of load probability (LOLP), cost/efficiency ratio, carbon emissions, power availability	GA, PSO, multiobjective programing (MOP), multiobjective genetic algorithm (MOGA)	Optimal sizing of HRES and maximum worth of the hybrid system occur when the LOLP/LPSP is in smaller level The reliability factor is determined for the PV-wind hybrid system	[52, 53]

6.6 SIMULATION SOFTWARE

The HRES is typically composed of two or more RESs for maintaining the continuous supply of electricity. For harnessing the optimized energy from the sources, different optimized techniques are applied in the system. Hence, simulation software is necessary for foreseeing the system performance before implementing them in real time. Several simulation software, as shown in Figure 6.6, are available for the size, financial, technical, and environmental analysis of the system. In this chapter, the most utilized, recent, and popular 14 software, namely, HOMER (Hybrid Optimization Model for Electrical Renewables), Hybrid2, RETScreen, HOGA (Hybrid Optimization by Genetic Algorithm), TRNSYS (Transient system), Bluesol, SAM (System Advisor Model), PVComplete, Aurora, HelioScope, PV*SOL, SOLARGIS, PVGIS, and SISIFO are compared.

A comparison among the 14 discussed simulation software for HRESs is presented in Table 6.4, which aims to help the researcher to select appropriate software for the specified research application.

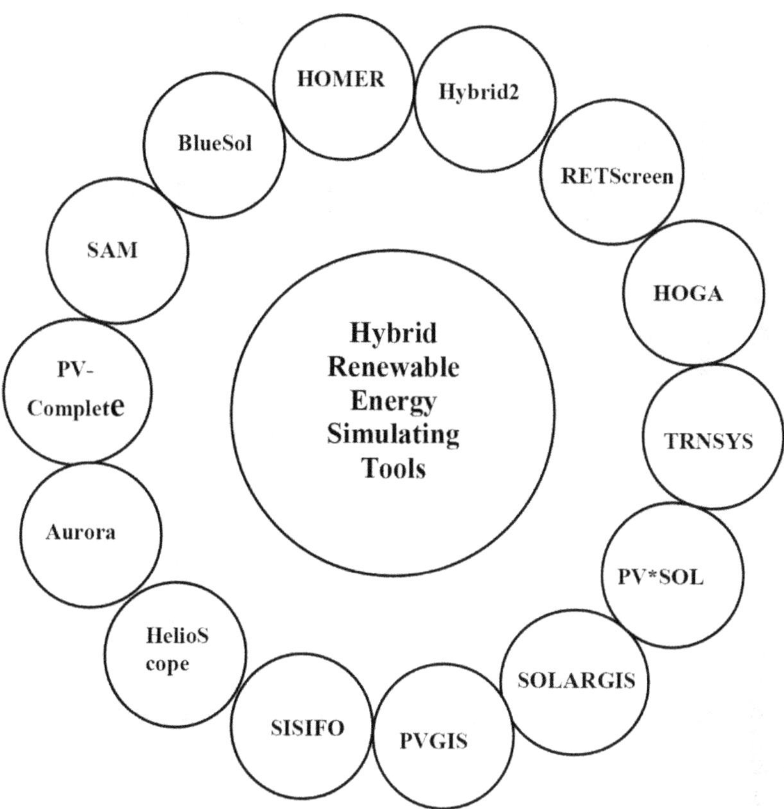

FIGURE 6.6 Fourteen recently available hybrid renewable energy-simulating tools.

TABLE 6.4
Comparison of 14 Simulation Software for HRESs [54, 55]

Software	Developed by	Latest Version	Computer Platform	Free/Prized	Analysis Facility	Availability
HOMER	NREL	HOMER Pro Version 3.13.8	Windows 10	Prized with a 21-day trial facility	Simulation, optimization, and sensitivity	www.homerenergy.com
Hybrid2	RERL	Version 1.0	Windows XP	Free	Performance and economics	https://www.umass.edu/windenergy/research/topics/tools/software/hybrid2
RETScreen	Government of Canada	Version 8	Windows 10	Prized	Assessment and optimization	http://www.retscreen.net/
HOGA	José L. Berna Agustín and Rodolfo Dufo-López, University of Zaragoza	iHOGA 2.3	Windows 7	Free	Simulation and optimization	http://www.unizar.es/rdufo/hoga-eng.htm
TRNSYS	University of Wisconsin	TRNSYS 18	Windows 10	Prized	Simulation on transient systems	http://www.trnsys.com/
PV*SOL	Berlin-based Valentin	PV*SOL premium 2020	Windows 10	PV*SOL premium is paid but PV*SOL is free	2D and 3D modeling and shading analysis	https://www.solardesign.co.uk/softwareindex.php
PVGIS	Joint Research Center (JRC)		Windows/Mac	An online free web application	Production and calculation of energy	https://ec.europa.eu/jrc/en/pvgis
HelioScope	Folsom Labs		Windows/Mac	Prized	Photovoltaic sizing, designing, and selling	https://www.helioscope.com/

Software	Developed by	Latest Version	Computer Platform	Free/Prized	Analysis Facility	Availability
BlueSol	CadWare	BlueSol V. 4.0.005	Windows XP, Vista, Windows 7, and Windows 8	Prized	Designing and optimizing photovoltaic (PV) systems	http://www.bluesolpv.com/dnnsite/Download.aspx
SOLARGIS	Marcel Suri and Tomas Cebecauer (founder)	Web-based	Windows and Mac	Prized	Weather data, online tools, and information for an effective solar energy assessment	https://solargis.com/pricing/products-and-plans
SISIFO	IES-UPM	Web-based	Windows and Mac	Free	Simulation of different kinds of PV systems and for increasing its bankability	https://www.sisifo.info/
Aurora	Sam Adeyemo (co-founder)	Aurora 2020	Windows and Mac	Prized	Shading analysis, simulating the performance of the PV system	https://www.aurorasolar.com/
SAM	NREL	SAM 2020.2.29	Windows 10/8 (64-bit)	Free	Model both renewable systems and financial strategy	https://sam.nrel.gov/download
PVComplete	Claudia Eyzaguirre and Daniel Sherwood (founders)	PVCAD, PVCAD MEGA, PVSKETCH	Windows and Mac	Prized	Precise solar engineering, intuitive design, and building multiple-megawatt layouts	https://pvcomplete.com/

6.7 CHALLENGES AND FUTURE TRENDS

The involvement of renewable energies to generate electrical power in hybridized form is strong evidence for the sustainability of the environment and healing factor to the damages of climate change, greenhouse effect, ozone layer degradation, etc. But still there are several hindrances that prevent these systems to be more efficient and optimal over conventional fuels.

Some of the challenges faced by the designers and the future scopes for the researchers are as follows:

i. The most efficient solar panels on the market available today have an efficiency of approximately 23% [18]. But the average efficiency of solar panels falls between 17 and 19% [10]. Again, fuel cells are 40–60% efficient [5].
ii. First of all, it is necessary to carry out vast research and investment to accelerate the efficiency of the modules or devices. Advanced technologies are needed to extract more power from these renewable sources.
iii. Some researchers have come up with the idea of completely rewiring homes to run on DC supply and proved the certainty of DC microgrid for localized loads because there has been a citable development on the field of equipment and household appliances that run on DC. Paying urgent heed to its technical and economic feasibility can depict a suitable future extent.
iv. High capital cost puts a significant increment of payback time. To lessen the payback time, there must be a collapse in the manufacturing cost.
v. Power loss is a factor in the percentage of efficiency. The less power will be lost, the more the performance will be.
vi. It is recommended to carry out transient analysis of the hybrid system by step changes in the varying parameters, such as solar radiation, wind speed, water flow, the temperature of the superheated fluid, and load demand for fixing stability issues of HRES.

6.8 CONCLUSIONS

This chapter mainly focuses on discussing the meticulous overview of the HRES, which is believed to be an indispensable source for the graduate and undergraduate students and researchers to learn a comprehensive knowledge about HRESs with a single article. The chapter recapitulates an overall brief of HRES, including the reasons for choosing HRES, an overview of different components in the design, unit sizing and optimization, storage and control over energy flow, the factors that make it optimal, and some examples. Overview of simulation software and optimization techniques of HRES is also elaborately discussed. The stability issues and maintenance (i.e., operation and maintenance of solar, wind, diesel set, battery, and so on) of HRES are briefly mentioned. Finally, the challenges and future trends of HRES are presented.

REFERENCES

1. ADB. (2017). Deployment of hybrid renewable energy systems in mini-grids. http://dx.doi.org/10.22617/TIM178889-2
2. Podder, A. K., Roy, N. K., & Pota, H. R. (2019). MPPT methods for solar PV systems: A critical review based on tracking nature. IET Renewable Power Generation, 13(10), 1615–1632.
3. Podder, A. K., Hasan, M. R., Roy, N. K., & Komol, M. M. R. (2019). Economic analysis of a grid-connected PV system: A case study in Khulna. European Journal of Engineering Research and Science, 3(7), 16–21.
4. Ibrahim, M., Khair, A., & Ansari, S. (2015, August). A review of hybrid renewable energy systems for electric power generation. International Journal of Engineering Research and Applications, 5(8), 42–48.
5. Podder, A. K., Ahmed, K., Roy, N. K., & Habibullah, M. (2019). Design and simulation of a photovoltaic and fuel cell-based micro-grid system. In: International Conference on Energy and Power Engineering: Power for Progress, ICEPE 2019, 1–6.
6. Lazarov, V. D., Notton, G., Zarkov, Z., & Bochev, I. (2005). Hybrid power systems with renewable energy sources types, structures, trends for research, and development. In: Eleventh International Conference on Electrical Machines, Drives and Power Systems ELMA 2005, 515–520.
7. Negi, S., & Mathew, L. (2014). Hybrid renewable energy system: A review. International Journal of Electronic and Electrical Engineering, 7, 535–542.
8. Bappy, F. I., Islam, M. J., Podder, A. K., Dipta, D. R., Faruque, H. M. R., & Hossain, E. (2019). Comparison of different hybrid renewable energy systems with optimized PV configuration to realize the effects of multiple schemes. In: First International Conference on Advances in Science, Engineering and Robotics Technology (ICASERT), 1–6.
9. Rekioua, D. (2020). Hybrid renewable energy system overview. In: Rekioua, D. (Ed.), Hybrid Renewable Energy Systems: Optimization and Power Management Control (pp. 1–37). Springer, New York.
10. Swarnakar, S. C., Podder, A. K., & Tariquzzaman, M. (2020). Solar, fuel cell and battery based hybrid energy solution for residential appliances. In: Fourth International Conference on Electrical Information and Communication Technology (EICT), 1–6.
11. Reddy, Y. J., Kumar, Y. V. P., Raju, K. P., & Ramsesh, A. K. (2012, December). Hybrid power system design with renewable energy sources for buildings. IEEE Transactions on Smart Grid, 3(4), 2174–2187.
12. Reddy, Y. J, Kumar, Y.V. P., & Raju, K. P. (2011, September). Real time and high fidelity simulation of hybrid power system dynamics. In: 2011 IEEE Recent Advances in Intelligent Computational Systems, 890–895.
13. Solanki, C. S. (2011). Solar Photovoltaics: Fundamentals, technologies and Applications. Second edition. Prentice Hall of India, Delhi.
14. Podder, A. K., & Habibullah, M. (2018). Model predictive based energy efficient control of grid-connected PV systems. In: 10th International Conference on Electrical and Computer Engineering (ICECE), 413–416.
15. Podder, A. K., Habibullah, M., & Roy, N. K. (2019). Current THD analysis of model predictive control based grid-connected PV inverter. In: International Conference on Electrical, Computer and Communication Engineering (ECCE), 1–6.
16. Brooks, W., & Dunlop, J. (2013). PV Installation Professional Resource Guide. NABCEP, Clifton Park, NY.
17. United States Agency for International Development. (2013). Solar PV System Maintenance Guide. Washington, DC.

18. Podder, A. K., Ahmed, K., Roy, N. K., & Biswas, P. C. (2017). Design and simulation of an independent solar home system with battery backup. In: 4th International Conference on Advances in Electrical Engineering (ICAEE), 427–431.
19. Clarke, S. (2003). Electricity generation using small wind turbines at your home or farm. Rural Environment/OMAFRA, 1–16.
20. Bhandari, B., Lee, K. T., Lee, G. Y, Cho, Y. M., & Ahn, S. H. (2015, January) Optimization of hybrid renewable energy power systems: A review. International Journal of Precision Engineering and Manufacturing-Green Technology, 2(1), 99–112.
21. Ghofrani, M., & Hosseini, N. N., (2016). Optimizing hybrid renewable energy systems: A review. IntechOpen.
22. Borowy, B. S., & Salameh, Z. M. (1996). Methodology for optimally sizing the combination of a battery bank and PV array in a wind/PV hybrid system. IEEE Transactions on Energy Conversion, 11(2), 367–375.
23. Markvart, T. (1996). Sizing of hybrid photovoltaic-wind energy systems. Solar Energy, 57(4), 277–281.
24. Markvart, T., Fragaki, A., & Ross, J. (2006) PV system sizing using observed time series of solar radiation. Solar Energy, 80(1), 46–50.
25. Karaki, S., Chedid, R., & Ramadan R. (1999) Probabilistic performance assessment of autonomous solar-wind energy conversion systems. IEEE Transactions on Energy Conversion, 14(3), 766–772.
26. Tina, G., Gagliano, S., & Raiti, S. (2006). Hybrid solar/wind power system probabilistic modeling for long-term performance assessment. Solar Energy, 80 (5), 578–588.
27. Posadillo, R., & López Luque, R. (2008). Approaches for developing a sizing method for stand-alone PV systems with variable demand. Renewable Energy, 33(5), 1037–1048.
28. Bhuiyan, M., & Ali, A. M. (2003). Sizing of a stand-alone photovoltaic power system at Dhaka. Renewable Energy, 28(6), 929–938.
29. Bhandari, R., & Stadler, I. (2011). Electrification using solar photovoltaic systems in Nepal. Applied Energy, 88(2), 458–465.
30. Siddaiah, R., & Saini, R. (2016). A review on planning, configurations, modeling, and optimization techniques of hybrid renewable energy systems for off-grid applications. Renewable and Sustainable Energy Reviews, 58, 376–396.
31. Kellogg, W., Nehrir, M., Venkataramanan, G., & Gerez, V. (1996). Optimal unit sizing for a hybrid wind/photovoltaic generating system. Electric Power Systems Research, 39 (1), 35–38.
32. Shahirinia, A. H., Tafreshi, S. M. M, Gastej, A. H., & Moghaddomjoo, A. R. (2005). Optimal sizing of hybrid power system using genetic algorithm. In: International Conference on Future Power Systems, Amsterdam, 1–6.
33. Hossam-Eldin, A., El-Nashar, A. M., & Ismaiel, A. (2012, August). Investigation into economical desalination using optimized hybrid renewable energy system. Electrical Power and Energy Systems, 43, 1393–1400.
34. Bai, Q. (2010, February). Analysis of particle swarm optimization algorithm. Computer and Information Science, 3(1), 180–184.
35. Huang Y., Xu Y., & Zhou X. (2011). Study on wind-solar hybrid generating system control strategy. In: International Conference on Multimedia Technology, Hangzhou, 773–776.
36. Petalas, Y. G., Parsopoulos, K. E., Papageorgiou, E. I., Groumpos, P. P., & Vrahatis, M. N. (2007). Enhanced learning in fuzzy simulation models using memetic particle swarm optimization. In: Proceedings of the 2007 IEEE Swarm Intelligence Symposium, Honolulu, HI, 16–22.
37. Pranav, M. S., Karunanithi, K., Akhil, M., Saravanan, S., Afsal, V. M., & Krishnan, A. (2017). Hybrid renewable energy sources (HRES) - A review. In: International Conference on Intelligent Computing Instrumentation and Control Technologies (ICICICT), 162–165.

38. Fung, C. C., Ho, S. C. Y., & Nayar, C. V. (1993). Optimization of a hybrid energy system using simulated annealing technique. In: Proceedings of the 1993 IEEE Region 10 Conference on Computer, Communication, Control and Power Engineering, Beijing, 235–238.
39. Wu, Y., Lee, C., Liu, L., & Tsai, S. (2010). Study of reconfiguration for the distribution system with distributed generators. IEEE Transactions on Power Delivery, 25(3), 1678–1685.
40. Sunanda, S., & Chandel, S. S. (2015). Review of recent trends in optimization techniques for solar photovoltaic–wind based hybrid energy systems. Renewable and Sustainable Energy Reviews, 50, 755–769.
41. Katsigiannis, Y. A., Georgilakis, P. S., & Karapidakis, E. S. (2012). Hybrid simulated annealing Tabu search method for optimal sizing of autonomous power systems with renewable. IEEE Transactions on Sustainable Energy, 3(3), 330–338.
42. Khatib, T., Mohameda, A., & Sopian, K. (2012). Optimization of a PV/wind micro-grid for rural housing electrification using a hybrid iterative/genetic algorithm: Case study of Kuala Terengganu, Malaysia. Energy and Buildings, 47, 321–333.
43. Abbes, D., Martinez, A., & Champions, G. (2014). Life cycle cost embodied energy and loss of power supply probability for the optimal design of hybrid power systems. Mathematics and Computers in Simulation, 98, 46–62.
44. Anayochukwu, A. V., & Nnene E. A. (2013). Simulation and optimization of photovoltaic/diesel hybrid power generation systems for health service facilities in rural environments. Electronic Journal of Energy & Environment, 1(1), 57–70.
45. Lal, D. K., Dash, B. B., & Akella, A. (2011). Optimization of PV/wind/micro-hydro/diesel hybrid power system in Homer for the study area. International Journal on Electrical Engineering and Informatics, 3(3), 307–325.
46. Amer, M., Namaane, A., & M'Sirdi, N. K. (2013). Optimization of hybrid renewable energy systems (HRES) using PSO for cost reduction. Energy Procedia, 42, 318–327.
47. Bilal, B. O., Sambou, V., Kebe, C. M. F., Ndiaye, P. A., & Ndongo, M. (2012). Methodology to size an optimal stand-alone PV/wind/diesel/battery system minimizing the levelized cost of energy and the CO_2 emissions. Energy Procedia, 14, 1636–1647.
48. Olatomiwa, L., Mekhilef, S., Huda, A. S. N., & Ohunakin, O. S. (2015). Economic evaluation of hybrid energy systems for rural electrification in six geopolitical zones of Nigeria. Renewable Energy, 83, 435–446.
49. Lujano-Rojas, J. M., Dufo-Lopez, R., & Bernal-Agustín, J. L. (2014). Technical and economic effects of charge controller operation and columbic efficiency on stand-alone hybrid power system. Energy Conversion and Management, 86, 709–716.
50. Hongxing, Y., Zhou, W., & Chengzhi, L. (2009) Optimal design and techno-economic analysis of a hybrid solar–wind power generation system. Applied Energy, 86, 163–169.
51. Kaabeche, A., Belhamel, M., & Ibtiouen, R. (2011) Sizing optimization of grid-independent hybrid photovoltaic/wind power generation system. Energy, 36, 1214–1222.
52. Shadmand, M. B., & Balog, R. S. (2014). Multi-objective optimization and design of photovoltaic wind hybrid system for community smart DC micro-grid. IEEE Transactions on Smart Grid, 5(5), 2635–2643.
53. Shivarama Krishna, K., & Sathish Kumar, K. (2015). A review on hybrid renewable energy systems. Renewable and Sustainable Energy Reviews, 52, 907–916.
54. Sinha, S., & Chandel, S. S. (2014). Review of software tools for hybrid renewable energy systems. Renewable and Sustainable Energy Reviews, 32, 192–205.
55. Kaur, D., & Cheema, P. S. (2017). Software tools for analyzing the hybrid renewable energy sources: a review. In: International Conference on Inventive Systems and Control (ICISC), 1–4.

7 Dynamic Modeling and Performance Analysis of Switched-Mode Controller for Hybrid Energy Systems

Linnet Jaya Savarimuthu and Kirubakaran Victor
Centre for Rural Energy, Gandhigram Rural Institute-Deemed to be University, Gandhigram, Tamil Nadu, India

CONTENTS

7.1 INTRODUCTION

Many rural and remote parts of the country still do not have a reliable electricity supply. Gradually, the numbers of electrical appliances are increasing and, correspondingly, the requirement of energy is also increasing. However, the power sources are not growing in the same ratio. This leads to the energy crisis in the country. With the availability of many renewable energy resources, like, solar photovoltaic (PV), wind, solar thermal, and biomass, there is a possibility of validating this energy deficiency, especially in rural and remote areas. Hence, a microgrid with one or more renewable energy sources can be developed to generate and consume power locally [1]. Hybridizing energy resources can definitely improve the system performance and can also enhance the system dynamics effectively. Therefore, hybrid renewable energy conversion systems are well matched for applications in which the average power demand is low, while the load dynamics are comparatively elevated [2]. The integration of the utility grid and renewable energy sources allows controlling the power flow efficiently. A lot of effort is going on to enhance the power efficiency of nonconventional sources and make them more reliable and beneficial [3]. When two or more resources of electricity are coupled to a common grid and operate together to deliver the desired load, then the system becomes a hybrid electric power system. Even though hybrid renewable energy systems have many advantages, there are few problems and issues associated with hybrid energy systems that have to be dealt with [4]. The majority of hybrid power systems necessitate a storage campaign where the usage of batteries is common. Such batteries need continuous monitoring and the cost is increasing rapidly, while the battery's lifetime is restricted to a few years. Therefore, it is to report that the practice of the battery usage is supposed to enhance for long time period for the economical consideration of hybrid renewable energy systems [5].

An eternally growing demand for conventional resources such as oil, natural gas, and coal is motivating the civilization towards the study, research, and development of renewable sources of energy. Renewable sources of energy are favored as they are environmentally friendly. The renewable energy-based distributed generators offer a leading task in electric power generation, for the aim of the cleaner energy solution [6]. The combination of renewable resources of energy to form a hybrid power system, which includes two or more renewable energy resources, is an incredible selection for the production of distributed energy systems [7]. Electric power ranges of renewable energy production are minimum; therefore, the renewable energy generation technologies are sited close to the load or associated with the utility grid [8]. The urban areas are distant away from the central grid network and the connection is achievable only by means of a weak transmission line. The perception of renewable energy integration is considered as a valuable solution to meet our energy demands [9]. Apart from unique mathematical models used in analyzing the behaviors of hybrid renewable energy systems, simulation and development software have also been established for a valuable variety of applications which comprises the control strategy, design, multiobjective optimization, and economic optimization. Also, the controller can be introduced to operate with reference to the state of charge and the probable input conditions [10].

7.2 PROPOSED HYBRID EB/PV MODELING

Electric power production from renewable energy resources has to enhance its significance to reach the goal of Sustainable Energy for All (SEforALL), where by 2030 the contribution of renewable energy should become double in the global energy mix-up. The emergent proportion of renewable energy resources (particularly wind and solar) in the power system promotes these challenges [11]. In various countries, the accesses of renewable energy in the utility are possible technically and economically; mainly, the growth of solar energy and wind energy technologies reaches the consistency of a utility while considering the economic conditions. The system makes use of locally existing renewable energy sources and end-use technologies for satisfying various needs, which include lightning domestic and communities, communication and educational mechanism, low-grade thermal energy and storage, compact water supply and cooking, and small-scale industries [12, 13]. The pathway of hybridization has to make suspicious and intentional planning for fulfilling the demands with existing energy sources in order to exploit benefits and better efficiencies to the end users [14].

Both electricity board (EB) and solar PV are utilized for the purpose of hybrid renewable energy power generation, although a battery technology may employ storage of electrical energy if needed. To harvest a maximum of the available energy, a hybrid EB–PV generation topology utilizes the outcomes of both solar PV and EB. The developed solar hybrid power system uses solar power from a PV system with an additional power-generating energy source. The flow of electric power can be monitored with the help of a load flow bus, which is connected to the utility buildings by means of a controlled switch. All the energy resources are modelled using MATLAB/Simulink library, a simulation software tool to analyze and predict their behavior. An effortless control technique tracks the maximum amount of power from the EB/solar energy resource to accomplish much higher generating power factors. The simulation results reveal the feasibility and consistency of this proposed system.

7.2.1 Proposed Block/System Diagram

The proposed system configuration permits the two power sources to provide the load individually or concurrently that depends on the energy resources available and also with the help of controlling the switching element. The hybrid renewable energy system (EB-PV) topology is represented in Figure 7.1. Energy from the solar PV is

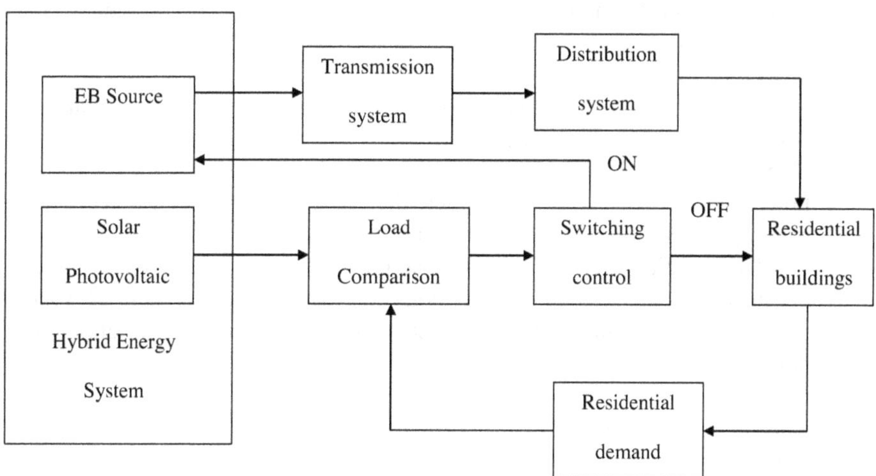

FIGURE 7.1 Hybrid renewable energy system (EB-PV) topology.

available here the whole time; however, the insolation levels of a solar PV system may vary because of the range of sunbeams and unpredictable dimness transmit by means of cloudiness and the natural world, etc. [15]. However, by combining the alternating resources, the entire system's power transmits can exist effectively and the consistency can be enhanced significantly. When the resources are not available or inadequate to meet the load demands, other energy resources are added to compensate for the variation.

7.2.2 Requirement-Based Control System

From a load flow bus, the requirement of residential demand will be sensed and a comparator is engaged to analyze the solar PV radiation intensity. If the available solar intensity is enough for pleasing the residential demand, the grid line will be bypassed and directly the solar PV system will provide the required load to the residential building. On the other hand, if the available solar intensity is not sufficient to satisfy the residential demand completely, the load comparator senses the given condition (the given condition is whenever the residential demand is greater than or equal to solar power) and turns ON the switching control. The switching control system helps us to get the required power from the EB source by switching over solar power to the EB source. To evaluate hybrid renewable energy system configurations techno-economically, it is necessary to develop a suitable model for optimization and simulation [16]. A demand-side assessment has to be analyzed and the model is designed through simulation software. Moreover, the assessment of roof space should be taken care of to simulate the model if incorporated with any other renewable energy sources [17].

7.3 MODELING OF THE DOMESTIC HYBRID RENEWABLE ENERGY SYSTEM

An electric power grid system comprises three segments which are to be considered for modeling. They are as follows:

i. Power generation sector where the generated power from the power station is transmitted with specific stepped-up voltage to the ultimate consumers by means of transmission lines.
ii. Power transmission sector at which the transmission lines connect the power generation sector and load distribution sector.
iii. Load distribution sector where the various categories of load, which include commercial load, residential load, domestic load, etc., are connected with the transmission lines which are established based on the variation of demand.

If the energy is not conserved and managed appropriately, huge trouble of energy crisis may occur [18]. Simulation, modeling, and implementation were executed in this study to build up a hybrid renewable energy system to meet the electric loads of residential buildings. For testing and analyzing, consider three residential buildings. The electric grid system was modelled along with a balanced three-phase source by holding Y grounded connection. The electrical cables and overhead lines were implemented in the three-phase transmission line section. The line provisions such as resistance (R), inductance (L), and capacitance (C) are defined by positive and zero-sequence stipulations per unit length, which resulted in coupling among the three-phase conductors and also the ground parameters. The three phases get balanced while defining the line parameters. The model of three-phase transformers with two windings (Y and delta) was utilized for the entire transformers of the network, where Y is grounded for high voltage and delta stands for medium voltage. The residential loads are designed with parallel RLC circuits which are represented in terms of active and reactive powers. Both the EB and solar PV are utilized for the hybrid renewable power generation system. The developed solar hybrid power system uses solar power from a PV system with an additional power-generating energy source. The modeling of a hybrid EB-solar power system, displayed in Figure 7.2, includes a solar PV and EB supply as an input source to satisfy the residential load according to the specified conditions. When any of the input resources are not sufficient to meet the load demands, the added energy resources compensate for the variation.

For modeling a solar power system, the irradiation of 1000 W/m^2 and temperature of 25°C have been considered as a standard testing condition and given as an input source for getting output power in terms of voltage and current.

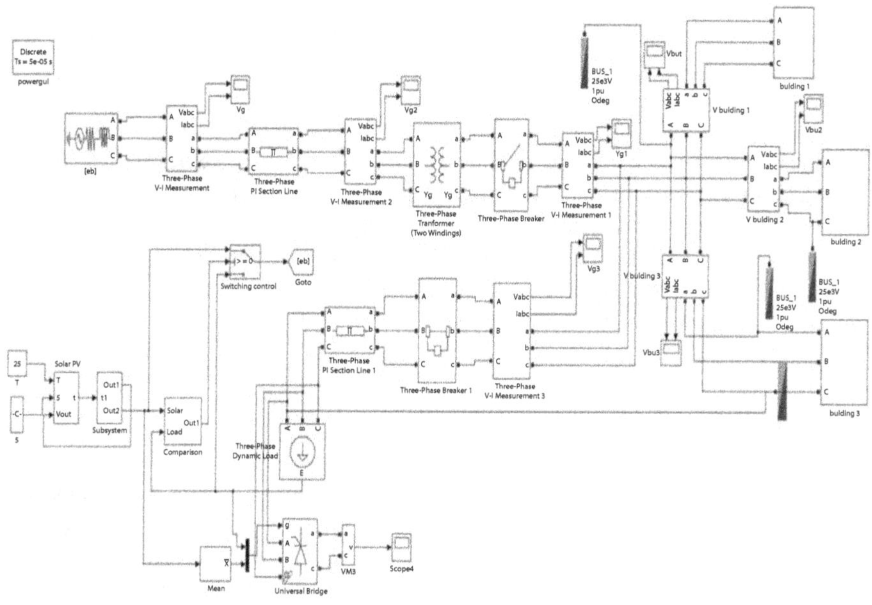

FIGURE 7.2 Modeling of an electric grid with solar power system (EB-solar).

The flow of electric power can be monitored with the help of a load flow bus, which is connected to the utility buildings by means of a controlled switch. The load flow bus senses the residential demand and a comparator is engaged to analyze existing solar power with the residential demand. Solar power can act as a primary source and the system permits the utility source to supply power by means of switching controller only if the available solar power is not enough to meet the residential demand. The switching control system is considered to switch over the supply of power processes to the EB source whenever solar power does not exist. The switching controller consists of both data inputs and control input. The switching process occurs based on the criteria of control input which is the output of the comparator. The defined criterion is whenever the control input is greater than or equal to the specified threshold value, then the switch passes through the data input (solar power) and the controller allows the solar energy system to deliver power to the residential buildings. If the control input fails to satisfy the defined criteria, then the output of the controller switches over to the EB source to meet the requirement of the residential buildings with remaining loads. Thus, the developed hybrid EB-solar power system satisfies all the residential demand through transmitting the required power.

7.4 PERFORMANCE OF AN ELECTRIC GRID WITH SOLAR POWER SYSTEM (EB-SOLAR)

The MATLAB/Simulink is a simulation software adopted to simulate the complete simulation of the hybrid renewable energy system architecture of the EB and a solar PV system. A grouping of EB source and solar PV energy system into a hybrid generation system possibly will enhance their efficiency by raising their whole energy output and by reducing the necessity of storing energy. The proposed model focuses on ensuring the load (residential) demand and managing the power flow from various energy sources such as solar PV and EB supply. Simulation results deliver the performance of the proposed hybrid control strategy for managing the power flow. Different waveforms revealed below are the output waveforms of an implemented simulation model of the hybrid system. The developed microgrid model has been tested under variable distributed power generation (solar PV and EB) and the residential buildings with a fixed load.

7.4.1 PERFORMANCE OF SOLAR POWER SYSTEM

The majority of geographical locations show the high potential of electric power production systems with solar PV because of their lesser impact on the environment and high accessibility of solar insolation [19]. The energy attained from the solar PV systems extremely depends on the ecological factors and ambient settings such as the solar insolation level, unpredictability in solar radiation (active weather state), and the module temperature. The important demerits of the solar PV systems are the process of efficiency conversion and the expenses of the PV system. Therefore, to facilitate minimizing the expenditure and to pick up the total effectiveness of the system, an extracted power from the solar PV panels should be maximized. To assure the maximum yield of energy at all environmental circumstances, the maximum power point tracking (MPPT) technique for solar PV systems is required [20].

The developed solar power system model includes a PV module, MPPT, and DC-DC converter. The MPPT delivers the required PV voltage to the boost converter all the way through variation in duty cycle *D*, which typically defines the rate of conversion. The perturbation and observation (P&O) algorithm is used to track the maximum power point. The P&O design evaluates the input of current and voltage; however, it randomly maximizes or minimizes the voltage by 0.001 step size. Subsequently, a new set of data is analyzed with the prior data as far as new readings attain lesser power than the prior data [21]. The flowchart for the P&O algorithm is represented in Figure 7.3.

The simulation was carried out to validate the developed model by verifying its functionality. Since the performance of a solar PV highly depends on two factors,

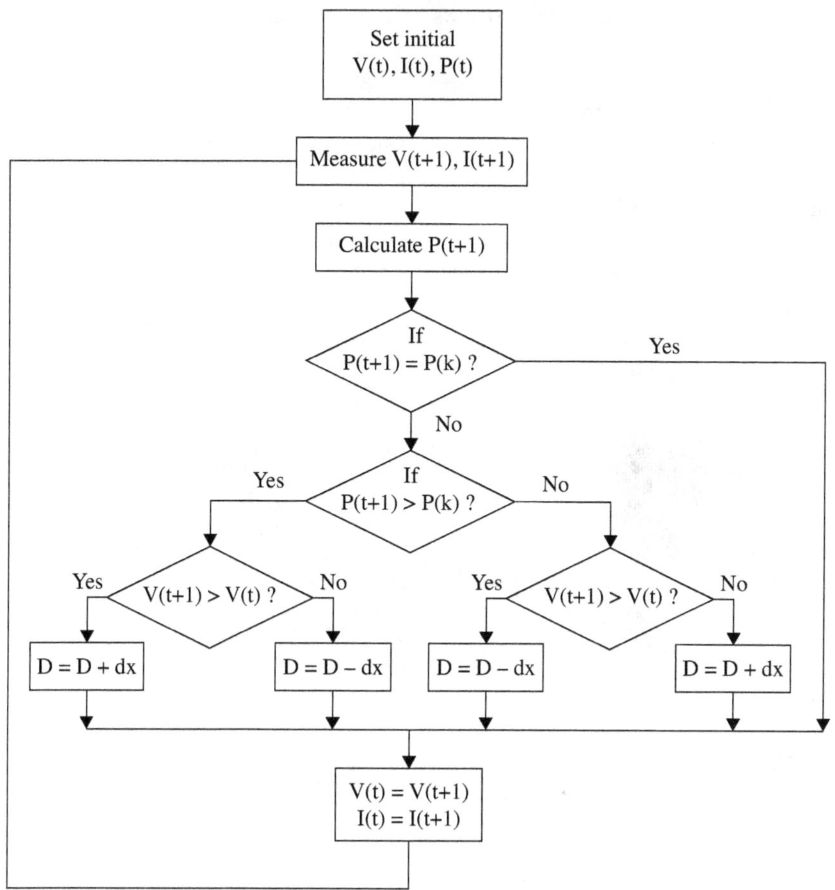

FIGURE 7.3 Flowchart for the P&O algorithm.

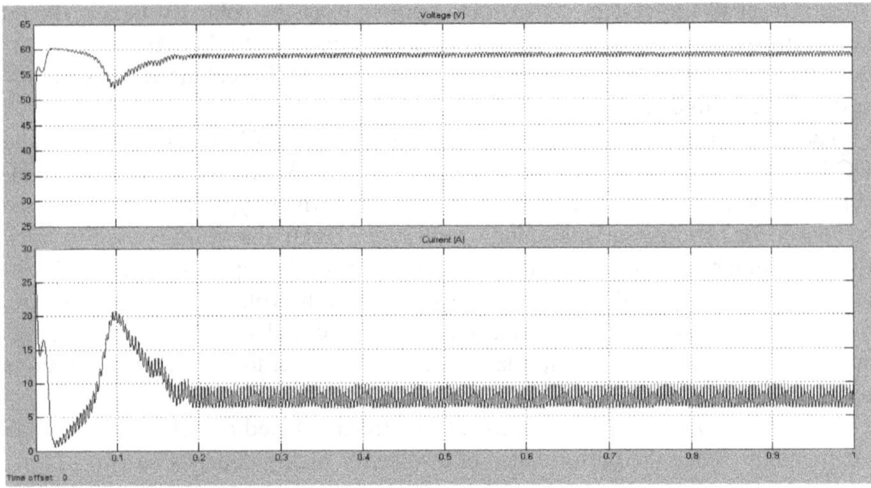

FIGURE 7.4 Plot of voltage and current waveforms of the solar cell.

namely, sun irradiance and temperature, the sun radiation of 1000 W/m^2 and temperature of 25°C have been considered as a standard testing condition and given as an input source for getting output power in terms of voltage and current. The resulted simulation plot delivers the voltage and current waveforms of the solar PV system, load power, and solar power output in watts over time. Figure 7.4 represents the plot of voltage and current waveforms of the solar cell.

The plot of gate pulse and output power of the developed solar PV system and the plot of load power and solar power from the developed solar PV system are depicted in Figure 7.5 and Figure 7.6, respectively.

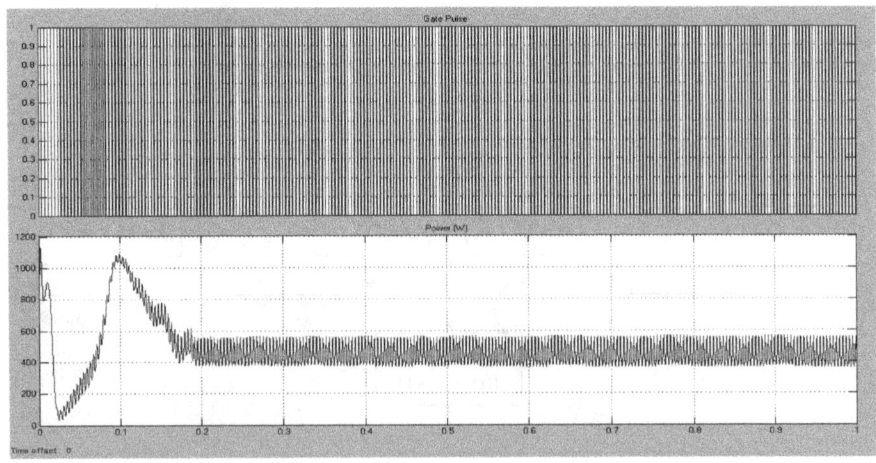

FIGURE 7.5 Plot of gate pulse and output power of the developed solar PV system.

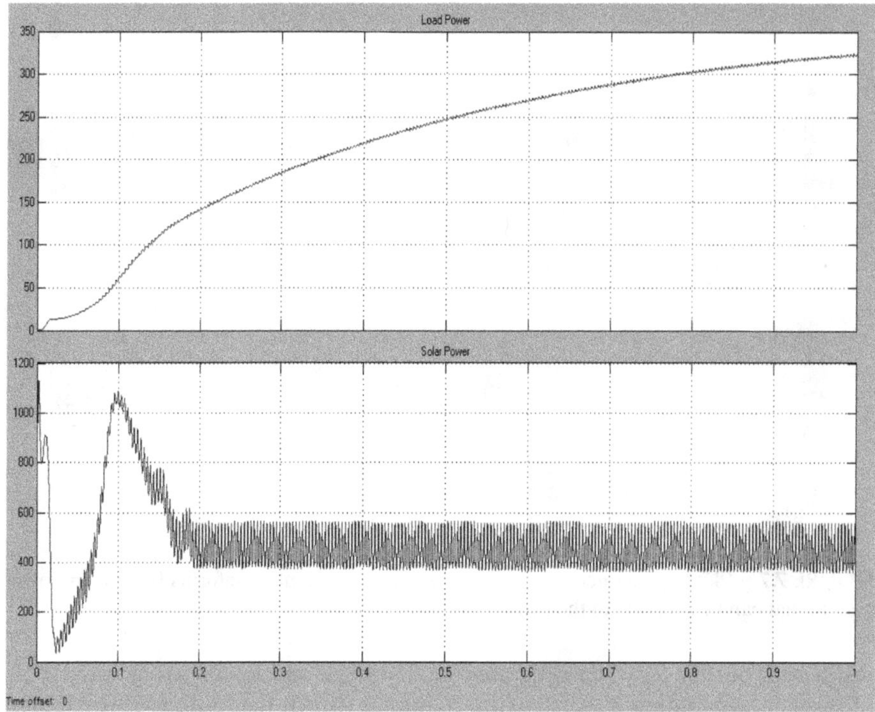

FIGURE 7.6 Plot of load power and solar power of the developed solar PV system.

From the resulted graphs, it is realized that the measure of current flow is influenced by the variation of sun irradiance. The solar power output enhances when the illumination of solar energy and sun irradiance goes up.

7.4.2 Performance of a Hybrid EB-Solar Power System

Once the developed solar PV is simulated and analyzed, the solar PV is connected with a developed electric grid system to achieve the performance of the corresponding system. The developed model is categorized into two operational modes.

7.4.2.1 Mode 1: When the Solar Power Transmits Residential Load

Initially, all the residential buildings get their corresponding load from the solar power system. The generated input voltage and input current waveforms from the hybrid solar-EB system where solar power transmits the required load to the residential buildings are represented in Figure 7.7. The generated load voltage and load current waveforms of residential building 1 are shown in Figure 7.8. Figure 7.9 represents the generated load voltage and load current waveforms of

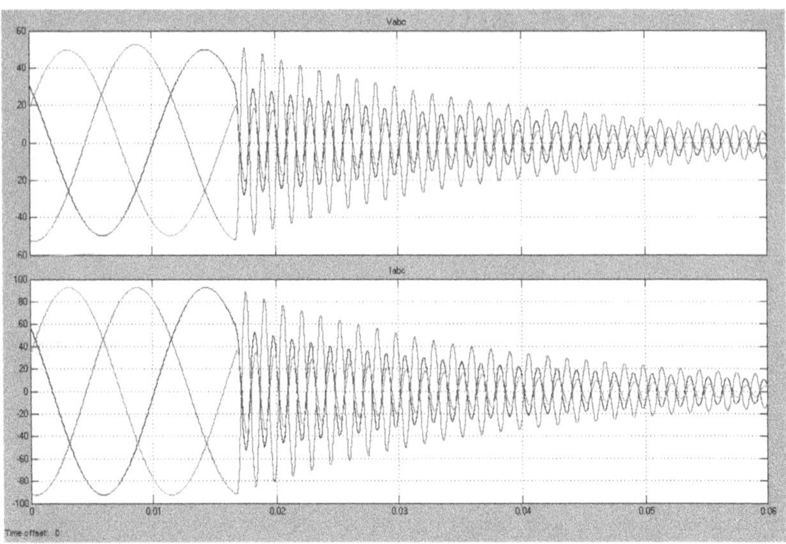

FIGURE 7.7 Plot of generated input voltage and input current waveforms from solar power system that can be supplied to the residential buildings.

residential building 2. The generated load voltage and load current waveforms of residential building 3 are shown in Figure 7.10. All the responses are plotted over time.

7.4.2.2 Mode 2: EB Source in Operating Mode

Whenever the available solar power system is not sufficient to meet all the required energy, then the switch controller allows the EB source to take charge of the power

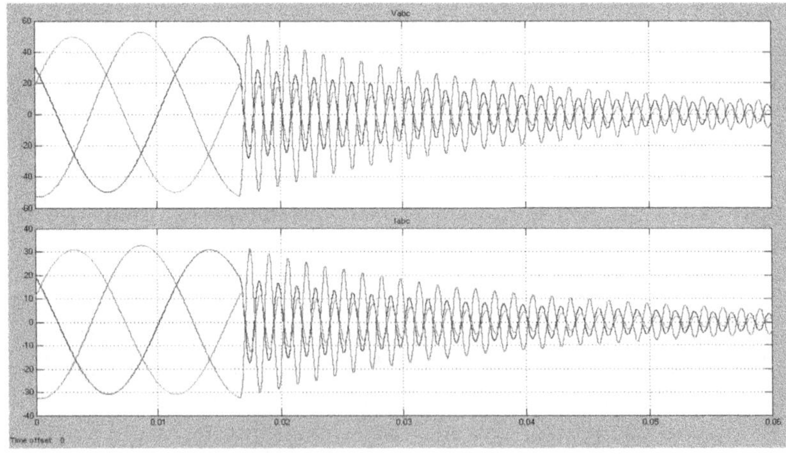

FIGURE 7.8 Plot of generated load voltage and load current waveforms of residential building 1.

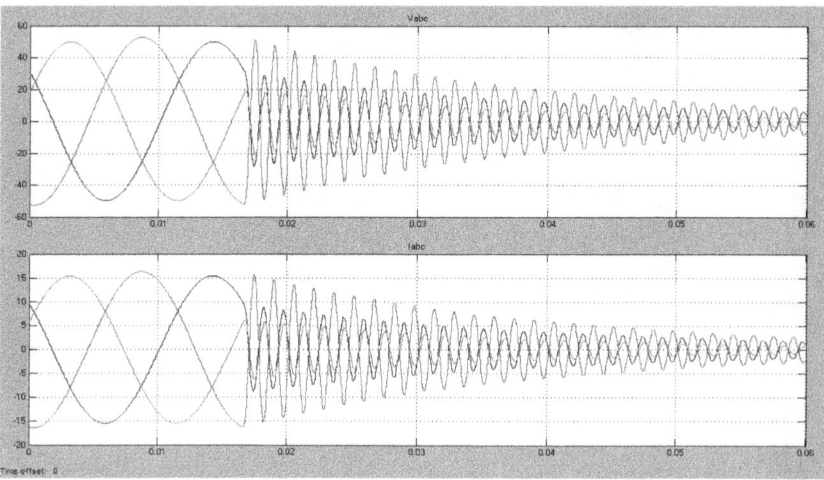

FIGURE 7.9 Plot of generated load voltage and load current waveforms of residential building 2.

delivering system. The switching control system is designed to switch over the supply of power processes to the EB source whenever solar power does not exist. The generated input voltage and input current waveforms from the hybrid solar-EB system where the EB source transmits the required load to the residential buildings after switching from a solar power system are represented in Figure 7.11.

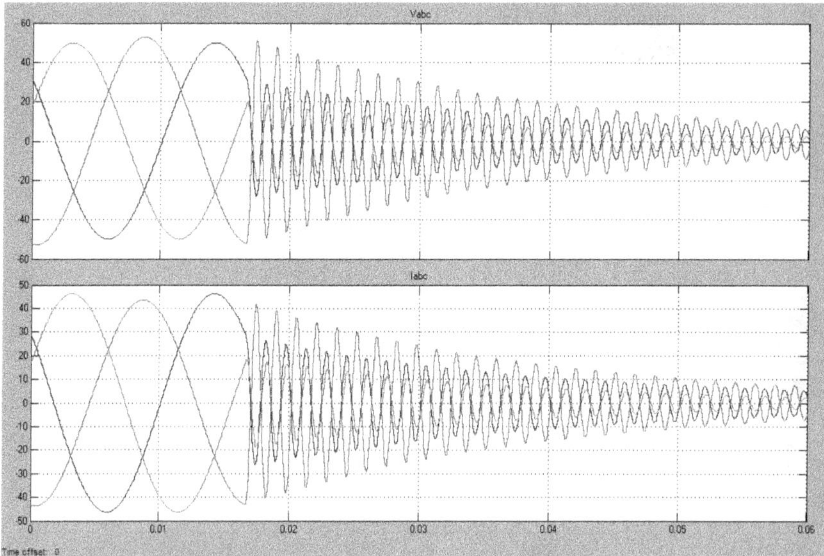

FIGURE 7.10 Plot of generated load voltage and load current waveforms of residential building 3.

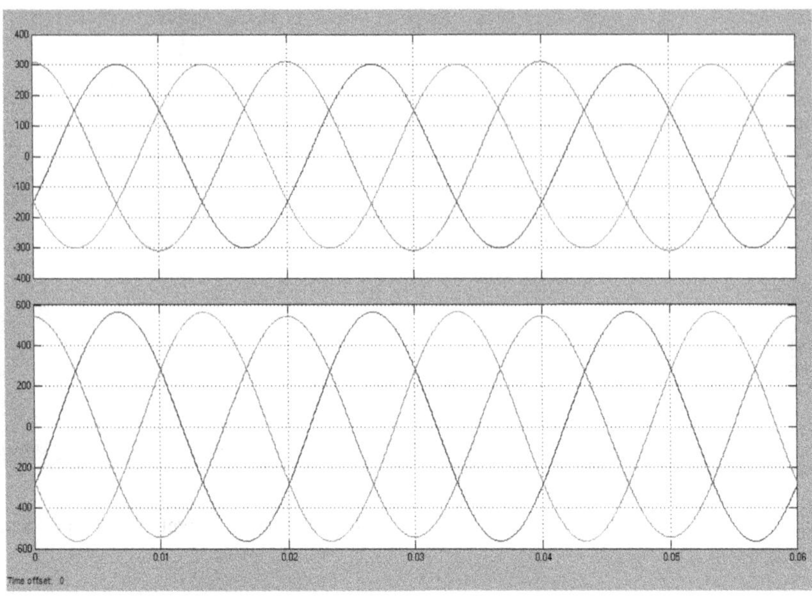

FIGURE 7.11 Plot of generated input voltage and input current waveforms from EB source that can be supplied to the residential buildings.

The generated load voltage and load current waveforms of residential building 1 are shown in Figure 7.12. Figure 7.13 represents the generated load voltage and load current waveforms of residential building 2. The generated load voltage and load current waveforms of residential building 3 are shown in Figure 7.14. All the responses are plotted over time.

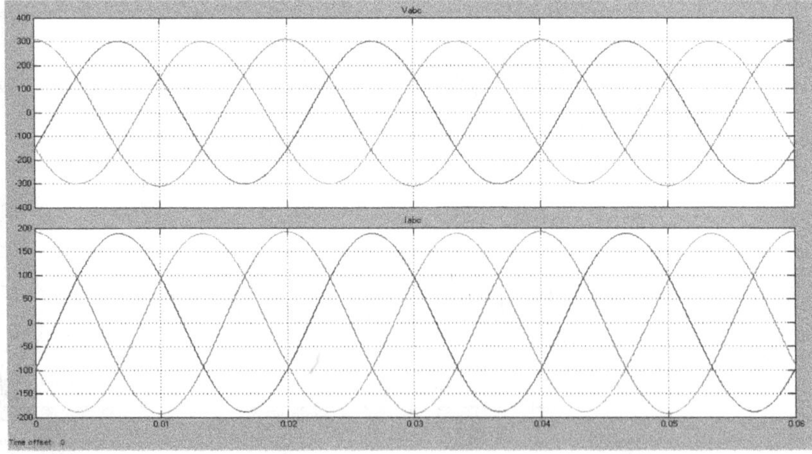

FIGURE 7.12 Plot of generated load voltage and load current waveforms of residential building 1.

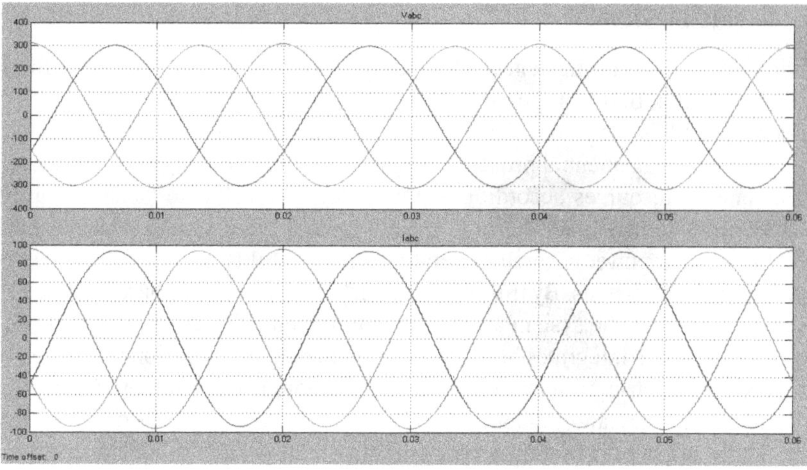

FIGURE 7.13 Plot of generated load voltage and load current waveforms of residential building 2.

The load voltage and grid voltage are constantly maintained with respect to time. The expected performance has been achieved from the system, and the imperfect sine waves that have been found in the resulted waveforms reveal that the system reports power losses caused by the switching process. Therefore, the developed system has to be considered to stabilize the system performance and to reduce the total harmonic distortions.

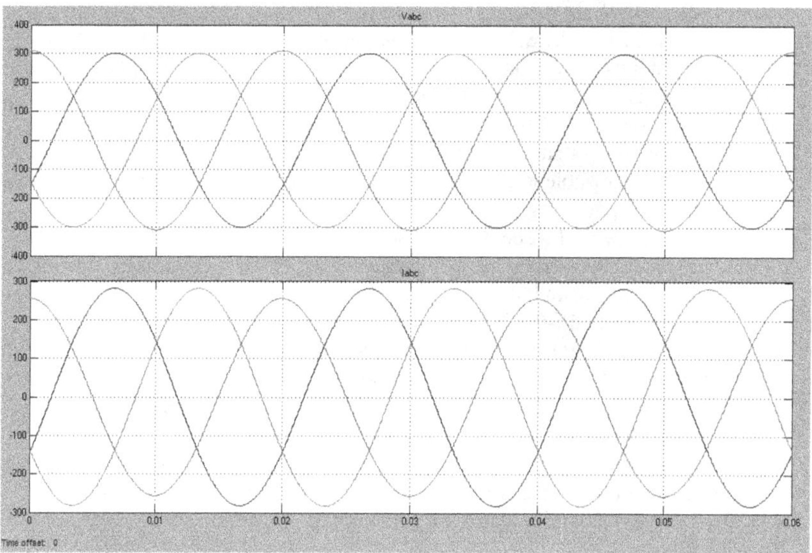

FIGURE 7.14 Plot of generated load voltage and load current waveforms of residential building 3.

7.5 CONCLUSION

Hybridizing or integrating renewable energy resources grants a sensible form of power production. In the proposed hybrid model, load demand is met from the combination of a PV system and the EB source. The obtained results reveal that the entire power management approach is efficient, and the flow of power between the various energy resources according to the meteorological state (solar energy) has the ability of automatic load sharing. Also, the results show the significance and the flexibility of solar power generator operation with the utility grid, which also reduces greenhouse gas emissions. By this, we can interface two or more sources of electrical power and optimize the system as per our requirement. The combination of renewable energy sources integrates both information and communication technologies in all features of electric power production, distribution, and consumption to increase the system efficiency and reliability and also to reduce the impact on the environment. Thus, the hybrid power system grants flexibility in terms of the successful exploitation of renewable energy sources.

REFERENCES

1. Abdallah, L., & El-Shennawy, T. (2013). Reducing carbon dioxide emissions from electricity sector using smart electric grid applications. Journal of Engineering. https://doi.org/10.1155/2013/845051
2. Abdelkader, H. I., Hatata, A. Y., & Hasan, M. S. (2015). Developing intelligent MPPT for PV systems based on ANN and P&O algorithms. International Journal of Scientific & Engineering Research, 6(2), 367–373.
3. Ahuja, R. K., & Kumar, R. (2014). Design and simulation of fuzzy logic controller based switched-mode power supply. IPASJ International Journal of Electrical Engineering, 2(5), 16–21.
4. Alward, Y. M. S., & Joshi, S. N. (2018). A review on control and automation based smart grid system and its impact on conventional grid. International Journal of Engineering Research & Technology, 7(04), 404–407.
5. Anoune, K., Bouya, M., Ghazouani, M., Astito, A., & Abdellah, A. B. (2016). Hybrid renewable energy system to maximize the electrical power production. In: 2016 International Renewable and Sustainable Energy Conference (IRSEC), Marrakech, pp. 533–539. https://doi.org/10.1109/IRSEC.2016.7983992
6. Bhandari, B., Lee, K. T., Lee, G. Y., Cho, Y. M., & Ahn, S. H. (2015). Optimization of hybrid renewable energy power systems: A review. International Journal of Precision Engineering and Manufacturing-Green Technology, 2(1), 99–112. https://doi.org/10.1007/s40684-015-0013-z
7. Diaf, S., Notton, G., Belhamel, M., Haddadi, M., & Louche, A. (2008). Design and techno-economical optimization for hybrid PV/wind system under various meteorological conditions. Applied Energy, 85(10), 968–987. https://doi.org/10.1016/j.apenergy.2008.02.012
8. Naik, M. D., Sreenivasulu, R. D., & Devaraju, T. (2014). Dynamic modeling, control and simulation of a wind and PV hybrid system for grid connected application using MATLAB. International Journal of Engineering Research and Applications, 4(7), 131–139.
9. Meziane, S., Feddaoui, O., Toufouti, R., & Ahras, S. (2013). Modeling and simulation of hybrid wind-diesel power generation system. International Journal of Renewable Energy, 8(2), 49–58. https://doi.org/10.14456/iire.2013.11

10. Ezema, L. S., Peter, B. U., & Harris, O. O. (2012). Design of automatic change over switch with generator control mechanism. Part 1. Natural and Applied Sciences, 3(3), 125–130.
11. Vera,Y. E. G., Dufo-López,R., & Bernal-Augustin, J. L. (2019). Energy management in microgrids with renewable energy sources: A literature review. Applied Sciences, 9(18), 3854.
12. Ramakumar, R. (2004). Role of renewable energy in the development and electrification of remote and rural areas. In: IEEE Power Engineering Society General Meeting, Denver, CO, pp. 2103–2105. https://doi.org/10.1109/pes.2004.1373253
13. Kaygusuz, K. (2009). Environmental impacts of the solar energy systems. Energy Sources, Part A: Recovery, Utilization, and Environmental Effects, 31(15), 1376–1386. https://doi.org/10.1080/15567030802089664
14. Kermadi, M., & Berkouk, E. M. (2017). Artificial intelligence-based maximum power point tracking controllers for photovoltaic systems: Comparative study. Renewable and Sustainable Energy Reviews, 69, 369–386. https://doi.org/10.1016/j.rser.2016.11.125
15. Negi, S., & Mathew, L. (2014). Hybrid renewable energy system: A review. International Journal of Electronic and Electrical Engineering, 7(5), 535–542.
16. Prajapati, V., Patel, S., Thakor, P., & Chaudhary, T. (2018). Modelling and simulation of solar PV and wind hybrid power system using Matlab/Simulink. International Research Journal of Engineering and Technology, 5(4), 619–623.
17. Su, W., & Wang, J. (2012). Energy management systems in microgrid operations. Electricity Journal, 25(8), 45–60. https://doi.org/10.1016/j.tej.2012.09.010
18. Tous, Y. E., Hafith S. A., & Arabia, S. (2014). Photovoltaic/wind hybrid off-grid simulation model using Matlab/Simulink. International Journal of Latest Research in Science and Technology, 3(2), 167–173.
19. Allani, M. Y., Jomaa, M., Mezghani, D., & Mami, A. (2018). Modelling and simulation of the hybrid system PV-wind with MATLAB/SIMULINK. In: 2018 9th International Renewable Energy Congress (IREC), Hammamet, pp. 1–6. https://doi.org/10.1109/IREC.2018.8362514
20. Sajadian, S., & Ahmadi, R. (2016). High performance model predictive technique for MPPT of gird-tied photovoltaic system using impedance-source inverter. In: 2016 IEEE Power and Energy Conference at Illinois (PECI), Urbana, IL, pp. 1–7. https://doi.org/10.1109/PECI.2016.7459236
21. Bakić, V., Pezo, M., Stevanović, Ž., Živković, M., & Grubor, B. (2012). Dynamical simulation of PV/wind hybrid energy conversion system. Energy, 45, 324–328. https://doi.org/10.1016/j.energy.2011.11.063

8 Artificial Intelligence and Machine Learning Methods for Renewable Energy

Sushila Palwe and Prerna Lahane
School of CET, MIT World Peace University, Pune, India

CONTENTS

8.1 INTRODUCTION

Global energy demands are growing rapidly. And, nonrenewable energy sources will not be able to meet our future energy requirement. Renewable energy, which has the benefit of minimum carbon emission, may be a feasible solution to make our planet safer and energy proficient. Over the past few years, many sorts of renewable energy resources such as wind, solar, geothermal, biomass, and tidal have been exploited. Computing and machine learning (ML) improves the potency and accessibility of renewable energy technology [1]. Computing (artificial intelligence [AI]), in conjunction with many AI advanced technologies, has incontestable immense potential to work on the renewable energy. AI and ML technologies will build a control by reducing emissions and increasing production potency. In this era, ML algorithms are applied to the information gathered from advanced sensors, good meters, intelligent device, and grid operators to estimate how individual appliances behave. Germany, as an example, uses advanced AI technology in the early warning system, which takes time period information from wind turbines and star panels around the country to predict the energy report for next two days. AI may also facilitate the trade to improve safety, responsibleness, and potency. It may also offer visibility into energy run, consumption patterns, and instrumentation health. As an example, prophetic analytics will take device knowledge from a turbine to watch wear and tear, and predict with a high degree of accuracy once it might want maintenance [2, 3].

8.2 RENEWABLE ENERGY DEFINITION

A renewable energy supply can be defined as consistent energy supply, which does not run out, or as energy supply that serves endless energy, such as the sun, wind, and sea tides. Renewable energy is also called alternative energy.

8.3 CHALLENGES OF RENEWABLE SOURCES

One of the major and most significant challenges of using renewable energy is the unreliable condition of the weather. These resources are prone to unpredictable weather vulnerabilities.

8.3.1 Availability of Power

One of the most important issues is power generation from alternative energy resources, reckoning on natural resources is out of control mankind. For example, star battery-powered electricity generation is possible only in the presence of sunlight and turns off at night due to unavailability of sunlight; similarly, wind energy generation depends on the availability of wind; thus, if the speed of wind is extremely slow, the rotary engine is not going to flip, which leads to no power flow to the electricity grid. On the other hand, an excessive amount of wind will damage the generator. Hence, a nonbreakable balance needs to be maintained for consistent power generation from wind energy [3].

8.3.2 Power Quality Issues

Persistently high power quality generation is key to maintain the stability and potential of the network. Power quality issues include disordered frequency, variation in voltage/current harmonics, low power issue, variation in transmission lines, etc. [3].

8.3.3 Resource Location

Many of the renewable energy plants supply their energy with the grid and thus need massive area. The location of a renewable energy plant is decided by considering the availability of natural power. Here the challenge starts, which means depending on geographical and environmental conditions location has to be decided. The second part includes the distance between energy supply and production in terms of cost and potential. Along with that, renewable energy completely depends on climate situation, location, and weather, which means that method used at one point may not be suitable at another point [3].

8.3.4 Information Barrier

It has been observed that many people are not aware about the advantages of renewable energy. Also, due to lack of knowledge many of them are not interested in these energies. A lot of expenditure and capital allowances are created out there for the implementation of renewable energies [3].

8.3.5 Cost Issue

Cost is considered as the big challenge in the field of renewable energy because the installation cost of power generation from natural sources is very high. This is the main hurdle in renewable power development era [3].

8.4 SOLUTIONS TO CHALLENGES

To overcome the challenges, some of the solutions are discussed in this section.

8.4.1 Smart and Central Control

The energy grids are interconnected with various actuators and sensor devices to gather a huge amount of information. This information using AI offers new insights on higher management operations. Such information is handled by grid operator for various operations [2].

8.4.2 Improvement in the Integration of Microgrids

AI in renewable energy is required to manage and improve the safety, reliability, efficiency, and intermittence. AI and advanced AI technologies will help us to find out the energy generation and consumption patterns. It also helps us to find out the health of the devices required for various power operations [2].

8.4.3 Expansion of the Market

AI and technologies of AI will facilitate renewable energy suppliers to enhance the market by suggesting new AI working models and also promote, for as much as possible, engagement. The AI-powered systems are ready to analyze the information associated with energy assortment and supply insights on energy consumption. Such knowledge will be facilitated by suppliers to reform the present service model and develop new service models. It is also useful to focus on new clients [2].

8.5 ARTIFICIAL INTELLIGENCE TECHNIQUES IN THE FIELD OF RENEWABLE ENERGY

In the field of renewable energy, there has been research on AI and advanced techniques from the past decades to achieve success. AI falls under various advanced technologies such as artificial neural network (ANN), fuzzy logic, deep learning, expert system, and many more. These many advanced technologies are useful for data analysis, modeling, and prediction of performance, failure, and control in the area of renewable energy processes. In the following subsections, short introduction to the AI techniques is presented.

8.5.1 Solar Power Generation Mechanism and AI Algorithms [4]

8.5.1.1 Smart Microgrid

One of the basic terms in the electricity generation operation is smart microgrid, which smartly does various tasks. It integrates the actions of users from power generation at generator side to consumption at user side to obtain an efficient, sustainable, economical, and safe source of electrical energy. It smartly coordinates the actions and actors in the field of electricity generation [5].

8.5.1.2 Maximum Power Point Tracking

With the help of maximum power point tracking (MPPT) method, the maximum voltage points and current output on the solar panel can be determined. MPPT is used to find out the most optimal power points on the photovoltaic (PV) solar panel. One of the components is DC-DC converter, which helps in generating maximum value of power, even though the irradiation varies due to various reasons. MPPT controls the duty cycle until maximum power generation points are captured by the system. Various AI technologies are being applied to obtain the maximum performance considering MPPT on PV solar panels [5, 6, 10].

8.5.1.3 Single-Ended Primary-Inductor Converter

The single-ended primary-inductor converter (SEPIC) is designed with different topology other than buck-boost type DC-to-DC converter. It does not change the polarity. The modification of SEPIC converter topology has been done by adding capacitors and inductors to minimize the output voltage [5, 6].

8.5.1.4 Perturb and Observe

Perturb and observe (P&O) algorithm is known as an optimization algorithm because it discovers the system's conditions much better than the previous system condition by changing the system parameter. In the case of MPPT, this optimization algorithm changes the duty cycle parameter of the system to obtain the best power output value of solar panel. P&O compares the power voltage values before change in duty cycle and after change in duty cycle, due to which precision and stability of the voltage sensor are very influential with the use of P&O algorithm [5–7].

8.5.1.5 Flower Pollination Algorithm

Flower pollination algorithm is an algorithm used for decision-making; it is also called optimization algorithm. This algorithm is used to find out points that give maximum function values. This algorithm uses the concept of pollination of flowers by insects and hence the name flower pollination algorithm. Insects always find out better flower for pollination; the same concept is applied here as well [5, 6].

8.5.1.6 Firefly

Swarm intelligence is a term used in the field of AI. Firefly algorithm (FA) is also a part of swarm intelligence. FA is also known as metaheuristic algorithm and is inspired by the flashing behavior of fireflies [5, 6].

8.5.2 AI Techniques in the Field of Solar Energy

8.5.2.1 ANN

ANN is an advanced technology of AI. This technique behaves like human brain, which has neurons, dendrites, etc. ANN is an architecture that consists of many artificial neurons, which act as simple processing units just like in normal animal species. It is basically a weighted directed graph that consists of neurons and interlinks among various neurons, and passes message or output from one neuron to another neuron. ANN is categorized into two parts, feed-forward network and feedback network, based on connections among nodes. In the case of renewable energy, to predict various results many hybrid statistical methods use ML algorithms along with ANN and support vector regression. Basically, to give more accurate prediction, hybrid statistical methods along with ML algorithms are used. If PV cells use physical configuration, then prediction could be possible using physical method and no ML model is needed.

8.5.2.1.1 Maximum Power Point Tracking Using ANN

Mostly, ANN prepares a model using nonlinear and complex functions. Majorly, the output of ANN depends on hidden layer and training of the model. In the ANN learning process, the learning model decides the weights at every neuron or node. After this, ANN updates the weights at every layer/network by considering adaption cycle until the model does not reach the state of equilibrium. To obtain accurate maximum power point of PV cell using ANN, weights (W_i) are determined appropriately by considering the relation of input and output values of PV cells. In an ANN-based MPPT, every neuron receives input value either from neighboring neurons or from the nonlinear system, which is associated with ANN input variables. The input variables for the ANN-based MPPT can be considered as PV array parameters such as voltage values and I_{SC}, climate information such as irradiance and temperature, or any combination of these [7]. The output of this ANN-based MPPT model is duty cycle signals, which control power conversion at the given or closed MPPT [4, 8, 9].

In the adaption cycle of training model, weights are adjusted as a part of the training process. During the training process, data patterns between input and output of ANN are recorded for longer period of time to obtain the result more accurately. Parameters such as temperature and irradiation are considered to be input variables to ANN, and at output side duty cycle ratio given to DC/DC boost regulator which converts the SPV voltage to its minimum value. The incoming layer gets an input value; consequently, the next layer, i.e., the hidden layer, comprises numerous neurons that obtain values after the first layer and direct the values to the output layer, i.e., the third layer. The output layer delivers the values, which in this case are duty ratio, to the system [10].

8.5.2.2 Fuzzy Logic

Fuzzy logic control is one of the most popular AI-based MPPT techniques and majorly used in enhancement in VLSI technology. There are three stages in the working of fuzzy logic controller, i.e., fuzzification, inference, and defuzzification. By considering the membership function, transfer of input variable to the linguistic variable has been done in the first stage of fuzzy logic, i.e., fuzzification process. Fuzzy values are assigned in the first stage of fuzzy logic. To get high accuracy, a number of membership functions are used by the controller. Fuzzification block assigns fuzzy values to input signals; these input signals are prepared by the same block. Fuzzy rules provide an explanation of unpredictable data that could be controlled using information given to the model. The inference mechanism works like an explanation facility of expert system. It generates a clarification of the data, which has considered the rules and the functions used for it. In the defuzzification stage, fuzzy information obtained from the inference mechanism is changed into nonfuzzy information. It is suitable for well-controlled mechanism [9, 4, 11].

Fuzzy logic controller gives output in the form of change in duty ratio of power converter or variable reference voltages of the DC link ΔV, which is called fuzzy rule algorithm. This algorithm associates the output of fuzzy logic with the fuzzy inputs depending on the power converter. Fuzzy logic controllers have many advantages: it deals with imprecise inputs, does not require accurate mathematical model, and also deals with nonlinearity. Fuzzy logic controller gives as minimum oscillations as possible around the MPPT. It gives better performance, which depends on the number of the fuzzy levels and membership functions [6, 10, 12].

8.5.2.3 Adaptive Neuro-Fuzzy Inference System

Adaptive neuro-fuzzy inference system (ANFIS) is a hybrid AI technology, which combines the neural network and fuzzy logic functionality of AI. ANFIS is designed by considering the nodes and guiding links among the nodes, which are interconnected to one another. ANFIS is a complex feed-forward network, which applies node function on input messages and the parameters related to the given node. ANFIS is divided into two adaptive network types, namely, circle node and square node. The difference between these nodes is that square node has parameters, whereas circle node does not contain any parameters. Square node is also known as "adaptive node." ANFIS is trained with data for initial membership function, and the same function is adjusted by using back-propagation algorithm or hybrid-learning algorithm. As it a hybrid technology, it consists of both functionality of decision-making, such as fuzzy logic, and training capability, such as neural network [13].

ANFIS architecture has been categorized into five layers [4, 9]:

1. In the first layer, by using membership function, every node represents their own new membership grade with the help of fuzzy set at every layer.
2. The second layer is denoted by PROD, which acts as a multiplier to multiply input signals.
3. At this layer, each node works as a static node and *j*th node is calculated as the ratio of *j*th firing strength to the sum of all rule strength.
4. The prediction of contribution of *j*th node toward complete output model is decided at this layer.
5. At this layer, overall result calculation of the ANFIS model has been done.

PV cells act as nonlinear sources, which are dependent on weather parameters. To obtain maximum power from the PV system, insolation and temperature factors play a major role. ANFIS uses the same training data as a neural-based system. After training, membership function at a given layer is generated using the tool named anfisedit in MATLAB/Simulink. The ANFIS model is depicted in Figure 8.1. ANFIS architecture uses layers as: two input (x_1 and x_2), one output (y), and two rules. In the first layer, also known as input layer, input signals are transferred from one node to all other layers. At the second layer, separation of input values from fuzzy set is done using a bell activation function. In the third layer, representation of fuzzy logic rules is done using the Sugeno fuzzy logic subtraction system. In the fourth layer, normalization of the input values and calculation of firing level occur.

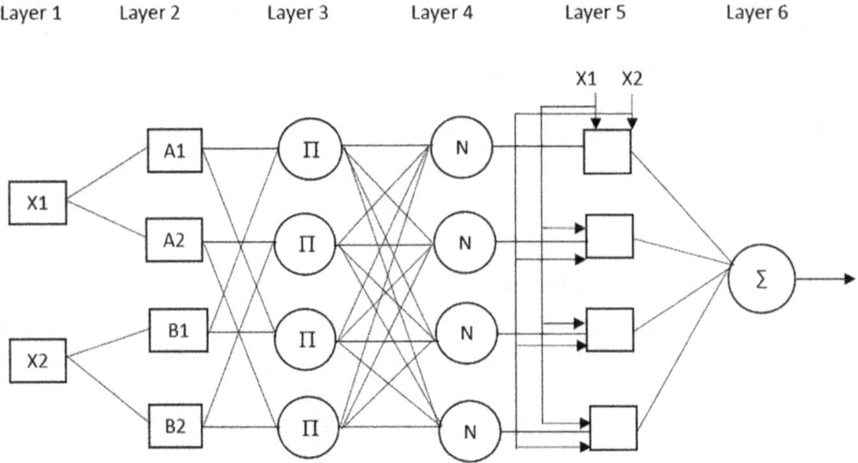

FIGURE 8.1 ANFIS architecture.

In the fifth layer, which is also called purification layer, calculation of weighted result values given by decontaminated layer is done. The last layer is known as summation layer and gives the last out value of the model. Solar energy prediction and radiation can be based on adaptive network-based fuzzy inference system [14].

8.5.2.4 Convolutional Neural Network

One of the important methods of deep learning is convolutional neural network (CNN), which works on image recognition and voice analysis. It is a type of biological neural network that has the weight-sharing functionality to minimize the complexity of the model. To recognize any type of image, it directly puts the image in the network instead of feature extraction or any other complex process. CNN can be efficiently used in the field of renewable energy to generate power. Using the CNN model, per day solar energy generation can be calculated easily and forecasting would be possible. Also, CNN is efficiently used for short-term prediction of wind power energy.

8.5.2.5 Deep Neural Network

Deep neural network (DNN) is one of the advanced technologies included in the field of ANN. It consists of many hidden layers between the input and output layers. DNN is capable of modeling and processing nonlinear relationships. There are two types of neural network: deep belief network and recurrent neural network. DNN accepts or processes a large amount of data; then, using specific deep learning algorithm, the data gets trained and then the model analyzes or predicts the future result. In the field of renewable energy, DNN plays a very important role: to predict the future situation of all types of energies, and according to that data can be forecasted for various objectives—for example, electricity generation, farming purpose, calculation of amount of energy in the future days, and many more.

8.6 SOLAR ENERGY ANALYSIS USING ML

Data analysis helps us to extract meaningful insights from data. These extracted insights are very useful to enhance the system as they provide useful strategies or future predictions. Data analysis can be done using two major techniques: exploratory data analysis (EDA) and predictive data analysis. EDA supports the insight extraction using data visualization techniques. Meaningful insights can be extracted by visualizing the graphs and charts in EDA.

Predictive data analysis aims to predict the data class, values, and hidden patterns, using intelligent ML algorithms.

8.6.1 Solar Energy Analysis Using EDA

EDA uses various data visualization techniques. Graphs and plots are very essential tools that allow us to visualize and learn the meaning relationship and patterns from data. Solar energy analysis is carried out using data visualization in the following section.

The dataset used in the following analysis contains various features, which contribute to solar energy prediction. The dataset demonstrated here for solar energy analysis is from open access data site [15]. Table 8.1 shows various features in this dataset.

All attributes of the data are continuous value. Following are the Features and Targets

Features: {Cloud coverage(%), Visibility(Miles), Temperature(C), Dew point(C), Relative humidity(%), Wind speed(mph), Station pressure(inch Hg), Altimeter(inch Hg)}
Target: {Solar Energy (watt)}

Various visualization tools and libraries are available for EDA analysis. These tool and libraries support various charts and plots graphs to express the insights from data. For example, with data mentioned in Table 8.1, day-wise data of all features with solar energy creation can be observed with scatter graph. By observing the scatter graph, the feature relationships with target (solar energy) and feature contribution for target can be observed. With scatter plot, positive and negative relationship of features can be observed, which is further useful to predict the amount of solar energy generation in certain environmental condition.

TABLE 8.1
Solar Energy Data Attribute

Cloud coverage	Visibility	Temperature
Dew point	Relative humidity	Wind speed
Station pressure	Altimeter	Solar energy

Correlation analysis using colormap is another example for understanding the correlation among the features and is further useful for discarding the correlated feature for dimensionality reduction.

8.6.2 Predictive Analysis for Solar Energy

Predictive data analysis aims to predict the data class, values, and hidden patterns, using intelligent ML algorithms.

ML enables a system to learn from the past experience. These experiences are collected in the form of historical data, which is treated as training data to make the ML model to learn. ML algorithms, once trained with an ample amount of training data, become capable of doing a variety of tasks, such as classification, clustering, and prediction. ML is a complex process. With the given training data, ML produces a suitable model based on that data to perform more precise task on that data. Every new sample of training data is useful to make model more precise toward accuracy.

8.6.2.1 Machine Learning Approaches

There are three different approaches of machine learning:

1. Supervised learning
2. Unsupervised learning
3. Reinforcement learning

Supervised learning starts with training the algorithm with labeled data. The purpose of this learning is to classify the given data into classes. Supervised learning methods use the labeled data form training which is always associated with class labels. The relationship between the features of data with each other and with class label is built as classification model. This classification model further predicts the class label of unseen data.

Unsupervised learning is useful to find hidden pattern from data. Unsupervised learning algorithms get trained with unlabeled data. Feature similarities are identified for modeling the data as patterns in the data.

Reinforcement learning is a behavioral learning model. These algorithms receive feedback while execution and then evolve the model to obtain the best outcome. Reinforcement learning is different from supervised learning and unsupervised learning as the training phase in not crucial part of this type of learning. Instead, the system learns through trial and error.

8.6.2.2 Machine Learning Techniques for Solar Energy Analysis

ML helps in prediction and pattern extraction from the data. The following are the areas where ML is helpful in solar energy analysis:

1. Analyzing the location and regions with similar weather condition and suggesting the suitable location to install solar plants
2. Predicting the units of solar energy based on past data
3. Analyzing important features for solar energy prediction and dimensionality reduction

8.6.2.3 Analyzing the Location and Regions with Similar Weather Condition and Suggesting the Suitable Location to Install Solar Plants Using Clustering

Clustering is an unsupervised learning technique that helps in identifying hidden patterns called clusters. With the given data, clustering algorithms identify the similarities among the features and cluster most similar data together. Clustered data further can be used to identify the type of unseen data based on similarity of the unseen data with the formed clusters.

There are various clustering algorithms which are researched by many authors. Most common clustering algorithms are K-means, DBSCAN, and spectral clustering. For solar energy analysis, K-means clustering is explained in this section.

In K-means clustering, the K clusters get formed based on similarity of the data. In solar energy analysis case, dataset with location-wise data is collected for various features as follows:

All attributes of the data are continuous value.

Features: {Cloud coverage(%), Visibility(Miles), Temperature(C), Dew point(C), Relative humidity(%), Wind speed(mph), Station pressure(inchHg), Altimeter(inchHg)}
Target: {Solar Energy (watt)}
K Means Clustering Algorithm for
Input: {K centroid (C_1,C_2),Features of Solar Dataset}
Output: {K cluster centroid}
Algorithm

Find Similarity Distance of Each Location Data with Initial Centroid
Assign the Location Data point to nearest centroid
Calculate new centroid
Repeat until converges
Label final Cluster
For unseen Data
  Find similarity of unseen data with cluster centers
  Label unseen data with closest cluster
  If Labeled cluster is cluster of locations with Good Solar Energy
    then recommend the unseen data point for solar plant
  Else
  unseen data point is not good for solar plant

8.6.2.4 Decision Tree to Identify the Class Label of Location as to Choose (Yes) or Not (No) for Solar Plant Implementation

Considering the solar data as discussed in the preceding sections, the different feature values, {Cloud coverage(%), Visibility(Miles), Temperature(C), Dew point(C), Relative humidity(%), Wind speed(mph), Station pressure(inchHg), Altimeter(inchHg),Solar Energy}, are associated with the class labels {Yes, No}, suggesting whether the location is good for solar plant or not.

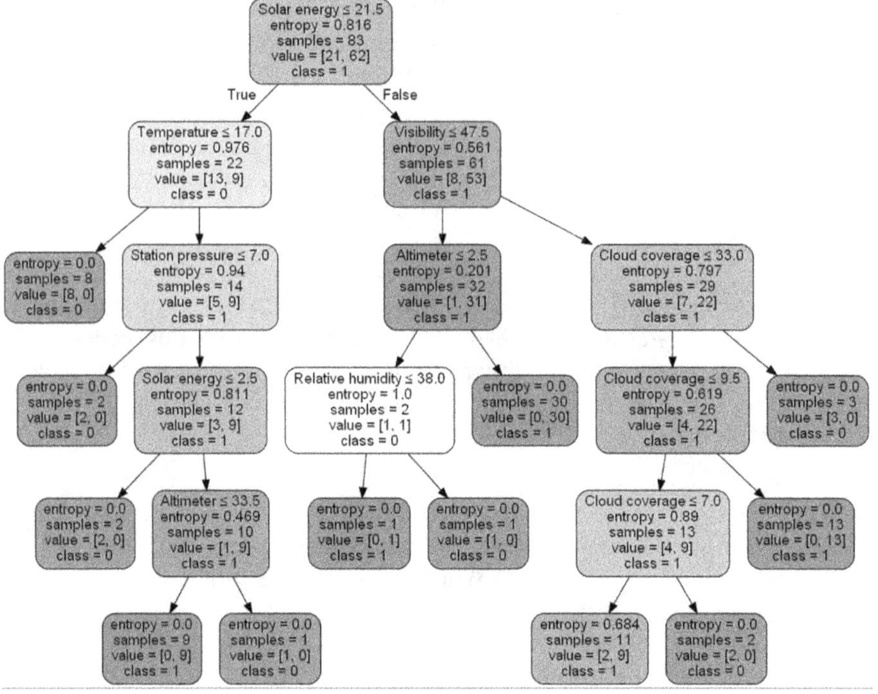

FIGURE 8.2 Decision tree for solar plant.

As the data is associated with class labels, this is supervised learning classification problem to predict whether the new given data of new location is to be used as solar plant (Yes) or not (No).

Figure 8.2 depicts the decision tree for solar plant, which shows the decision tree for solar data. Each feature and its values are used to construct the decision tree model.

Decision tree algorithm is a classification algorithm that models the classifier as a tree structure. Each node in this decision tree is decision variable (feature) and each branch is outcome values of that feature. While predicting the class of unseen data, the decision tree model is traversed from root to leaf node following outcome of branch. This traversed path leads to predict the class label of unseen data.

8.7 CONCLUSION

In this chapter, we have discussed the various AI techniques and ML algorithms and how these approaches have been applied to different types of renewable energy, especially solar energy. The use of ML is important for renewable energy sectors nowadays. Various AI algorithms, such as ANN, are discussed with dataset description. ML methods such as EDA and predictive analytics are discussed with various techniques such as clustering, classification, and dimensionality reduction.

With these discussed AI and ML techniques, the optimization, optimal utilization, and accurate prediction are certainly possible to make real-time use of renewable energy as a major energy source and these will certainly help to make lives easier in rural areas.

REFERENCES

1. Artificial intelligence in renewable energy. Available online at: https://www.frontiersin.org/research-topics/12362/artificial-intelligence-in-renewable-energy
2. The role of artificial intelligence in renewable energy. Available online at: https://www.imaginovation.net/blog/artificial-intelligence-in-renewable-energy/
3. Meet A. Moradiya (2019). The challenges renewable energy face. Available online at: https://www.azocleantech.com/article.aspx?ArticleID=836
4. M. Hanan, X. Ai, M. Y. Javed, M. Majid Gulzar and S. Ahmad (2018). A two-stage algorithm to harvest maximum power from photovoltaic system. In: Energy Internet and Energy System Integration (EI2) IEEE Conference, pp. 1–6.
5. S. Suyanto, L. Mohammad, I. C. Setiadi and R. Roekmono (2019). Analysis and evaluation performance of MPPT algorithms: Perturb & observe (P&O), firefly, and flower pollination (FPA) in smart microgrid solar panel systems. In: 2019 International Conference on Technologies and Policies in Electric Power & Energy, pp. 1–6.
6. P. Dimitroulis and M. Alamaniotis (2020). Residential energy management system utilizing fuzzy based decision-making. In: 2020 IEEE Texas Power and Energy Conference (TPEC), pp. 1–6.
7. Naveen and A. K. Dahiya (2018). Implementation and comparison of perturb & observe, ANN and ANFIS based MPPT techniques. In: Inventive Research in Computing Applications International Conference, pp. 1–5.
8. M. A. Samadhan and S. S. Kamble (2020). Introduction of different maximum power point tracking method using photovoltaic systems. In: Advanced Computing and Communication Systems International Conference (ICACCS), pp. 752–755.
9. E. H. M. Ndiaye, A. Ndiaye, M. A. Tankari and G. Lefebvre (2018). Adaptive neuro-fuzzy inference system application for the identification of a photovoltaic system and the forecasting of its maximum power point. In: Renewable Energy Research and Applications Conference, pp. 1061–1067.
10. P. Vinay and M. A. Mathews (2014). Modelling and analysis of artificial intelligence based MPPT techniques for PV applications. In: Advances in Green Energy International Conference, pp. 56–65.
11. M. Aurangzeb, X. Ai, M. Hanan, M. U. Jan, H. U. Rehman and S. Iqbal (2019). Single algorithm Mpso depend solar and wind MPPT control and integrated with fuzzy controller for grid integration. In: 2019 IEEE 3rd Conference on Energy Internet and Energy System Integration (EI2), pp. 583–588.
12. S. B. Raha and D. Biswas (2020). Fuzzy controlled demand response energy management for economic micro grid planning. In: Power Electronics, Smart Grid and Renewable Energy IEEE Conference (PESGRE), pp. 1–6.

13. T. M. Sanjeev Kumar, S. Sharma, C. P. Kurian, S. M. Varghese and A. M. George (2020). Adaptive neuro-fuzzy control of solar-powered building integrated with daylight-artificial light system. In: 2020 IEEE International Conference on Power Electronics, Smart Grid and Renewable Energy Conference (PESGRE2020).
14. K. Geetha and B. Sangeetha (2013). Application of artificial intelligent technique to power quality issues in renewable energy sources. In: Renewable Energy and Sustainable Energy International Conference, pp. 233–237.
15. Renewable energy data set. Available online at: http://s39624.mini.alsoenergy.com/Dashboard/2a5669735064572f4342554b772b71513d

9 Artificial Neural Network-Based Power Optimizer for Solar Photovoltaic System

An Integrated Approach with Genetic Algorithm

Revathy Subbiah Rajaram and Kirubakaran Victor
Centre for Rural Energy, The Gandhigram Rural Institute-Deemed to be University, Gandhigram, Tamil Nadu, India

CONTENTS

9.1 INTRODUCTION

There is an exponential growth of photovoltaic (PV) power production and the cost of electricity from solar energy is falling. Various renewable energy sources have been compared by Khalil in 1981 [1], and it is proposed that solar photovoltaic (SPV)based energy production has more potential and promising future compared to other forms such as wind and hydropower. As a growing economy, the power demand of India is increasing at an exponential rate, which presses the country to move towards renewable energy. The rise in the number of utility-scale solar power plants and the residential rooftop power generation units is contributing well to the power need of the country. But a good control over the generated electricity is required for better distribution and consumption. Solar-based energy production is gaining importance on a global level. As a part of clean energy utilization, manufacturing industries are moving towards renewable energy-based production to achieve carbon-neutral operations [2].

The major challenge in renewable energy sources is the fluctuation of the input sources such as SPVs and wind turbines. A maximum power point tracker (MPPT) can be employed to extract maximum power output at uncertain conditions. It can also be described as an electronic converter used to match the SPV power output to the utility or battery input. Either way, it performs an optimization function and helps in enhancing the overall PV system efficiency, increases battery life, and decreases grid utility failure. Pulse width modulation (PWM) is often confused with the MPPT: although they are both intended to act as DC-DC converters, the PWM controllers regulate only the output voltage, whereas the MPPT regulates both output current and voltage, making it more efficient, accurate, and suitable for large-scale PV power production.

There are many technologies for tracking the maximum power point. Conventional techniques such as perturb and observe (P&O) algorithm and incremental conductance algorithm are employed widely and these techniques are also called the hill-climbing algorithms because of the way they work. However, they have their disadvantages such as oscillation around the maximum power point and false MPPT when there is a rapid change in irradiance. Soft computing-based techniques such as artificial neural networks (ANNs), fuzzy logic, and swarm-based algorithms offer high precision and accuracy in a very low time frame [3]. ANNs emulate the functioning of brain neurons to solve a problem, whereas fuzzy logic is a computational approach based on different degrees of truth and swarm-based algorithms mimic the natural events such as bird flocking and ant colony organization to arrive at solutions for a given problem. The ease of implementation of these algorithms is another perk. The objective of this work is to explore the possibilities of ANNs for optimizing the output power of an SPV system. It also considers the advantages offered by the genetic algorithm (GA) in improving the performance of the proposed ANN-based power optimizer.

9.2 MODELING OF THE POWER OPTIMIZER

The design of a power optimizer includes the modeling of a PV array, an MPPT algorithm, and a DC-DC converter. Figure 9.1 depicts the structure of a power optimizer system with the PV array.

9.2.1 Solar Photovoltaic System

The solar array designed for the study has five Canadian Solar CS5C-80M modules connected in series. The datasheet for the module provided by the manufacturer indicates that the maximum power of a single module is 80.15 W with an open-circuit voltage of 21.8 V and a short-circuit current of 4.97 A. The array of five Canadian Solar CS5C-80M modules in series is simulated in MATLAB/Simulink using equation (9.1) and the values obtained from the datasheet.

$$I = N_{\mathrm{p}} I_{\mathrm{ph}} - N_{\mathrm{p}} I_{\mathrm{o}} \left[\exp\left\{ \frac{q\left(V_{\mathrm{pv}} + I_{\mathrm{pv}} R_{\mathrm{s}}\right)}{N_{\mathrm{s}} AKT} \right\} - 1 \right] \tag{9.1}$$

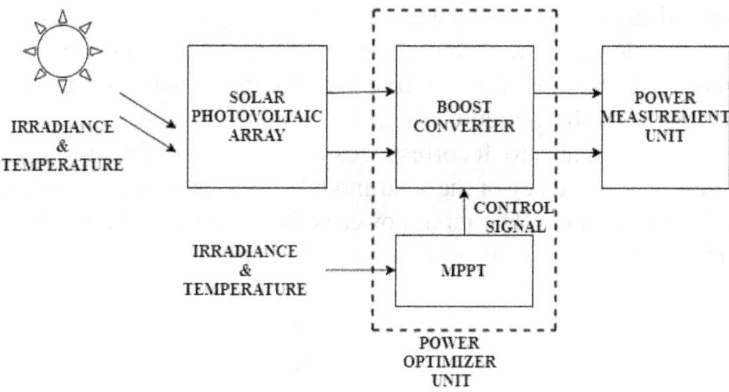

FIGURE 9.1 Block diagram of solar photovoltaic system with power optimizer.

where I_{ph} is the photocurrent of the module, N_p is the number of modules connected in parallel, N_s is the number of modules connected in series, I_o is the diode saturation current, K is the Boltzmann constant, and q is the elementary charge of the electron. A is the quality factor of the diode, R_s is the series resistance, and V_{pv} and I_{pv} refer to module voltage and current, respectively.

The *V-I* characteristics of the curve in Figure 9.2 describe the relationship between the current and the voltage of the solar module at standard testing conditions. P_M refers to the maximum power that can be delivered by the module, and V_M and I_M are

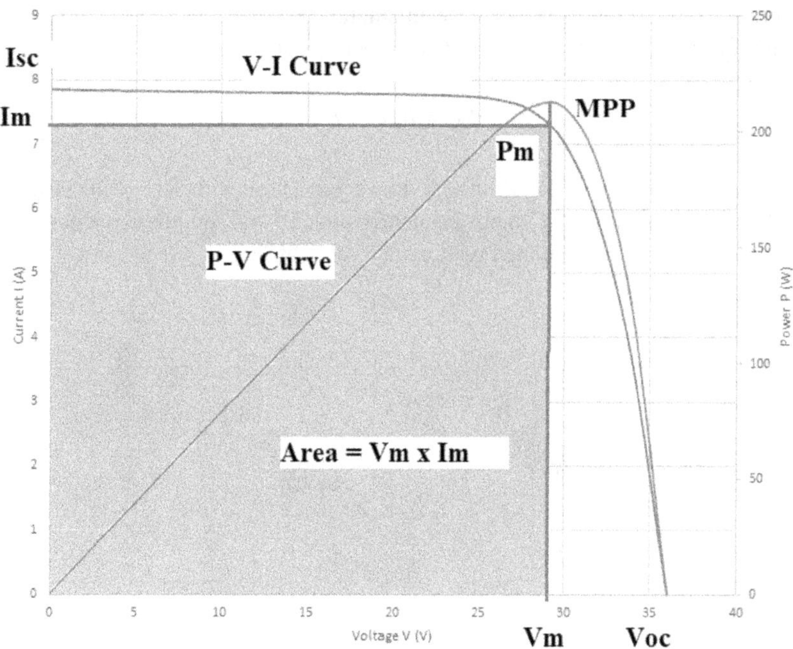

FIGURE 9.2 *V-I* and *P-V* characteristics of the PV array.

the voltage and current associated with the maximum power. The maximum voltage of the solar module when the current is zero is called the open-circuit voltage V_{OC}. It corresponds to the amount of forwarding bias due to the photogenerated carriers at the PN junction. The short-circuit current I_{SC} is obtained when the voltage across the module is maintained at zero. It corresponds to the number of photogenerated carriers generated. The efficiency of the solar module is the ratio of the maximum power output to the input power. The input power is the product of the incident radiation falling on the module G (W/m^2) and the area of module A (m^2).

$$n = \frac{P_M}{P_{IN}} = \frac{V_M I_M}{GA} \tag{9.2}$$

9.2.2 Boost Converter

The boost converter is employed in this study to step up the DC voltage produced by the PV array. The circuit of the boost converter is given in Figure 9.3.

The rate of conversion is usually determined by the duty cycle D, which is generated by the MPPT [4]. The relationship between the input voltage V_i and output voltage V_o of the boost converter is given in equation (9.3)

$$\frac{V_o}{V_i} = \frac{1}{1-D} \tag{9.3}$$

The optimum D value to remove the mismatch between the resistance of the PV module R_{PV} and the load resistance R_L is given by equation (9.4)

$$R_{PV} = R_L (1-D)^2 \tag{9.4}$$

The parameters of the 1 KHZ, 100 V boost capacitor with RL of 10 Ω includes a 0.625-mF inductor, 0.01-C input capacitor, and 2.5-mC output capacitor. All the component values are calculated with a duty value of 0.5.

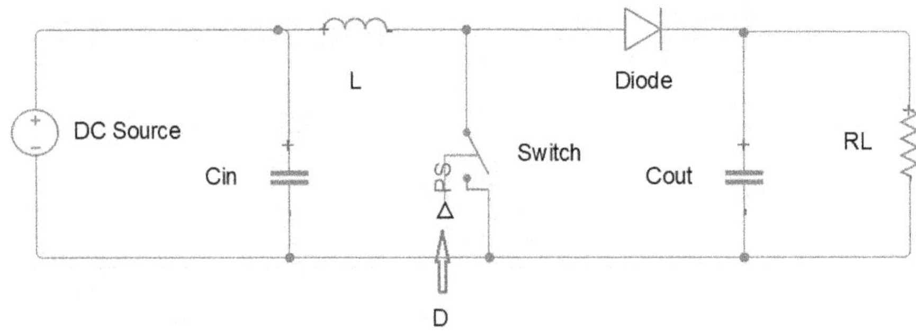

FIGURE 9.3 Boost converter.

9.3 MAXIMUM POWER POINT TRACKER

9.3.1 Perturb and Observe MPPT

P&O algorithm is a continuous process, which observes the power values and keeps perturbing until the maximum operational value of the power is reached. It is commonly referred to as a hill-climbing technique of tracking the maximum power point. This algorithm periodically perturbs the panel voltage and compares the output power with the previous value of the perturbation cycle [5]. When the power increases, the perturbation is kept in the same direction until maximum power is reached. Perturbation is reversed when there is a decrease in power, thus oscillating around the MPP [6]. The P&O algorithm is depicted as a flowchart in Figure 9.4.

The perturbations are usually of very low value (+/– 0.5), so that nothing can be missed in the PV curve and the entire area of the curve is traced with continuous observations and perturbations for accuracy. But this results in increased response time. The size of perturbations can be increased for faster performance but it will cause steady-state oscillations around the maximum power point [7].

9.3.2 Artificial Neural Networks-Based MPPT

ANNs emulate the working principle of brain neurons to solve real-time nonlinear, system-independent, stochastic problems. Neurons are the basic units of the network; they are interconnected and process the inflowing information. The interconnection space between neurons is referred to as synapse, which acts like a valve controlling the rate of flow of information from one neuron to the other. A simple neural network has three layers: an input layer, a hidden layer, and an output layer. However, the

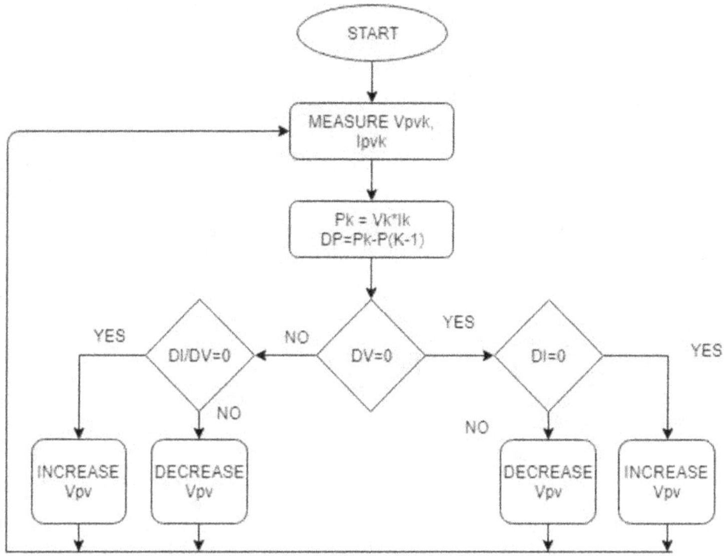

FIGURE 9.4 Perturb and observe algorithm.

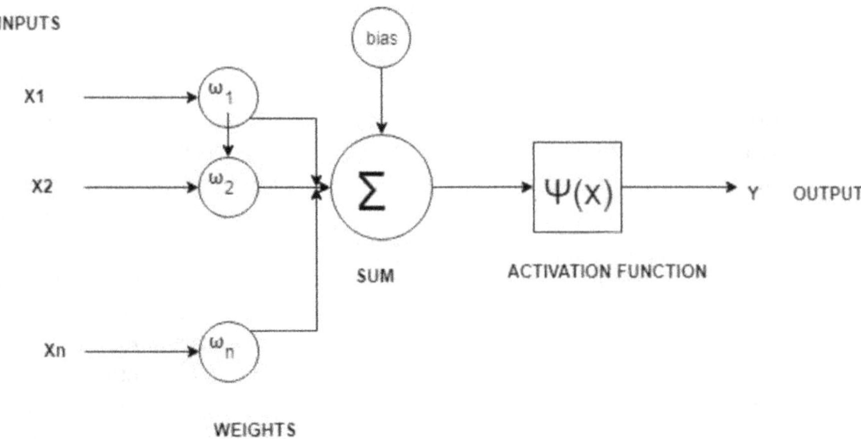

FIGURE 9.5 Architecture of ANN.

number of hidden layers can be designed as per the design and application requirements. The number of neurons in each layer is variable and problem-dependent [8]. The basic architecture of the network is given in Figure 9.5.

The information flows from the input nodes forward, while the errors associated with the process learning flows in the reverse direction until the weights of all neurons are adjusted accordingly. The network is trained for a particular application through training algorithms. A network can be trained multiple times until the desired performance is achieved [9]. In Levenberg-Marquardt (LM) algorithm, the training process is terminated when there is no improvement in generalization. It requires only less time but more memory. Scaled conjugate gradient algorithm has more similarity to the LM algorithm but requires less memory. The Bayesian regularization algorithm is suitable for difficult and noisy datasets. The training is based on adaptive weight minimization; it offers good generalization but is time-consuming. The performance of the network is determined based on the mean square error (MSE) and the regression value [10].

ANNs are employed as MPPT systems to predict the maximum voltage or power that is produced at a particular instance of time. This reference value is then compared with the instantaneous values to determine the duty cycle. The atmospheric parameters such as ambient temperature and solar irradiance are also considered as input variables apart from the module parameters such as short-circuit current, open-circuit voltage, temperature coefficient of voltage, temperature coefficient of current, voltage, and current. These input variables are processed by the hidden layers and the overall performance of the system depends on factors such as the number of neurons in the hidden layer, activation function chosen, and training process. The accuracy of the network depends on the quality of the training datasets. To achieve more accuracy, a large number of datasets are required [11]. In this method, the atmospheric parameters such as solar irradiance and temperature are considered as inputs to predict the maximum reference voltage for the PV array at a given weather condition. This reference voltage is compared with the instantaneous voltage of the PV array to generate the duty cycle for the boost converter. The proposed ANN has

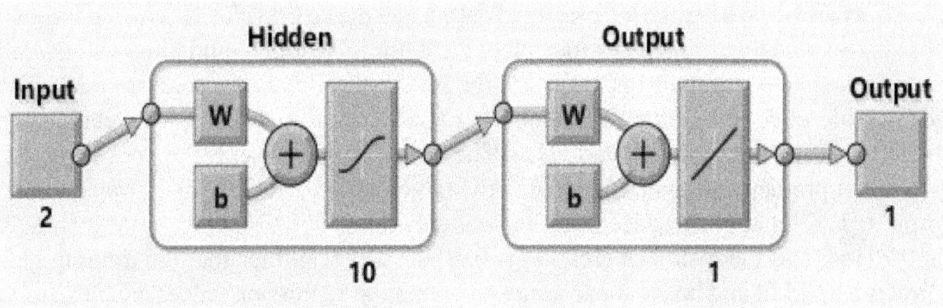

FIGURE 9.6 Structure of the proposed ANN.

an input layer with two input nodes, a hidden layer with 10 neurons, and an output layer with one output node and is given in Figure 9.6.

The best data are selected by simulating the PV array in MATLAB and employed for training purposes. A total of 231 datasets were obtained, among which 70% are used for training, 15% for testing, and 15% for validation. LM algorithm was used to train the network, where the back-propagation method is employed for error correction. The MSE is used to validate the performance of the ANN in all three stages – training, testing, and validation. The best validation is indicated by the dotted line in the plot. Figure 9.7 indicates that the best validation is achieved at epoch 6.

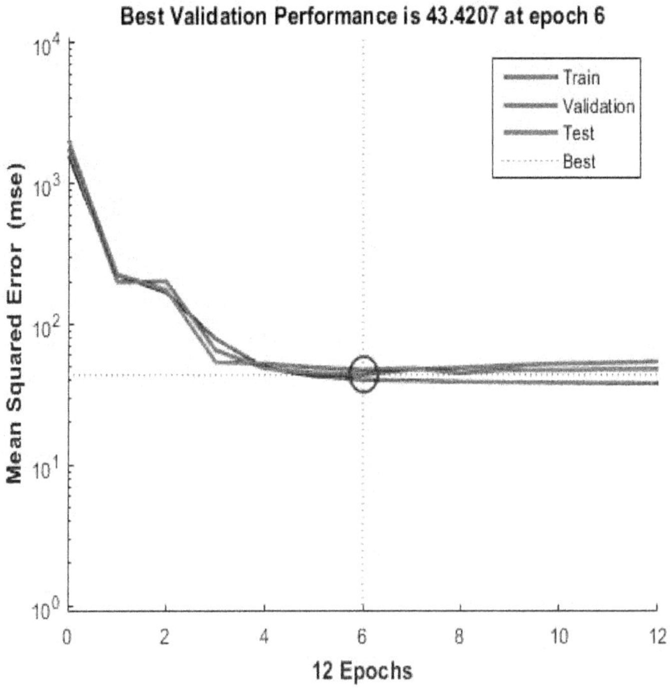

FIGURE 9.7 MSE of the ANN.

The correlation between the predicted output and the target value is depicted through the regression plots. It includes four plots for training, testing, validating, and overall data each. The continuous line indicates the best fit through linear regression (*R*). The ideal value of *R* is 1, and any value close to 1 signifies a perfect linear relationship between the target and predicted value. The dotted line represents the perfect results, where the predicted outputs are equal to the target values. The regression value for the proposed ANN at training is 0.91083, at validation it is 0.91676, at the testing phase it is 0.90364, and the overall regression is 0.91097. This signifies that the training data directs a good fit and hence the testing and validation regression values are close to 1.

9.3.3 A Genetic Algorithm Trained ANN-Based MPPT

GA is a method for optimizing discontinuous, highly nonlinear problems through evolutionary ideas of natural selection and genetics processes such as selection, crossover, and mutation. It randomly selects individuals from the given population as parents and produces children for the next generation. The population evolves towards an optimized solution over successive generations. The flowchart of the GA is given in Figure 9.8. This technique can be employed to generate an optimized reference value to generate a control pulse for the DC-DC converter to achieve MPP [12].

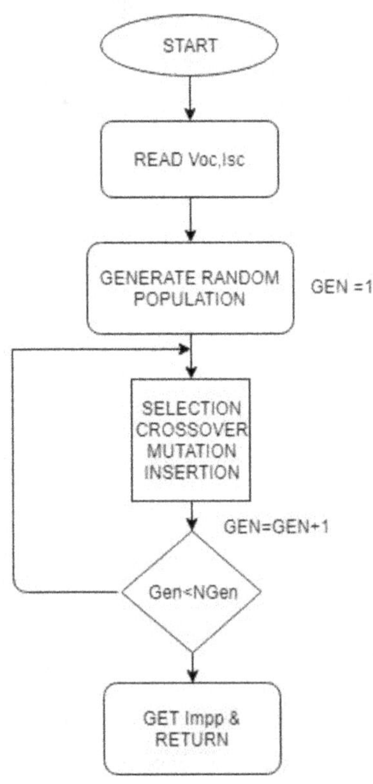

FIGURE 9.8 Genetic algorithm-based MPPT.

TABLE 9.1
Steps for Implementing Genetic Algorithm

Step No.	Functions
Step 1	Define objective function and the optimization parameters
Step 2	Initialization of the population size
Step 3	Evaluate the population with the objective function
Step 4	Test for convergence. If converged, stop the process or go to Step 5
Step 5	Start the reproduction process through genetic operators such as selection, crossover, and mutation
Step 6	Evolve new generation and go to Step 3

The most popular form of GA-ANN hybrid is to train the ANN with the GA-optimized datasets. GA is used to select the best value or the global maxima of V_m or I_m from a given population [13]. The optimized datasets increase the efficiency of the ANN even under rapidly varying weather conditions (VWCs) [14]. The authors in [15] have developed a neural network, where network learning is implemented with an online adaptation algorithm to inherit robustness and high speed of learning is achieved through GA optimization for predicting maximum power values under shading effects but it involves complex calculations which makes the PV plant scaling complex.

The role of the GA-based optimization technique is to determine the maximum value of power under certain ambient weather conditions, the output of which is used to train the customized ANN to control the boost converter [16]. The steps to implement GA are described in Table 9.1 [13].

The objective function of the GA is to detect the optimum value of $x = (x_1, x_2, x_3, \ldots, x_n)$. $F(x)$ is the array power output that is to be maximized [16],

$$F(x) = V(x) * I(x) \tag{9.5}$$

$$F(x) = N_s\left[V_o - \frac{R_s}{N_p} I(x) + \left(\frac{nK(T+273)}{q}\right) \times \ln\left(I_{pv} - \frac{I(x)}{N_p} + I_o\right) / I_o\right] * I(x), \tag{9.6}$$

bounded to $0 < I(x) < I_{sc}$.

The GA parameters used for the optimization of the nonlinear problem is given in Table 9.2.

Figure 9.9 indicates the best fitness identified by the GA through various generations. The straight line indicates the mean fitness and the dotted line indicates the best fitness. The best fitness is identified at –158,179 at a mean value of –156,600.

The design setup of the GA-based optimization in MATLAB window is shown in Figure 9.10. The lower and upper bounds for the three design variables are set in the problem setup window. These optimized values are applied to train the ANN, which acts as an MPPT.

TABLE 9.2
Parameters of the Genetic Algorithm

Parameters	Value
Number of design variables	3
Crossover function	0.8
Population size	50
Mutation rate	10%
Selection type	Roulette
Maximum generations	20

The proposed ANN has an input layer with two input nodes, a hidden layer with 24 neurons, and an output layer with one output node and is given in Figure 9.11.

A total of 224 datasets were obtained, among which 60% are used for training, 20% for testing, and 20% for validation. Scaled conjugate gradient algorithm was used to train the network. The scaled conjugate algorithm is a variation of conjugate gradient method but avoids the line search per learning iteration and offers supervised learning with a superlinear convergence rate.

MSE is the average squared difference between outputs and targets, and it is used to validate the performance of the ANN in all the three stages – training, testing, and

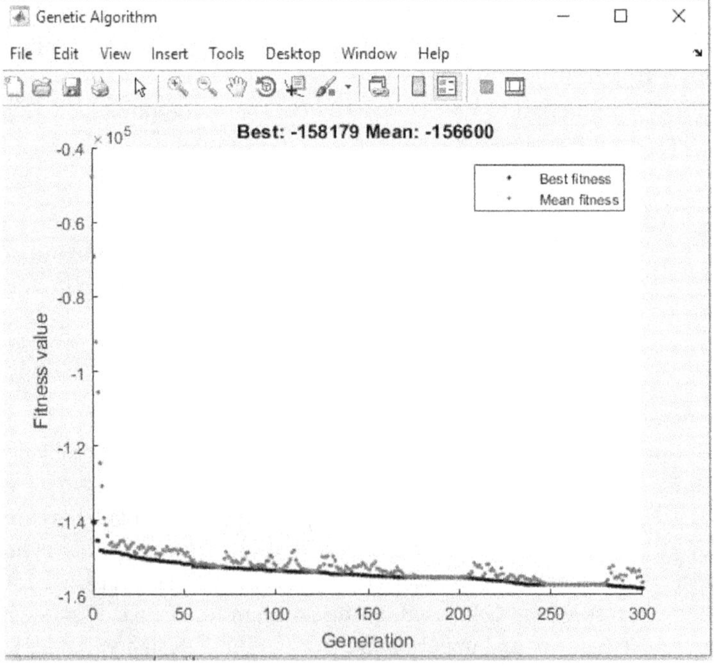

FIGURE 9.9 Fitness of the genetic algorithm.

Problem Setup and Results

Solver: ga - Genetic Algorithm

Problem

Fitness function: @reva

Number of variables: 3

Constraints:

Linear inequalities: A: b:

Linear equalities: Aeq: beq:

Bounds: Lower: [0 200 15] Upper: [4.97 1000 45]

Nonlinear constraint function:

Integer variable indices:

Run solver and view results

Use random states from previous run

Start Pause Stop

Current iteration: 300 Clear Results

Final point:

1	2	3
4.97	1,000	45

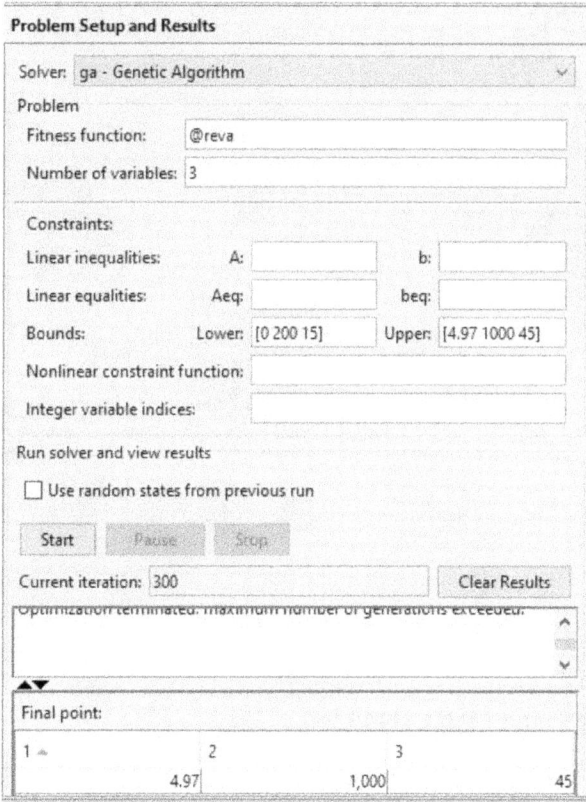

FIGURE 9.10 Genetic algorithm optimization window.

validation. Lower values close to zero indicate fewer errors. The best validation is indicated by the dotted line in the plot. Figure 9.12 indicates that the best validation achieved is 0.01553 at epoch 30.

The regression value at training is 0.99605, at validation it is 0.98634, at the testing it is 0.99436, and the overall regression is 0.99422. This signifies that the training data directs a good fit and hence the testing and validation regression values are close to 1.

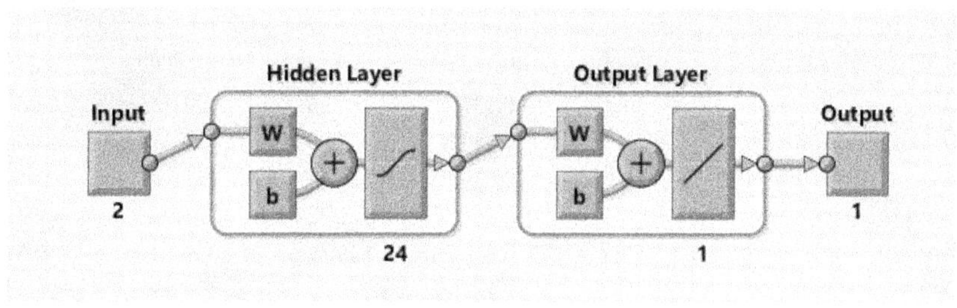

FIGURE 9.11 Structure of the ANN trained with GA.

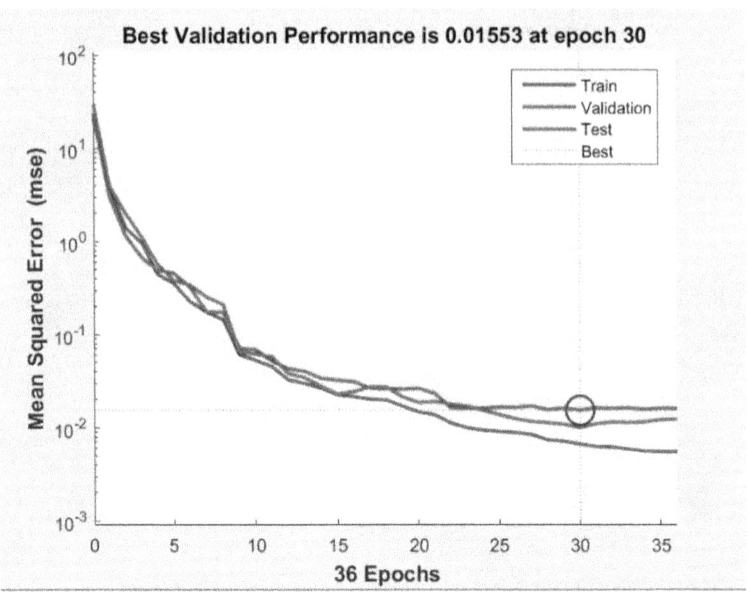

FIGURE 9.12 MSE of the ANN trained using GA.

9.4 RESULTS AND DISCUSSION

The performance of the proposed ANN is determined through the accuracy of the predicted values. The overall performance of the power optimizer is studied under both normal and shaded conditions to prove the strength of the proposed method. The performance of the ANN is interpreted through the error percentage calculated from the actual value and the predicted value. Table 9.3 displays

TABLE 9.3
Error Analysis

		ANN			GA-Trained ANN		
G	*T*	Trained *V*	Predicted *V*	Error (%)	Trained *V*	Predicted *V*	Error (%)
200	30	77.28082	77.35958	0.099364	78.112	78.152	0.05121
375	25	65.16298	64.98423	−0.27431	140.263	140.312	0.03502
450	25	107.0495	106.9432	−0.13792	200.037	200.097	0.03
500	15	144.1296	144.2467	0.139235	214.808	214.681	−0.0593
625	35	101.8984	101.2466	−0.63973	256.75	256.799	0.01908
775	40	199.2141	199.0246	−0.191	301.924	302.024	0.03312
850	20	306.8923	307.0654	0.161998	386.321	386.521	0.05177
900	45	296.6097	295.8976	−0.73705	390.393	390.835	0.11325
1000	35	400.1474	400.4663	0.318423	400.75	399.998	−0.1878

TABLE 9.4
Comparison of ANNs

	Training			Performance		
MPPT Technologies	**Training Datasets**	**Number of Epochs**	**Error (%)**	**Efficiency (%)**	**Tracking Speed (ms)**	**Oscillations at MPP**
ANN	231	10,000	1.7%	28.2	5	No
ANN-GA	225	30	0.1%	30.21	3	No

the inputs, actual values, predicted values, and the error percentage calculated by equation (9.7)

$$\text{Error } (\%) = \frac{\text{Predicted } V_{max} - \text{Actual } V_{max}}{\text{Actual } V_{max}} \times 100 \quad (9.7)$$

Table 9.3 depicts the error in the predicted values of the voltage. The error percentage ranges between −0.05% and 0.1% for the ANN trained using the GA predicted values, whereas the error percentage of the ANN trained using the PV simulated values ranges from 0.09% to 0.31%. The calculated values of error are very low and make no big difference in the duty cycle values; hence, it is negligible.

The two neural networks are compared in terms of the different parameters under both the training phase of the network and the performance aftermath in Table 9.4. The training datasets optimized by the GA mark a greater difference in both the phases.

The proposed maximum power point techniques are tested under five conditions. They are standard test conditions (STCs), VWCs, and three different partial shading conditions (PSC-I, PSC-II, PSC-III). The STCs indicate that the SPV array works with an irradiance of 1000 W/m^2 and a temperature of 25°C. The output power of various maximum power point techniques at STCs is given in Figure 9.13.

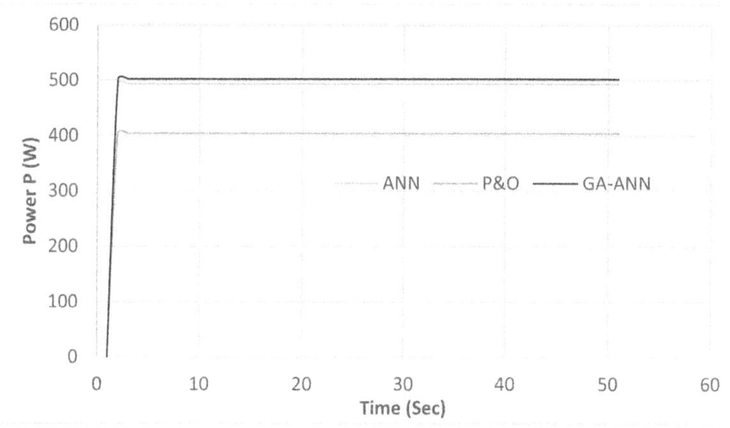

FIGURE 9.13 Output power under STC.

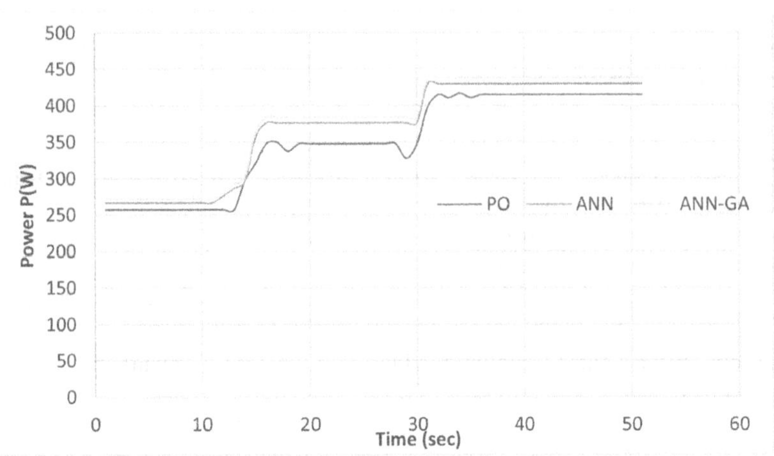

FIGURE 9.14 Output power under varying weather conditions.

Figure 9.14 depicts the results of different MPPTs under VWCs. Under this condition, the PV array was tested for rapidly varying atmospheric conditions, which were achieved by varying the irradiance and temperature pattern. G (W/m^2) = [700, 800, 900]; T(°C) = [25, 30, 35].

PSC-I is influenced in the PV array by reducing the irradiance input of one module by half (500 W/m^2). In PSC-II, two modules out of the five in the PV array are partially shaded with an irradiance value of 400 W/m^2 and 500 W/m^2, respectively. In PSC-III, three modules out of the five in the PV array are partially shaded with an irradiance value of 400 W/m^2, 500 W/m^2, and 600 W/m^2, respectively. Figure 9.15 describes the effect of the three PSCs on the power-voltage characteristics of the array.

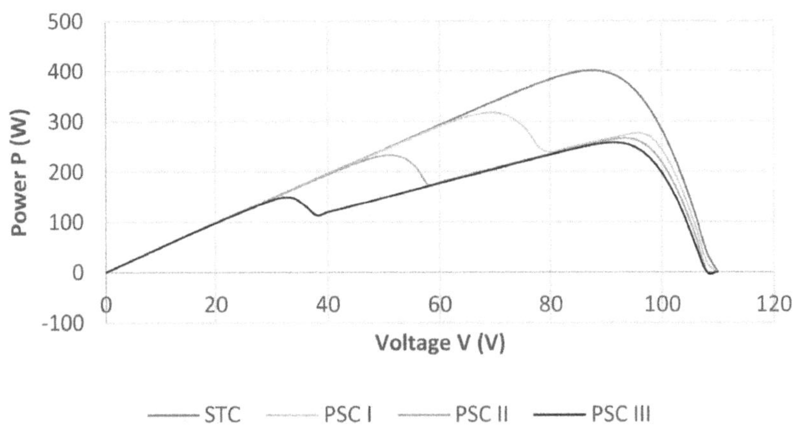

FIGURE 9.15 PV curve under PSCs.

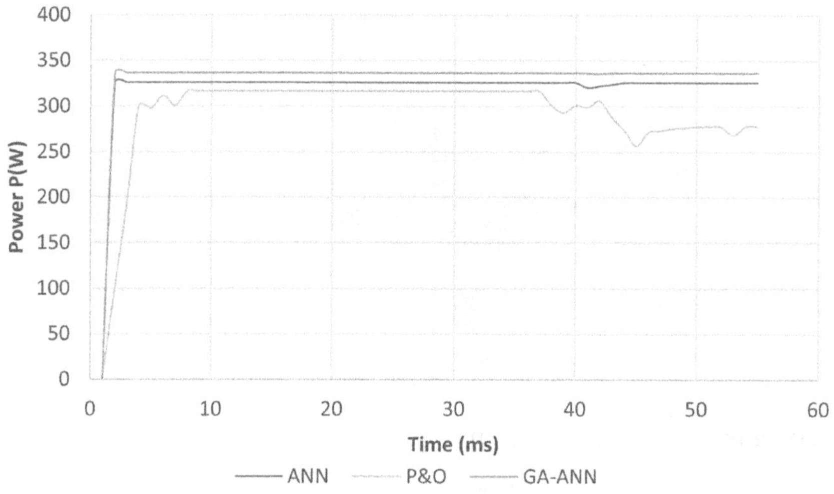

FIGURE 9.16 Output power at PSC-I.

The performance of the different MPPTs under PSC-I is shown in Figure 9.16. The ANNs converge faster than the conventional P&O algorithm. The maximum power of the ANN optimizer is 325 W and it is higher than the PV array maximum power by 2.8%. The GA-trained ANN outperforms the ANN trained with the PV simulated values by 3.51%.

Figure 9.17 represents the performance of the MPPTs under PSC-II. The neural network-based optimization comes with the P_{max} value of 269 W, which is 15% greater than the array value. The GA-trained ANN outperforms the ANN trained

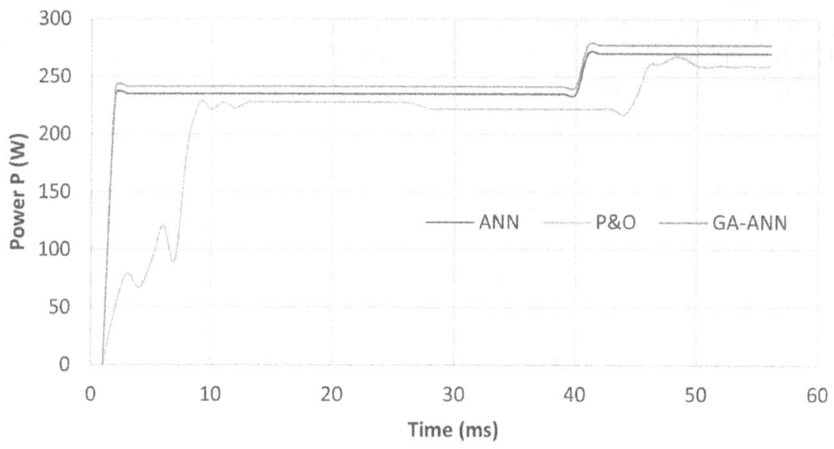

FIGURE 9.17 Output power at PSC-II.

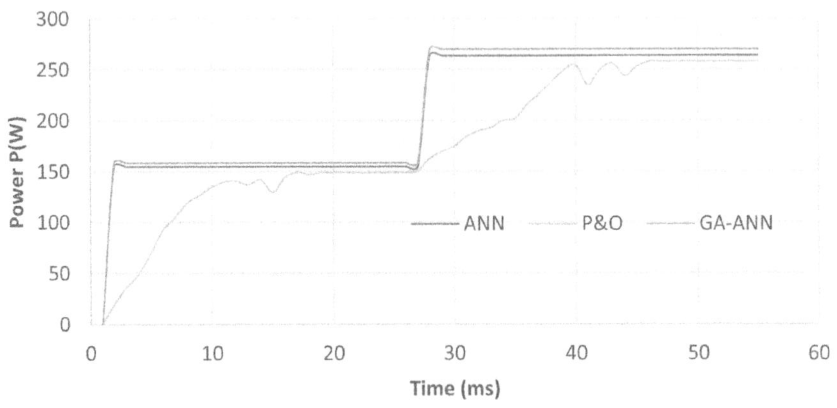

FIGURE 9.18 Output power at PSC-III.

with the PV simulated values by 4%. In addition to the maximum power value, the response time is also less than the ANN response time under the PSC-II.

The performance of the MPPTs under PSC-III is represented in Figure 9.18. The neural network-based optimization comes with the P_{max} value of 263 W, which is 7.7% greater than the array value. In addition to this, there are no fluctuations and the response time is higher than the conventional P&O MPPT when three modules are partially shaded in the array of five modules. The GA-trained ANN outperforms the ANN trained with the PV simulated values by 2.34%. In addition to the maximum power value, the response time is also less than the ANN response time under the PSC-III.

Table 9.5 presents the comparison of the three maximum power point techniques discussed in this work. It is evident from the results that the ANN-based MPPT works at a good speed and reduced oscillations compared to the conventional P&O algorithm. However, the role of a GA in optimizing the training datasets for the ANN has increased the efficiency of the ANN by 2% and at a decreased tracking time of 3 seconds. Hence, integrating GA with ANN has a positive impact on the overall performance of the power optimizer.

TABLE 9.5
Comparison of the MPPTs

Evaluation Parameters	P_{max} (KW)	Tr (ms)	η (%)	Oscillations	Sensors
P&O	0.414	15	24	High	*V*, *I*
ANN	0.493	5	28.2	No	*G*, *T*
ANN-GA	0.502	3	30.21	No	*G*, *T*

9.5 CONCLUSION

In general, the MPPT design depends on the type of PV system, its application, cost, and geographic location. The diverse conditions make it difficult to evaluate the MPPT techniques with some preset evaluation parameters. For a better and fair assessment of the hybrid technologies, the knowledge of the independent technologies is significant. ANNs depict pattern-based learning, which makes it faster and independent of any system or process it is applied for. However, their performance depends on the training process, so an ANN trained for a particular PV system cannot be employed for another, which makes it exclusive. The ANN technology demonstrates better convergence and efficiency, but the training datasets needed some fine-tuning. Hence, a GA was applied to generate highly optimized datasets for training the ANN. The GA was applied offline due to memory constraints. In the proposed work, ANN improves the performance of the power optimizer over the conventional P&O algorithm and the GA enhances the training part of the ANN. Thus, both the technologies involved are complementing one another, resulting in good tracking speed, no oscillations when MPP is tracked, very low convergence speed, and high reliability under changing irradiances.

REFERENCES

1. Khalil, T. M. (1981). Comparative analysis of energy resources. *International Journal of Production Research*, *19*(4), 401–409.
2. Jin, T., Shi, T., & Park, T. (2018). The quest for carbon-neutral industrial operations: Renewable power purchase versus distributed generation. *International Journal of Production Research*, *56*(17), 5723–5735. DOI: 10.1080/00207543.2017.1394593
3. Sedaghati, F., Nahavandi, A., Badamchizadeh, M. A., Ghaemi, S., & Fallah, M. A. (2012). PV maximum power-point tracking by using artificial neural network. *Mathematical Problems in Engineering*. https://doi.org/10.1155/2012/506709.
4. Motahhir, S., El Ghzizal, A., Sebti, S., & Derouich, A. (2018). Modeling of photovoltaic system with modified incremental conductance algorithm for fast changes of irradiance. *International Journal of Photoenergy*. https://doi.org/10.1155/2018/3286479.
5. Christopher, I. W., & Ramesh, R. (2013). Comparative study of P&O and InC MPPT algorithms. *American Journal of Engineering Research*, *2*(12), 402–408.
6. Sreedhar, S., & Jagadeesh, D. (2016). A review on optimization algorithms for MPPT in solar PV system under partially shaded conditions. *IOSR Journal of Electrical and Electronics Engineering*, *1*, 23–32 (Two day National Conference on "SMart grid And Renewable Technologies" (SMART-2016).
7. Ezinwanne, O., Zhongwen, F., & Zhijun, L. (2017). Energy performance and cost comparison of MPPT techniques for photovoltaics and other applications. *Energy Procedia*, *107*, 297–303. https://doi.org/10.1016/j.egypro.2016.12.156
8. Babaie, M., Sebaaly, F., Sharifzadeh, M., Kannan, H. Y., & Al-Haddad, K. (2019). Design of an artificial neural network control based on Levenberg-Marquart algorithm for grid connected packed U-Cell Inverter. In: *2019 IEEE International Conference on Industrial Technology (ICIT)*, Melbourne, 1202–1207.
9. Rai, A. K., Kaushika, N. D., Singh, B., & Agarwal, N. (2011). Simulation model of ANN based maximum power point tracking controller for solar PV system. *Solar Energy Materials and Solar Cells*, *95*(2), 773–778. https://doi.org/10.1016/j.solmat.2010.10.022

10. Hadji, S., Gaubert, J. P., & Krim, F. (2015). Theoretical and experimental analysis of genetic algorithms based MPPT for PV systems. *Energy Procedia*, *74*, 772–787. https://doi.org/10.1016/j.egypro.2015.07.813
11. Bouselham, L., Hajji, M., Hajji, B., & Bouali, H. (2017). A new MPPT-based ANN for photovoltaic system under partial shading conditions. *Energy Procedia*, *111*, 924–933. https://doi.org/10.1016/j.egypro.2017.03.255
12. Balasubramanian, G., & Singaravelu, S. (2012). Fuzzy logic controller for the maximum power point tracking in photovoltaic system. *International Journal of Computer Applications*, *41*(12), 22–28. https://doi.org/10.5120/5594-7840
13. Rezvani, A., Izadbakhsh, M., Gandomkar, M., & Vafaei, S. (2015). Implementing GA-ANFIS for maximum power point tracking in PV system. *Indian Journal of Science and Technology*, *8*(10), 982–991. https://doi.org/10.17485/ijst/2015/v8i10/51832
14. Paul, S., & Thomas, J. (2014). Comparison of MPPT using GA optimized ANN employing PI controller for solar PV system with MPPT using incremental conductance. In: *2014 International Conference on Power Signals Control and Computations, EPSCICON 2014*, Thrissur, 1–5. https://doi.org/10.1109/EPSCICON.2014.6887518
15. Sarenyadhevi, K., & Rajasekaran, N. (2015). An intelligent approach for MPP tracking in solar panel using artificial neural network and genetic algorithm. *International Journal of Research and Engineering*, *2*(4), 37–40.
16. Prasad, L. B., Sahu, S., Gupta, M., Srivastava, R., Mozhui, L., & Asthana, D. N. (2017). An improved method for MPPT using ANN and GA with maximum power comparison through perturb & observe technique. In: *2016 IEEE Uttar Pradesh Section International Conference on Electrical, Computer and Electronics Engineering, UPCON 2016*, Varanasi, 206–211. https://doi.org/10.1109/UPCON.2016.7894653

10 Predictive Maintenance

AI Behind Equipment Failure Prediction

S. Sharanya[1], Revathi Venkataraman[1], and G. Murali[2]
[1]Department of Computer Science and Engineering, SRM Institute of Science and Technology, Tamil Nadu, Chennai
[2]Department of Mechatronics Engineering, SRM Institute of Science and Technology, Tamil Nadu, Chennai

CONTENTS

10.1 INTRODUCTION

Artificial intelligence (AI) has acquired a profound place in almost all fields. The ongoing changes in the energy sector demand the need for dynamic and robust system that could very well adapt to changes. As the world has shifted its paradigm to explore renewable energy resources, more capital investments are made in diverse geographical locations to cater the needs of power-thirsty world, in deploying and installing the renewable power generation equipment. As a matter of fact, the equipment is prone to failures due to incidental or accidental damages and normal degradation (aging) due to wear and tear. The restoration or maintenance activity is accompanied by two serious consequences: (1) disruption of power generation and (2) partial or complete replacement of equipment. Hence, predicting the onset of equipment failures so as to schedule planned, predictive maintenance activity is the optimal solution to mitigate the effect of failures.

Condition monitoring (CM), fault diagnosis, and failure prediction are related terminologies habituated in almost all industrial sectors for health surveillance of equipment. CM is the process of unceasing surveillance of the system, consisting of activities such as system monitoring, fault detection, fault diagnostics, and fault prognostics. Quantitative CM of any equipment is performed by assessing the critical variables involved in the system's function. The reactive detection of fault in the machinery activates the diagnostic module to identify and characterize the fault for further investigation. Fault prognostics, a proactive fault prediction strategy, is gaining more significance, since it uses CM tools to predict the possible faults from their early signs, which in turn lowers the maintenance frequency to avoid unplanned reactive maintenance. Predicting equipment failure is crucial for building and maintaining renewable energy infrastructure. Intelligent fault prediction strategies will revive the perception of countries to invest in renewable energy sources and to harvest clean energy.[1]

The twin challenges confronted by the renewable energy sector are the complexity of the equipment and location of its deployment. Predicting equipment failure through AI-based techniques will help the plant to achieve its highest energy spec by:

- Reducing the overall equipment maintenance time
- Increasing the working hours of the equipment
- Minimizing the cost of spare parts and supplies
- Ensuring operator safety

Thus, a comprehensive vision of predictive equipment failure strategies will confirm uninterrupted power generation in the long run by learning the system/equipment behavior through the critical parameters of the equipment. The AI-based techniques naturally fit into equipment failure prediction as they can forecast the system behavior from the past and present conditions/data to plan or schedule predictive maintenance. Intelligent robotic probe, automatic plant monitoring through sensors, remote monitoring and control, and reflexive activation of prognostic modules are some of the preexisting failure-handling mechanisms backed by the power of AI. This chapter throws light on the various AI-based failure prediction strategies that are deployed on renewable source infrastructure.

10.2 FAILURE MODES AND EFFECT ANALYSIS

Failure modes and effect analysis (FMEA) is a proactive, systematic and structured approach for analyzing reliability concerns in equipment/systems to identify the failure modes and its effect on the product/operation and hence to uncover the possible recovery actions.[2] The term *failure modes* is an extensive list of possible ways in which a failure or fault can occur and *effect* is the damage or harmful consequence of the occurrence of failure. The vital factors in deciding the impact of failures are frequency of its occurrence, damage intensity, and their isolation.[3] *Risk priority number* (RPN) is a quantitative metric that assesses the impact of the failure on any equipment and is given by Eq. 10.1:

$$RPN = Severity \times Occurrence \times Isolation \tag{10.1}$$

The failure modes are effectively ranked based on the RPN value. Some of the common types of FMEA are as follows:

- *Design failure mode and effect analysis (DFMEA)*: This delves into the area of design-related failure prevention, malfunctioning of the product and other safety and reliability aspects of the equipment. DFMEA attracts significance since it is one of the effective tools to detect failures at a very early stage, thus optimizing the entire process of FMEA by mitigating the negative effects of the failure.
- *Process failure mode and effect analysis (PFMEA)*: This explores the faults and failures that may appear during assembling, fabricating and manufacturing the equipment. The common faults encountered are missing machinery parts, misalignment, disproportionate dimensions, etc.

10.2.1 Process of FMEA

The disciplined engineering of FMEA is disseminated into stringent modules to investigate the failure modes to focus on reliable equipment development process.[4]

i. Identify possible failure modes of the equipment
ii. Rank the failure modes based on RPN

iii. Determine the total risk estimate (TRE) as mentioned in Eq. 10.2.

$$TRE = \frac{\sum_{i=1}^{n} RPN_i}{n \times 1000} \times 100\% \tag{10.2}$$

Here, the RPN_i is the actual RPN value of the ith cause among n causes listed in step (i).

iv. Isolate critical components/equipment which are of high priority.

v. Formulate guidelines for optimal corrective and preventive actions: the impact of the corrective action is determined by the difference in RPN values before and after the corrective action is implemented and is mentioned in Eq. 10.3.

$$\frac{RPN}{F_i} = \frac{RPN_{i\_before} - RPN_{i\_after}}{F_i} \tag{10.3}$$

where F_i is the feasibility ranking in the range [1, 10].

vi. Quantize the effectiveness of applying FMEA (ΔRPN_R) procedure through Eq. 10.4, which is the difference between the sum total of the RPN before ($\sum RPN_{i\_before}$) and after ($\sum RPN_{i\_after}$) the corrective actions.

$$\Delta RPN_R = \frac{\sum RPN_{i\_before} - \sum RPN_{i\_after}}{\sum RPN_{i\_before}} \times 100 \tag{10.4}$$

This score is a direct indication of mitigated impact of the failures that may occur at various stages of product development and deployment.

10.2.2 Frequently Occurring Failures in Renewable Power Generation

The infrastructure costs for renewable power generating units such as wind turbines and hydroelectric power plants are high. Some of the common failures that are more prone to occur in various power-generating systems are discussed in brief.

- *Wind turbines*: Heavy, inconsistent winds and unpredictable weather induce different kinds of failure modes in wind turbines. Some of the notable ones are: surface delamination and cracks in blades; faults in main bearing, gear boxes, and generators; defects in pitch systems, convertor units, yaw systems, and braking systems; and buckling of wind towers due to geometric imperfections. The failure modes in wind turbines are mainly due to lack of competent core technology, tradeoff in cost, and material quality, absence of design standards and quality checks, and climatic differences.[5]
- *Hydroelectric power plants*: Violation of safety standards in hydroelectric power units may cause devastation and sometimes casualties. The frequent

failure modes in hydraulic turbines are cavitation, fatigue, material defects, corrosion, shaft misalignment, rotor generator defects, and faults in transformer and switchgears.[6]

- *Solar power plants*: Degradation in solar panels can occur due to short-circuiting, development of thermal stress or hotspots, assembly damage, snail trail contamination, PID (Potential Induced Degradation) effect, encapsulant disorders, corrosion, thermal cycling, damp heat, microcracks, shading of photovoltaic (PV) cells, dirt and debris accumulation, delamination, and busbar soldering faults.[7] In addition to these, inherent upper bound power limitation of inverter, low operating voltage of PV generator, undersized inverter, isolation faults, gear faults, wrecking of electronic components, and circuit breaking also impart failures in solar power plants. The effect of environmental factors such as ambient temperature, snowfall, dust, sandstorms, wind velocity, and humidity should not be undermined in failure analysis.[8]
- *Geothermal systems*: This versatile renewable energy has a very complex system, which may be prone to various failure modes. The defects may occur at cooling towers, gas extraction system, generator and other electrical components, turbines, steam, and transmission stations.[9]
- *Biomass and biogas plants*: Failure modes in biomass and biogas plants may be bifurcated into structural and chemical failure. Defects in biogas utilization components, pipelines, production systems, and disposal system are structural failures.[10] The extensive chemical reactions in the biogas plants generate unfavorable byproducts such as volatile fatty acids, oxygen, struvite (compound of magnesium ammonium phosphate), and carbon dioxide. This renewable energy source has faults and failures specific to its kind such as alkalinity, imbalanced pH, foaming, variation in temperature, heavy metal, and effective usage of residue.

To summarize, the renewable energy generation has diverse failure modes, which demands continuous surveillance to render uninterrupted power generation. The integration of AI methods with the classical CM promises a safer and economical way to tamper the renewable power generation without getting trapped into any unprecedented crisis.

10.3 CONDITION MONITORING STRATEGIES

CM is continuous surveillance of the system by assessing the critical variables that are vital for functioning of the system. A failure in the equipment is identified when the values of the variables deviate from the normal operational profile.[11] The course of CM is disintegrated into the phases of fault detection, fault identification or isolation, fault diagnosis, and finally prognostics, as shown in Figure 10.1.

CM targets to smoothen the impact of failures on the equipment, thereby increasing its healthy operation time which is quantized in terms of availability (Eq. 10.5). The factors that arbitrate equipment availability are operational time and downtime. The duration in which the equipment meets its expected performance is *operational*

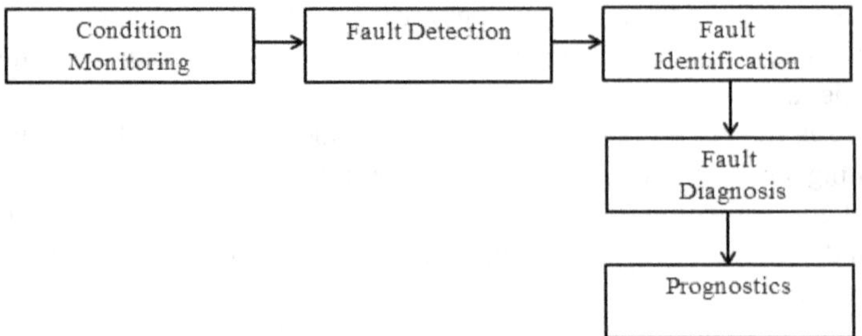

FIGURE 10.1 Condition monitoring and fault diagnosis.

time. On the other hand, *downtime* is the duration in which the equipment fails to give its conferred performance.

$$Availability = \frac{Operational\_time}{Operational\_time + Down\_time} \tag{10.5}$$

It is evident that all the renewable energy generators are expected to work to their full functionality to cater the power demands of the country, which implies that they should be ensured with high degree of availability. The maintenance strategies for monitoring equipment failure can be classified as reactive and proactive maintenance. As the name suggests, reactive maintenance policies are event triggered. They are invoked after the occurrence of equipment failure, which is of little use in critical systems like power generators. The proactive policies are farsighted; they detect or anticipate the failures from the data collected through CM. This makes the proactive strategies to be more proficient in handling maintenance activities of mission critical power plants. Proactive CM policies come in two flavors, namely, preventive and predictive maintenance.

10.3.1 Preventive Maintenance

Periodic and time-scaled maintenance activities are the characteristic features of preventive maintenance. The power plant is inspected for the occurrence of any failures in fixed time intervals (time-based) or after fixed running time (usage-based). Both these methods will prevent the equipment failure or breakdown before it occurs, thus saving time and costs.

10.3.2 Predictive Maintenance

Predictive maintenance is powered by CM tools for early equipment failure detection with manifold advantages[12]:

- Reducing the overall equipment maintenance time
- Increasing the working hours of the equipment
- Minimizing the cost of spare parts and supplies
- Ensuring operator safety

The notion of predictive maintenance is fueled by predictive analytics, machine learning, deep learning and other AI methods.

10.4 NICHE OF ARTIFICIAL INTELLIGENCE IN PREDICTIVE MAINTENANCE

The quest for proactive equipment failure prediction in renewable energy plants has seen renowned intensification due to prominent factors such as Industry 4.0, extensive usage of sensors, advancements in data acquisition and handling tools, progression in data analytics and competent predictive power of machine learning algorithms.[13] Predictive maintenance, under the canopy of prognostics and health management, has become a recent research attraction.

The failure of equipment in power plants can be detected through diverse data such as: acoustic variation; vibration analysis; temperature, pressure and humidity transitions; and debris, desalination and snow formation in the components. The classic analytical failure detection mechanisms deployed in power plants demand physical investigation of equipment, fatigue, quality and capacity of generated power and electrical effects. All the above-mentioned strategies incur cost in terms of human labor and impaired accuracy. Also, none of the strategies could fore-warn about the occurrence of failures. All of these factors envision the AI-based equipment failure prediction methodologies that could monitor the health of power plants deployed in remote locations without intensive human interactivity and with improved accuracy.

The entire work plan of building an AI-based system for equipment failure prediction begins at data acquisition from sensors, which is followed by calculating the metric for performance parameter monitoring. The deviations from the normal operational profile are matched for the predefined patterns to diagnose the faults. The trend analysis shows future behavior which can further project the exact operational failure time. This is depicted in Figure 10.2.

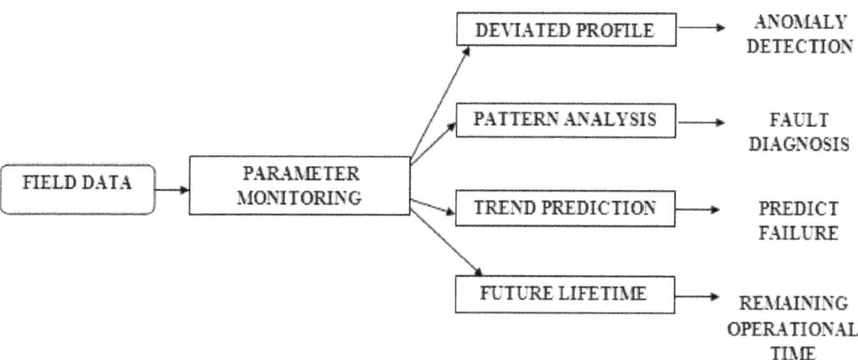

FIGURE 10.2 AI-based equipment failure prediction.

10.5 FAILURE PREDICTION IN SOLAR POWER SYSTEMS

The solar energy is sourced in two ways, namely, thermal energy and PV energy. The latter is the major source of solar power generation and directly converts the solar energy to electricity, which is in contrast to the former that converts the sunlight to heat energy; this heat drives a heat engine to finally produce electricity. Predictive maintenance of solar power generators using AI augmented with Big Data and IoT (Internet of Things) has become more common. The strategy of equipment failure can be generalized as *learning phase*, where the AI model will be trained to learn the faults, and *prediction phase*, where the model will predict the occurrence of failures from the learned knowledge. Both the phases may use same or different techniques to build the entire failure prediction model.

10.5.1 Approach Using Big Data

This uses supervisory control and data acquisition (SCADA) tool to collect the data evidence to predict the incipient faults in two modules, namely, supervision diagnosis module (SDM) and short-term fault prediction model (FPM).[14]

- *SDM*: This employs self-organizing map (SOM) for failure prediction of inverters from the deviant values of power. The topology-preserving property of SOM enables the diagnosis of incipient faults through the variant output mappings. This module is capable of predicting only generic failures but the specific failure class prediction is done in the next phase.
- *Short-term FPM*: The generic failures diagnosed in the previous phase are classified into specific fault classes using pattern recognition-based feed-forward artificial neural networks (ANNs). This module is capable of predicting inverter faults proactively, in an advance of 7 days. The failure classes considered are AC switch open, thermal fault, input overcurrent and DC ground fault.

10.5.2 Approach Using Deep Learning

Unshared convolution neural network is deployed to predict the failures in PV panels from the 2D power curves.[15] The architecture has two convolution layers: one that shares the convolution parameters to the next layer and another that does not share the parameters. The unshared convolution layer is capable of predicting the shadowing effects, thus obtaining performance gain from it. This is a simple architecture that effectively interpolates the input data to predict the failures in the PV panels.

10.5.3 Approach Using ANN

Deploying models based on ANN could also be a promising solution for failure prediction in PV systems. The primary tasks involved in building ANN are deciding the block structure, mode of data acquisition, categorizing fault classes and testing the network.[16] A simple neural network that may take three variables, namely, voltage, current, and power, as an input with a reasonably bigger hidden layer will output fault classes. The failures can be diagnosed based on the heat flow density obtained in Watt/m^2.

Alternatively, ANN can also be deployed in an advanced scenario to make more accurate prediction. The fault prediction can be done by determining the difference between normal and anomaly values of solar irradiance and temperature, expressed as residue vector.[17] The failure prediction alerts are generated after observing any degradation pattern from the normal operational profile.

Studies indicate that the above-mentioned methods can forecast failures in PV systems for about a period of 7 days in advance, which may help the operator to schedule the maintenance activity.

10.6 FAILURE PREDICTION IN WIND TURBINES

The wind turbines are complex machinery where minor faults propagate and may cause failures in other parts also. The deployment of wind turbines is done in two modes, onshore and offshore, which makes them more susceptible to unpredictable extreme weather conditions. To alleviate the adverse effects of wind turbine failures, it is very essential to predict the faults to facilitate the scheduling of maintenance activity. The common parameters used in CM of wind turbines, as shown in Figure 10.3, are[20] as follows:

- Wind parameters: speed and deviations
- Performance: output power, pitch angle, rotor speed
- Vibration: tower and drivetrain acceleration
- Temperature: oil, bearing and gearbox temperature

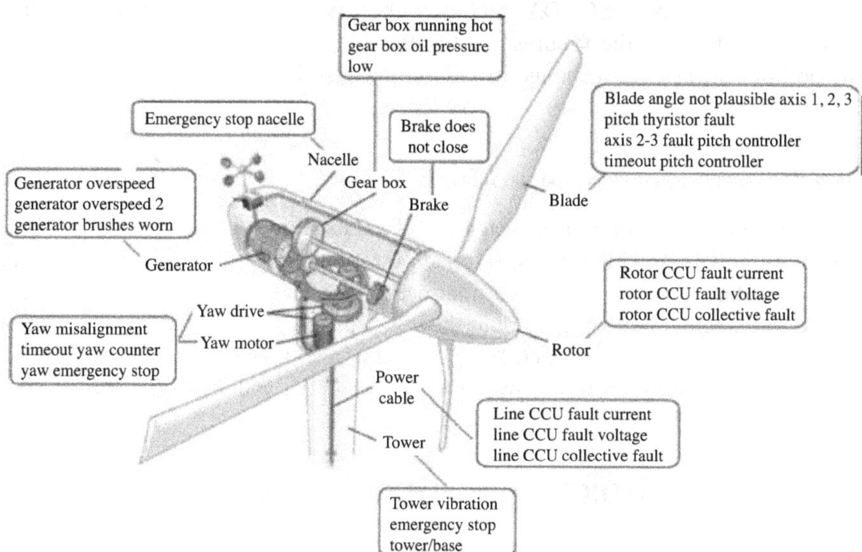

FIGURE 10.3 Provisioning of condition monitoring system in wind turbines.

A variety of AI-based approaches that fall under the umbrella of clustering, normal and abnormal behavior modeling, trend and pattern analysis, and assessment of expert systems are used in failure prediction of components in wind turbines.[18]

10.6.1 Approaches Using Unsupervised Learning

The temperature SCADA data from the wind turbines is preprocessed to eliminate redundancy and noise. The nonlinear principal component analysis (PCA) is implemented using autoassociative neural networks.[19] The vibration data of drivetrain and tower obtained from the SCADA also serves as best predictor metric in predicting failures. The abundance of vibration data measured through the installed accelerometers is scaled down through dimensionality reduction techniques such as PCA or unsupervised clustering techniques such as K-means.[20]

10.6.2 Approaches Using ANN

The ANNs have become very robust and are extensively used in failure prediction of all components. Gearbox faults in wind turbines are effectively predicted by using a three-layered ANN by monitoring temperature, thermal difference and oil temperature.[21] This can be augmented with knowledge-based expert system with fuzzy rules to diagnose and predict the gearbox failure time, recommendations to maintenance and optimal time for maintenance.

10.6.3 Approaches Using Bayesian Network

The Bayesian network functions based on the predefined user posterior probabilities to arrive the unobserved probabilities of the future events. A Venn diagram-based Bayesian network over SCADA data uses the evidence and their corresponding inferences to forecast the failures. As the Bayesian networks can derive the exact probabilities of failures, a change in failure probabilities can be noticed as the evidence accumulates over time.[22]

10.6.4 Approach Using Neuro-Fuzzy Systems

Adaptive neuro-fuzzy inference system (ANFIS) is deployed to find the pitch faults in wind turbines using the wind speed and power output from the SCADA data.[23,24] The high adaptivity can model the nonlinear relationships between the input and output. The detailed knowledge about the wind turbines is encoded into the system using a priori knowledge transfer. A more common approach to predict the failures is to count the alerts generated by the rules on a day, and when the threshold is breached, then failures can be forecasted.

10.7 FAILURE PREDICTION IN HYDROELECTRIC SYSTEMS

The habitual failures that are more prone to occur in hydroelectric power plants are vibrations, fluctuations in pressure, cavitation, and declined turbine efficiency. The schematic representation of CM in hydroelectric power plant is given in

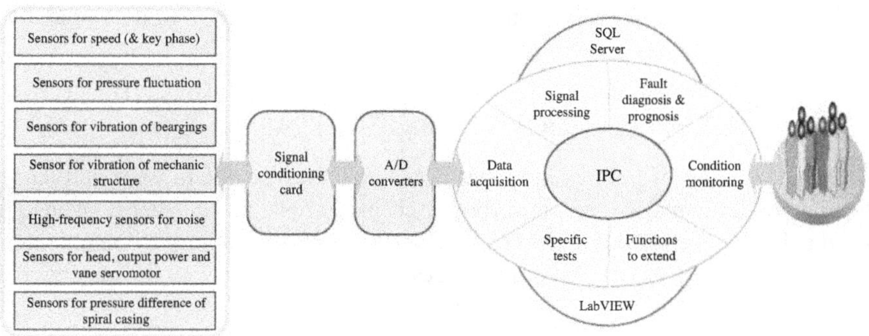

FIGURE 10.4 Condition monitoring framework for hydroelectric turbines based on LABVIEW.

Figure 10.4.[25,26] AI-based techniques are predominantly used in failure prediction in various components of hydroelectric power generators.

10.7.1 Deep Learning Approaches

A closed-loop recurrent neural network (RNN) can manipulate time-based sequential data acquired from the planetary gear train to predict the deterioration of gear box. RNN is immensely useful in manipulating temporal signals and is capable of predicting the failure rate of components, since it is very sensitive even to minor deviations.[27,28]

The RNN can forecast the failures one step ahead based on values of the previous state variables. Deep forest is yet another deep learning technique that is gaining more significance in failure prediction. The CM information based on the rotational speed of hydraulic turbine is used to diagnose coupled parallel misalignment and rub impact faults in hydroelectric turbines.[29]

10.7.2 ANN-Based Hybrid Approaches

The process of quantizing the qualitative failure analysis has inherent uncertainty and fuzziness. Cloud neural network is featured as $C(E_x, E_n, H_e)$, where the fuzziness is decoded as entropy (E_n) and E_x is the expectation of point x. The hyperparameter (H_e) is the membership degree of E_n that induces randomness to avoid overfitting. Each tuple C represents qualitative concept and superposition of these tuples forms discrete-valued quantitative results, through cloud transformation.[30]

10.7.3 Clustering Approaches

The signals from magnetoelectricity vibration sensor are disintegrated into multiple intrinsic mode functions (IMFs) to detect the power line failures in hydroelectric plant.[31] Multidimension permutation entropy technique is employed to extract the useful features from the IMF. The obtained entropy is processed using Shannon's entropy function to sense the deviations.

Thus, it is evident that AI techniques are predominantly used in failure prediction of hydroelectric power plants.

10.8 FAILURE PREDICTIONS IN BIOGAS PLANTS

The chief sources of failures in biogas production may be due to wear and tear in equipment, fermentation, and unfavorable chemical reactions inside the plant. The solid oxide fuel cell (SOFC), the chief convertor deployed in biogas system, is more susceptible to failures. AI-based techniques are extensively used to forecast the failures in biogas plants.

10.8.1 Role of Expert Systems in Biogas Failures

An IF-THEN-ELSE rule-based expert system with comprehensive modules is used for pretreatment, installation of biogas components, manure utilization, fault analysis, and management of knowledge base.[32] The diagnostic fuzzy rules are framed by experts and are realized by the inference engine by consulting FACT-ID-based RULE NAME. The fuzzy rules are established based on the current operative condition, and result in accurate prediction. The outcomes of this hybrid model are promising in predicting failures by monitoring the temperature, pH and oxidation rate of the system.

Alternatively, assignment of fuzzy numbers for monitoring the biogas parameters also proves to be an effective technique.[33] The input parameters expressed as fuzzy numbers are relative yield of the biogas, hydraulic retention time, methane production and capacity of the digester. The rule-based expert system categorizes the plant safety from the fuzzy numbers, in terms of fuzzy classes expressed in linguistic statements. This system is successful in predicting failures irrespective of the seasons and it recommends the practices to improve the efficacy of the plant.

10.8.2 SVM in Biogas Plant Failure Prediction

The classification of performance of SOFC into normal, cathode humidification and anode poisoning can be efficiently done using least-squares support vector machine (LS-SVM).[34] The voltage data from the SOFC can be retrieved and trained using hidden semi-Markov model, which can model the health condition as well as estimate the remaining useful life of the cell. This model gives higher accuracy of 97%, which is better than most of the state-of-the-art techniques.

10.9 AI-BASED APPROACHES FOR FAILURE PREDICTION IN GEOTHERMAL ENERGY SYSTEMS

Currently, failure prediction of geothermal energy systems is done through visualizing graphs and patterns of online data. The operator manually predicts the occurrence of failures and issue alerts, which may delay the prognostic module to act upon the failure. Hence, automatic failure prediction using AI-based approaches will meliorate the predictive maintenance.

10.9.1 Fault Detection of Sensors in Geothermal Plants

It is a well-known fact that the temperature inside the ground level is dramatically higher, making it a potential cause for the failure of heat exchanger in the geothermal system. The heat pump thrusts the thermal energy from the ground. Various sensors are installed in the geothermal system to monitor the ground temperature and energy absorbed from the ground. The machine learning techniques such as ANN, random decision forests, randomized trees, and gradient boosting are deployed to study the occurrence of failures at the sensor level.[35] The results of the model based on extremely randomized trees showed promising results.

10.9.2 ANN in Heat Exchanger Fault Prognostics

Building ANN with optimal number of layers proves to be very effective in modeling components of geothermal heat exchanger systems such as heat recovery steam generator, gas turbines, boilers, and steam turbines.[36] The critical predictors are output poser, inlet and outlet pressure, temperature, fuel, and gas flow. The modeled ANN acts as a tool for prediction of faults and failures, thus enabling the scheduling of predictive maintenance.

10.9.3 Failure Prediction in Heat Pumps

Prediction of failure of heat pumps housed in a heat exchanger of geothermal systems is done using multilayer perceptron and decision trees (J48).[37] The temperature deviation in the monitoring parameters is considered as predictor variable to detect the anomalies such as leak in pumps. Both the algorithms depicted astounding results in predicting failures in heat exchanger systems.

10.9.4 Classifiers on Failure Prediction of Geothermal Plant Engines

Electrical currents, resistance, heat, and pressure values of the engines play a vital role in forecasting failures in engines of geothermal plants.[38] Machine learning algorithms such as XGBoost, SVM, and back-propagation ANN exhibit better performance in fault prediction using the data that is preprocessed by the K-means clustering model.

10.10 CHALLENGES IN EQUIPMENT FAILURE PREDICTION

The equipment failure prediction evolves from manual diagnostics leading through FMEA landing in AI techniques. The accuracy in fault prognosis comes with the cost of time and data availability. The following are the major hurdles and challenges faced by the domain of failure forecasting:

- Lack of expertise in the field of renewable energy
- Limited deployment of state-of-the-art techniques
- Varied geographic locations emanating diverse problems and failures

- Nonavailability of universally acceptable robust methods
- Unplanned plant construction
- Higher initial installation costs that limit the provisions for setting up predictive maintenance strategies
- Political and geographical variations across countries, which hinder the formulation of universally acceptable guidelines for renewable energy plants

The above-mentioned list is nonexhaustive. All these challenges offer potential scope for investigation and stand chances for research problems.

10.11 FUTURE DIRECTIONS

The renewable energy is looked upon with valor by the future power-thirsty world. To ensure uninterrupted and quality power supply, the renewable energy plants must be maintained in healthier state. Any unforeseen maintenance activity will disrupt the power generation. AI-based techniques have contributed the lion's share in CM of critical components. Although many machine learning and deep learning models are currently deployed to predict failures, there is definitely scope for development of robust and comprehensive models, since the current state-of-the-art techniques are limited to model individual components and not the entire plant.[39] Hybridization of multiple techniques is emerging as a powerful failure prediction solution, which when geared can catalyst more AI-based models with good predictive power.

REFERENCES

1. Yaguo Lei, Feng Jia, Jing Lin, Saibo Xing, Steven X. Ding. (2016). An intelligent fault diagnosis method using unsupervised feature learning towards mechanical big data. IEEE Transactions on Industrial Electronics, 63(5), 3137–3147.
2. Jia Huang, Zhaojun Li, Hu-Chen Liu. (2017). New approach for failure mode and effect analysis using linguistic distribution assessments and TODIM method. Reliability Engineering & System Safety, 167, 302–309.
3. K. D. Sharma, S. Srivastava. (2018). Failure mode and effect analysis (FMEA) implementation: A literature review. Journal of Advance Research in Aeronautics and Space Science, 5(1), 1–17.
4. Z. Bluvband, P. Grabov. (2009). Failure Analysis of FMEA. 2009 Annual Reliability and Maintainability Symposium, Fort Worth, Texas, 344–347. doi: 10.1109/RAMS.2009.4914700.
5. Jui-Sheng Chou, Chien-Kuo Chiu, I-Kui Huang, Kai-Ning Chi. (2013), Failure analysis of wind turbine blade under critical wind loads. Engineering Failure Analysis, 27, 99–118.
6. Evrencan Özcan, Rabia Yumuşak, Tamer Eren. (2019). Risk based maintenance in the hydroelectric power plants. Energies, 12, 1502; doi:10.3390/en1208150.
7. S. Karthikeyan, M. Subramaniyam, A. Ghosh, M. Prashanth, R. Karunanithi. (2019). Micro hardness and corrosion properties of A390 alloy + x vol.% zirconium dioxide composites processed by P/M method. International Journal of Microstructure and Material Properties, 14(6), 511–523.
8. Manju Santhakumari, Netramani Sagar. (2019). A review of the environmental factors degrading the performance of silicon wafer-based photovoltaic modules: Failure detection methods and essential mitigation techniques. Renewable and Sustainable Energy Reviews, 110, 83–100.

9. Hamid Reza Feili, Navid Akar, Hossein Lotfizadeh, Mohammad Bairampour, Sina Nasiri. (2013). Risk analysis of geothermal power plants using failure modes and effects analysis (FMEA) technique. Energy Conversion and Management, 72, 69–76.
10. Shikun Cheng, Zifu Li, Heinz-Peter Mang, Kalidas Neupane, Marc Wauthelet, Elisabeth-Maria Huba. (2014). Application of fault tree approach for technical assessment of small-sized biogas systems in Nepal. Applied Energy, 113, 1372–1381.
11. E. Zio. (2012). Diagnostics and Prognostics of Engineering Systems: Methods and Techniques. IGI Global, Chap. 17.
12. Sule Selcuk. (2017). Predictive maintenance, its implementation and latest trends. Journal of Engineering Manufacture, 231(9), 1–10.
13. Thyago P. Carvalho, Fabrízzio A. A. M. N. Soaresa, Roberto Vita, Roberto da P. Francisco, João P. Basto, Symone G. S. Alcalá. (2019). A systematic literature review of machine learning methods applied to predictive maintenance. Computers & Industrial Engineering, 137, 106024.
14. Alessandro Betti, Maria Luisa Lo Trovato, Fabio Salvatore Leonardi, Giuseppe Leotta, Fabrizio Ruffini, Ciro Lanzetta. (2017). Predictive Maintenance in Photovoltaic Plants with a Big Data Approach. 33rd European Photovoltaic Solar Energy Conference and Exhibition (EUPVSEC), 1895–1900.
15. Timo Huuhtanen, Alexander Jung. (2018). Predictive Maintenance of Photovoltaic Panels via Deep Learning. IEEE Data Science Workshop (DSW), Lausanne, 66–70. doi: 10.1109/DSW.2018.8439898.
16. Aicha Djalab, Ahmed Hafaifa, Mohamed Mounir Rezaoui, Ali Teta, Nassim Sabri. (2019). An Intelligent Faults Diagnosis and Detection Method Based an Artificial Neural Networks for Photovoltaic Array. The 3rd International Conference on Applied Automation and Industrial Diagnostics, Elazig, Turkey, 1–6.
17. Massimiliano De Benedetti, Fabio Leonardi, Fabrizio Messina, Corrado Santoro, Athanasios Vasilakos. (2018). Anomaly detection and predictive maintenance for photovoltaic systems. Neurocomputing, 310, 59–68.
18. Thomas Kenbeek, Stella Kapodistria, Alessandro Di Bucchianico. (2019). Data-Driven Online Monitoring of Wind Turbines. Proceedings of the 12th EAI International Conference on Performance Evaluation Methodologies and Tools, 143–150.
19. K. Kim, G. Parthasarathy, O. Uluyol, W. Foslien. (2011). Use of SCADA Data for Failure Detection in Wind Turbines. Energy Sustainability Conference and Fuel Cell Conference, 1–9.
20. K. Wang, V. S. Sharma, Zhang, Z. (2014). SCADA data based condition monitoring of wind turbines. Advanced Manufacturing, 2, 61–69.
21. Mari Cruz Garcia, Miguel A. Sanz-Bobi, Javier del Pico. (2006). SIMAP: Intelligent system for predictive maintenance application to the health condition monitoring of a windturbine gearbox. Computers in Industries, 57(6), 552–568.
22. Bindi Chen, Peter J. Tavner, Yanhui Feng, William W. Song, Yingning Qiu. (2012). Bayesian Networks for Wind Turbine Fault Diagnosis. EWEA 2012, Copenhagen, 16–19 April.
23. Bindi Chen, Peter C. Matthews, Peter J. Tavner. (2013). Wind turbine pitch faults prognosis using a-priori knowledge-based ANFIS. Expert Systems with Applications, 40, 6863–6876.
24. Nassim Laouti, Nida Sheibat-Othman, Sami Othman. (2011). Support Vector Machines for Fault Detection in Wind Turbines. Proceedings of the 18th IFAC World Congress, Milan, 7067–7072.
25. Z. Liu, S. Zou, L. Zhou. (2012). Condition Monitoring System for Hydro Turbines Based on LabVIEW. Asia-Pacific Power and Energy Engineering Conference, Shanghai, 1–4.

26. W. Jiang. (2008). Research on Predictive Maintenance for Hydropower Plant Based on MAS and NN. Third International Conference on Pervasive Computing and Applications, Alexandria, 604–609.
27. R. Yam, P. Tse, L. Li. (2001). Intelligent predictive decision support system for condition-based maintenance. International Journal of Advanced Manufacturing Technology, 17, 383–391.
28. E. J. Amaya, A. J. Alvares. (2010). SIMPREBAL: An expert system for real-time fault diagnosis of hydrogenerators machinery. IEEE 15th Conference on Emerging Technologies & Factory Automation, 1–8.
29. Xiaolian Liu, Yu Tian, Xiaohui Lei, Mei Liu, Xin Wen, Haocheng Huang, Hao Wang. (2019). Deep forest based intelligent fault diagnosis of hydraulic turbine. Journal of Mechanical Science and Technology, 33(5), 2049–2058.
30. L. Han, Z. Li. (2009). Research of Cloud Neural Network Based on Cloud Transformation and Its Application on Vibration Fault Diagnosis of Hydro-Turbine Generating Unit. Asia-Pacific Power and Energy Engineering Conference, 1–4.
31. Dong Liang, HuaShan Guo, ZeWei Guo, Tao Zheng. (2019). Fault diagnosis technique for hydroelectric generators using variational mode decomposition and power line communications. Journal of Physics: Conference Series, 1176(6), 1–6.
32. Man Zhou, Zhiyong Zou. (2018). Design of an Intelligent Control System for Rural Biogas Engineering. 2nd IEEE Advanced Information Management, Communicates, Electronic and Automation Control Conference (IMCEC), Xi'an, 1636–1639.
33. Djordje Djatkov, Mathias Effenberger, Milan Martinov. (2014). Method for assessing and improving the efficiency of agricultural biogas plants based on fuzzy logic and expert systems. Applied Energy, 134, 163–175.
34. XiaoJuan Wu, Qianwen Ye. (2016). Fault diagnosis and prognostic of solid oxide fuel cells. Journal of Power Sources, 321, 47–56.
35. Hector Alaiz-Moreton, Manuel Castejón-Limas, Jose-Luis Casteleiro-Roca, Esteban Jove, Laura Fernandez Robles, Jose Luis Calvo-Rolle. (2019). A fault detection system for a geothermal heat exchanger sensor based on intelligent techniques. Sensors, 19, 1–16.
36. M. Fast, T. Palme. (2010). Application of artificial neural networks to the condition monitoring and diagnosis of a combined heat and power plant. Energy, 35, 1114–1120.
37. José Luis Casteleiro-Roca, Héctor Quintián, José Luis Calvo-Rolle, Emilio Corchado, María del Carmen Meizoso-López, Andrés Piñón-Pazos. (2015). An intelligent fault detection system for a heat pump installation based on a geothermal heat exchanger. Journal of Applied Logic, 17, 36–47.
38. Zulkarnain, I. Surjandari, R. R. Bramasta, E. Laoh. (2019). Fault Detection System Using Machine Learning on Geothermal Power Plant. 16th International Conference on Service Systems and Service Management (ICSSSM), Shenzhen, 1–5.
39. S. Sharanya, Revathi Venkataraman. (2020). An intelligent context based multi-layered Bayesian inferential predictive analytic framework for classifying machine states. Journal of Ambient Intelligence and Humanized Computing. https://doi.org/10.1007/s12652-020-02411-2.

11 AI Techniques for the Challenges in Smart Energy Systems

S. Dwivedi
S.S. Jain Subodh P.G. (Autonomous) College, Jaipur, India

CONTENTS

11.1 INTRODUCTION

Artificial intelligence (AI) is an emerging technology envisaged for upfront changes in energy systems existing currently around the world for cleaner, greener, and highly efficient energy production [1, 2]. Carbon-free economy is highly desirable for the preservation of naturally occurring systems, environment, flora and fauna, better living style, high health level, and wellness [3]. The traditional systems of energy production are inclusive of fossil fuels, renewable energy systems, solar energy, wood-based fuel materials, coal, and other similar systems. Specific problems of carbon emission are associated with all these traditional sources of energy whether renewable or non-renewable. There is a need to make these materials more efficient for producing energy in a specific advanced manner with automated systems. AI is an emerging technology with a wide collection of computational methods that collect and collate all the information together from an efficient conclusive standpoint to assist in implementation of technologically advanced tools for effective harnessing of energy [4–7]. This can have far-sighted social, economic, technological, environmental, and political implications for sustainable development as it has a direct bearing on the life of the people.

AI can be an advanced aid system for providing powerful support in the form of neural networks, expert systems, optimized complex renewable energy systems, big data management systems, prediction of renewable energy potential, and fuzzy logic (FL), for example, to solve technical problems of energy management, management and security systems for smartly energized homes, data-security management system of smart grids, fault-detection of smart power grids, energy-related evaluations and predictions, supply chain smart energy management systems, and predictions of cost estimations of energy systems [2, 7, 8].

Nowadays, advanced intelligence systems are capable of performing human mind-based activities, such as, writing computer programs and mathematical equations, driving a machine-handled automobile having automation capabilities, diagnosing diseases using automated machines with AI capabilities, automated synthesis of complex organic chemical compounds, analysis of electronic circuits, and developing small programs for specific types of works. The main objectives of AI research are to develop the understanding of human cognition, automation, automated machines with AI capabilities for intelligent actions, amplification for building intelligent systems for improving the thought process of humans, intelligence-based problem-solving capabilities, superhuman intelligence, establishing communication with people in a phased manner, employing common language, intelligent systems for storing information, automated capabilities of the system to collect and collate the data together, and intrinsic system capabilities to act on self-intelligence. Intelligent technologies can be integrated with renewable energies and support systems to offer efficient energy transformation. Smart grid technologies integrated with intelligent systems make it possible to use the renewable energy sources efficiently and sustainably. They also help in cutting down the costs drastically for both front- and end users of energy, and also provide a decentralized open-ended system and design for energy systems.

AI is an emerging advanced technological field with numerous opportunities in different sectors of the economies and industries globally. AI forms a wide collection of computational techniques for digging out information from numerous data sources and processes an action based on the data. It also includes the data called "small data," which is produced by the algorithm itself. AI forms a technology of common use that has substantial technological, socio-economic, and political implications. AI is slated to generate transformations in the energy sector radically. AI-induced energy transformations in the energy sector directly influence the stability factor and financial affluence. Neural networks, FL, and expert systems included in AI technologies have been applied successfully to solve different technological challenges. Other focus areas include smart grid fault detection, forecasting, technological, socio-economic, and political predictions in the energy sector, management of demand-oriented masses and their requirements, and security of smart grid data management with the latest AI technologies and blockchain. The highest potential of AI is its ability in forecasting and prediction of renewable energy, optimization of complex renewable energy architectures and system designs, big data management, and other systems and methods for accelerating the energy transition process.

11.1.1 AI Techniques

AI techniques consist of many branches that include FL, artificial neural network (ANN), data mining (DM), and adaptive network-based fuzzy inference system (ANFIS) [2].

11.1.1.1 Fuzzy Logic

FL can be defined as a rational system that is an addition of multivalued logic in the narrow sense [2, 9]. However, in a broader sense, FL sounds in consonance with the theoretical concept of fuzzy sets. Fuzzy set is a theoretical concept related to classification of articles with not-so-intense peripheries that forms a set consisting of elements with closed membership [10–12]. However, as a matter of fact, FL belongs to a larger classification of fuzzy theory; in the narrow perspective, it differs both in theoretical concepts and subject-matter from orthodox multivalued logical architectures. FL is easy to grasp, flexible, having features such as non-precision data management, high tolerance level, modeling of non-linear functions with random complexities, and is built up on platforms designed as a result of experiences of users; it has flexible features to make a blend with traditional control techniques and is based on languages of common use. Mathematical concepts contained in FL are very simple and do not possess outstanding complexities. It can be added arbitrarily at any point of operation with any given system without requiring any added functionality. FL gives an impression that data precision is not granted for the process, on having a closer look. Fuzzy system can be created to tie up closely with any set of input-output datasets, which is performed by adaptive techniques, for example, ANFIS [9, 13–15]. The platform of FL is based on the realistic expertise of experienced users who have a deep understanding of the system. This is opposed to the neural networks that exploit training data to produce opaque and non-penetrable models. FL has the property to substitute orthodox control systems and can also augment them in certain cases along with simplification of the implementation methodology [9]. FL relies on the concept of human communication by exploiting common-mode language systems [9]. Normally, an FL-based model forms an efficient relation between two multi-dimensional spaces. Connection between output and input fuzzy spaces is called fuzzy associative memories (FAMs) [16, 17]. Linguistic variables and features [18, 19] are well-defined and collective rules are well-defined between different fuzzy sets to set up a system. Following steps are required for implementation of an FL-based system [9, 17]:

1. Fuzzification is the name of coding methodology that belongs to every numerical input of a linguistic variable being changed into membership functional principles of attributable values.
2. Inference is the methodology of coding that belongs to the process of computational rule that is formed by intersection of lone surroundings, which can be applied to the fuzzy operator and a number of rules are combined together to reach a conclusion, which means joining the singular confidence levels by use of fuzzy operator (OR).
3. Defuzzification is the decoding operational methodology of the information that forms the resultant fuzzy sets derived from inference process that constitutes the most suitable crisp resultant value.

11.1.1.2 Adaptive Network-Based Fuzzy Inference System

ANFIS is a complex model that is attained as a result of combination of fundamentals of FL and neural networking in a singular integrated platform [9, 20, 21]. This model is formed by the combination of a fuzzy inference system typified in the form of an adaptive framework for identification of system methodologies, and a prognostic instrument that shapes a provided input to the related output universe that forms the basis of illustrative training sets of data. ANFIS inference system forms the dataset that combines the fuzzified human knowledge and a set of input-output pair of datasets or patterns. Fuzzified human knowledge is modeled in the typology of fuzzy "if-then" sets of patterns to perform input-output methodology of the mapping process [22, 23]. ANFIS coding is broadly applicable in systems involving imprecision or uncertainties in defining the variables that constitute the behavior of the system. This methodology has the capabilities to model and identify the nonlinear systems and prediction of chaotic time-dependent behavior. Two processes of fuzzy inference systems are Mamdani [24] and Sugeno [16] in which difference is generated from consequential parts that include fuzzy membership functionalization in Mamdani and constant or linear functions in Sugeno. Inference system related to ANFIS relies on both human content of fuzzified knowledge, which is modeled in "if-then" defined typology of sets and the patterns of classified elements of input-output set elements for completion of methodology of input-output diagramming [25]. ANFIS modeling is broadly applicable to systems involving fuzzy or messy definitions of variables that constitute behavior of the system.

11.1.1.2.1 Fuzzy If-Then Set of Rules Combined with Fuzzy Inference Systems

11.1.1.2.1.1 Fuzzy If-Then Rules These conditional proclamations are expressed in typology of IF X THEN Y, where X and Y are labels of sets of variable elements declared by proper functions defining those member elements [9, 20, 21]. Fuzzy if-then sets of rules are applied in concise form to capture messy configurations of reasoning, playing an important role in human capability to take proper decisions where uncertainty in environment or unpredictable conditions prevail. Jang et al. provided an example that when volume is small, pressure becomes high; in this case, both volume and pressure are linguistic labels or values that are defined by functions entailing the membership elements [9]. Takagi and Sugeno defined another type of messy "if-then rule" that involves fuzzy sets intricacy in the premise part only [16]. Jang et al. defined this rule with the help of an example of resistive force acting on an accelerating object as follows [9]:

$$\text{Force} = \text{k} * (\text{Velocity})^2$$

The above equation fits in case when velocity is high in the premise part characteristic of a properly functionalized membership element. The consequent domain of the equation is well-defined by a nonfuzzy or precise equation of the changing feedback, which is velocity. Both modeling and control employ the two types of fuzzy if-then set of rules.

Fuzzy inference methodologies can also be recalled as fuzzy models, fuzzy-based systems, fuzzy controllers or even FAM process shown schematically in Figure 11.1. A fuzzy inference-based system is composed of five functional basic elements:

1. A fundamental rule that consists of numerous fuzzy elements of set of if-then rules.
2. A dataset stating the functions that define the membership properties.

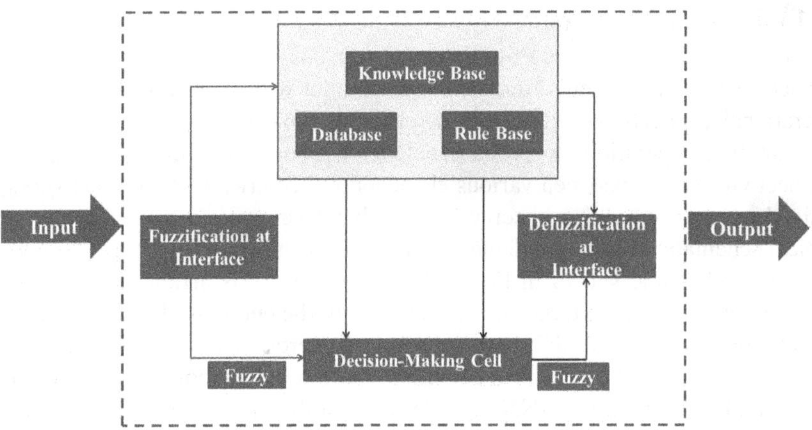

FIGURE 11.1 Fuzzy controllers or FAM or fuzzy inference system.

3. A decision-making cell that performs the inference.
4. A complex interface or fuzzified surface-interface transforming capabilities of crusty feeds.
5. An interface with defuzzification characteristics with transformation capabilities of fuzzy elements.

Normally, knowledge base is the name given to the joint platform of the rule base and the database:

1. Fuzzy sets applied in the fuzzy rules.
2. Different operations performed on the rules.
3. Fuzzy sets of elements defining the fuzzy rules.
4. Points of match with certain linguistic values.
5. Output collated as a result of drawings from fuzzy inference systems.

Different steps involved in the process of fuzzy reasoning as a result of application of if-then rules to perform inference operations in the fuzzy inference systems are defined as follows: [9]

i. This step is called fuzzification which involves comparison of input variables with membership functionalization on the side of premise part for obtaining membership values or measuring elements of compatibility of assigned individual linguistic label.
ii. This basic step is called a particular T-norm operator obtained as a result of combination of membership functionalization values on the premise part to attain the firing strength of weightage of individual rule.
iii. Defuzzification is the step involving aggregation of qualified consequent of either crisp or fuzzy element of each rule that depends on the strength of firing.

11.1.1.3 Artificial Neural Networks

Based on the input provided by biological nervous systems, neural networks are formed of normal elements functioning in parallel with each other that define the operational connectivity between these elements [26–29]. A neural network is fed with artificial operations to produce output of a particular function by adjusting connectivity values between various elements of the network. Neural networks are trained or arranged with a specific feed so that it generates a particular output. A typical schematic of basic fundamental of ANNs is shown in Figure 11.2.

In the schematic shown in Figure 11.2, the network is arranged that relies on comparison between the target and its output until the output is almost similar to the target. Fundamentals of ANNs are applied in numerous areas that include function approximation, FAMs, pattern recognition, pattern association, and generation of novel significant patterns. ANNs can be successfully applied in various fields that include neurology, meteorology, psychology, medicine, sound, pattern and speech recognition, in electromyographs, at the interface of human-robotics machinery, military targets, and in load predictions related to thermal and electrical forecasts [26–29]. ANNs are also useful in process control as they can be applied to build up process models derived from multi-dimensional data collected and collated from sensors. Neural networks are generally composed of an input layer, few hidden layers, and an output layer. Each single neuron of every layer is attached to other neurons of the preceding layer through adaptable encumbrances. These connected encumbrances generally store knowledge ostensibly that corresponds to synapse efficiency in neural systems. In training or feed process, connection encumbrances are applied in an ordered manner by a suitable methodology. ANNs use a learning methodology in which an input is fed back into the network along with the required outcome and the encumbrances are arranged such that the neural network is programmed for producing the required output. Encumbrances rearrange themselves in an ordered manner after feeding. Before training, these encumbrances are arranged in a disordered manner, which does not constitute any significant meaning. Information processing methodology in a neural network unit can illustrate about the procedure through

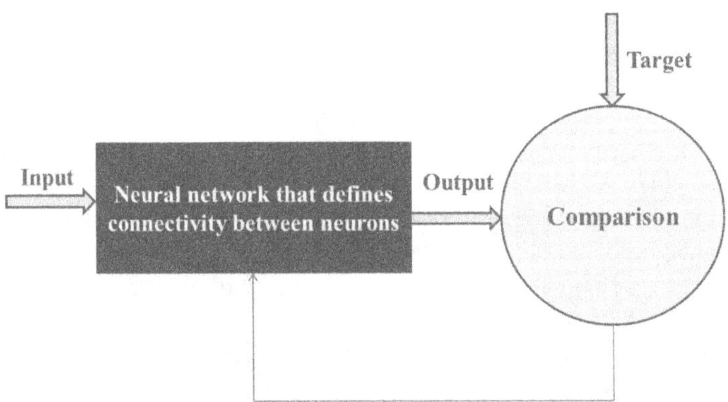

FIGURE 11.2 Basic fundamentals of ANNs.

a single node. The node accepts loads of information from other nodes via the input leads. Initially, the nodes are summed up and the outcome is then processed through an activation functionalization so that the output generates an activated node. The generated activation lead gets multiplied by a particular encumbrance for each of the output leads, which are transferred to the next node.

11.1.1.4 Genetic Algorithms

Genetic algorithms (GAs) are driven by the methodology of adaption of living organisms to the punitive veracities of life in the harsh evolution and inherited world [30–33]. The evolution process is best optimized by an algorithm that replicates the best-fit individuals for population multiplication. This generation of an optimized algorithm is truly based on the fundamentals of survival of the fittest and natural selection. It can be applied by selecting a limited number of possible solutions to the proposed problem that evolves as a function of time called individuals. GA utilizes three major genetic operators: selection, crossover, and mutation. In each of the reproduction process called generation, characters involved in the current generation can be assessed by generation of a process methodology of fitness that is equipped with technological capabilities to solve the underlying problem. Subsequently, each of the individual is regenerated proportionately according to its fitness. The higher is the fitness, the higher is its chance for participation in reproduction to generate an offspring. Only a handful of newly generated offspring undergo the transmutation process. After many reproduction cycles, only those individuals possessing the best genetics with optimum fitness function are able to survive. GA makes use of fitness function to produce results of individuals who will survive in the population and take part in reproduction. The function of a GA is the breeding of a number of individuals that will make the optimized solution to the concerned problem.

11.1.1.5 Data Mining (DM)

DM offers a powerful methodology of automated analysis for extraction of prognostic information from huge databases [34–36]. It is based on first principles theoretical calculations that take lesser time to solve problems which are otherwise time-taking. Hoffmann and Apostolakis have specified that DM relies on digging out information from databases and reveals information in a particular manner, which is otherwise too much complex to decode [37]. DM finds application in fields as diverse as production control, share market, customer relationship management, medical, and astronomical data analysis. DM process relies on collection of data initially, followed by processing methodologies and identification of problems related to data-processing quality, and gaining first-hand acumenship into the data. In the next step, pre-processing of data is performed to construct data in a phased manner to build up a well-defined dataset. This step involves data cleaning by removal of inconsistent and noise factors, and extraction of embedded features in data transformation. Data is refined utilizing various tools and techniques to smoothen loads of data to make them significant and understandable. The modeling step involves determination of optimized modeling parameters to get the desired outputs. In the evaluation phase, model is determined in the final stage for validation so as to adequately set it into the needed model of the problem. Pre-process cleaning of data is a big step in the way to

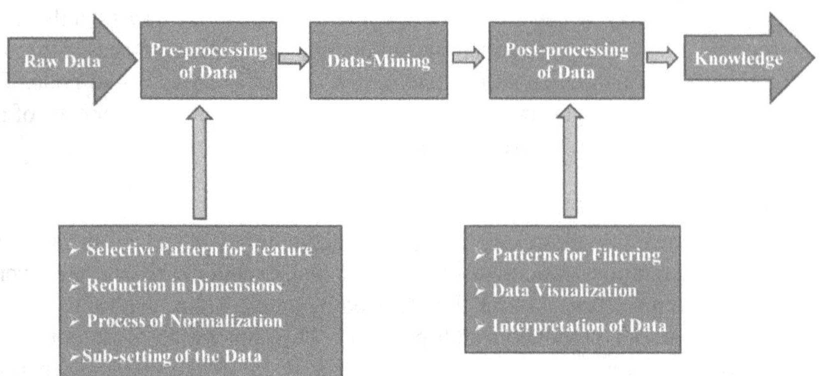

FIGURE 11.3 Schematic for the process of discovery of knowledge in databases.

gain information about knowledge discovery. A schematic of process of knowledge is shown in Figure 11.3, which shows pre-process and post-processes involved in the steps of DM.

There are many applications of aforementioned artificial intelligence techniques that include FL, ANFIS, ANN, GA, and DM. Fuzzy logic has been applied successfully to a variety of applications in solar energy systems that include sun-tracking system, control of solar buildings, estimation of solar radiation, control system of a solar energy-based air-conditioning system and photovoltaic (PV) solar energy-based systems. Salah et al. fabricated a fuzzy algorithm that codes a process of energy management for a domestic PV panel, and validation is performed on a 1-kilowatt peak (kWp) PV panel on different days in various seasons [38]. The knowledge discovery acquired through the algorithm provides an insight into energy saving in the daytime. Atlas and Sharaf proposed a PV energy system that relies on a complex of electric encumbrance automated by an FL-based algorithmic dynamic system, recognition, and automated controller system for enhanced tracking ensuring maximum powerpoint (MPP) operative methodology in the area of concern that receives solar radiation, atmospheric temperature, and fluctuations in electric encumbrances [39]. Apart from MPP detector and double-oriented FL MPP controller system for tracking, this process methodology also includes two control units. These control units consist of voltage control units for common DC load bus and other for controlling velocity of the permanent magnet DC motor by applying DC or DC choppers. This tracking system generates results with minimized error percentages with changes in solar irradiation levels that generate various maximized power generation output points. Sugeno proposed a fuzzy inference system for design and modeling of the controller system [16]. Fuzzy if-then rules are followed to account for determination of the area under direct irradiation from the sun. Simulation tools along with features of 3D augmented reality are the powerful instruments to ascertain systematic methodology before installation making it possible to develop real-time applications. In the design of multifarious sun-tracking system, step tracking is performed in each of the four minutes for 1° movement by the sun requiring lesser energy to drive these systems. Gomez and Casanovas performed a study on physical fuzzy modeling of a climate variable that defines ambiguity of solar irradiance in the form of fuzzy uncertainty [40, 41]. A right approach is presented by comparison of fuzzy models with nonfuzzy models of solar irradiance; an improved performance can be observed by uncertainties considering the fuzziness of the data and output obtained from the model. Paulescu et al. developed FL algorithms to forecast the solar radiation transmission that can be applied in determination of solar energy [42]. In this regard, the first model makes a crossover to self-determined fuzzy modeling of each individual characteristic transmission of solar radiation. The second model is an appropriate FL model for stream light and diffuse transmission encompassing over the whole atmosphere. Outcome of the fuzzy algorithms concluded that development of parametric models parallel to the parameters of FL is a sustainable option of classical parameterization methodologies. This classical parameterization has developed higher orders of flexibility in acclimatizing to local meteorological conditions owed to the experiential nature of the input-output characteristics of the fuzzy model. In 1998, Şen developed an FL-based algorithm for determination of solar irradiation from measurements of duration of sunlight [43]. In 2006, Gouda et al. developed a quasi-adaptive FL-based controller for regulation of space heating in solar buildings with chief objective of reduction in overheat lagging due to passive solar heat gain in a room space [44]. In 2007, Lygouras et al. closely observed the outcome of a variable FL-based controller for implementation in a solar-empowered air-conditioning system [45].

Similarly, there are many applications for ANFIS-based solar-empowered energy systems. In 2008, Mellit et al. set up a novel neuro-fuzzy model for forecasting the sequential clearness index on a monthly basis for generation of solar radiation used for size determination of the PV system [46]. In 2008, Chaabbene and Ammar applied a neuro-fuzzy dynamic model for forecasting of solar radiation and atmospheric temperature relying on medium-term forecasting (MTF) for routine meteorological changes [38]. In case of short-term forecasting (STF), it determines meteorological parameters evolution process five minutes ahead.

ANNs-based algorithms have been developed for application in a lot of renewable energy systems. In 2008, Rehman and Mohandes applied atmospheric temperature, comparative humidity values, and day of the year as feedstock in the neural network for forecasting of global solar radiation (GSR) on surfaces lying parallel to the horizon [47]. In 2010, Mellit and Pavan produced an algorithm specifying multi-layer perception (MLP) network for prediction of solar irradiance 24 hours in advance with mean daily ambient temperature and daily solar irradiance as input feeds [48]. In 2009, Benghanem et al. synthesized ANN-based fundamental models for determination of routine global solar radiation [49]. Six ANN models were developed incorporating air temperature, sunshine timings, day of the year, and ambient temperature. In 2008, Mubiru applied ANN for estimation of average daily global solar irradiation on a monthly basis falling over a horizontal area [50]. This proposed ANN-based model was superior in functioning as compared to the empirical model due to its strength of capturing the non-linear behavior of solar radiation.

GAs can be applied to a variety of solar energy systems that are inclusive of solar hot water systems, complex solar-wind systems, solar cells, PV solar energy systems, flat plate solar air heating systems, and in determination of Angstrom equation coefficients. In 2010, Zagrouba et al. applied GA for identification of electrical parameters of PV solar cells and modules [51]. GA-based algorithm is successfully applied to tackle problems of local minima related to criteria of non-convex optimization. In 2002, Loomans and Visser applied GA for calculating the yield and costs of large solar-based hot water systems relying on financial and technical data of the system constituents [52]. The GA optimizes separate variables such as number of collectors, collector heat exchange area, and mass of the heat storage. The choice of number of variables is unlimited, as the GA is a discrete optimization instrument and makes use of databases for implementation in the architecture. In 2010, Varun applied GA for determination of optimized thermal performance of a flat plate solar-based air heater with capabilities of handling various systems and operating parameters [53].

11.1.1.5.1 AI in Photovoltaic Electricity Production

AI can provide effective algorithmic solutions to technological problems in cases in which it is not becoming possible to develop a feasible theoretical model. Feedforward neural networks have proved efficient for modeling of renewable energy resources [9, 49, 54]. It can be successfully applied for modeling applications in effective classification, data predictions, and pattern recognition. Feedforward artificial neural networks (FANNs)-based algorithms are capable of tackling non-linearity in data and data uncertainties in models or in the already applied datasets. This specific capability is inherited in its architecture culminating from collective elementary neuron units, which is structured in three interconnected layers of input, hidden, and output layers. A real-time algorithmic model is drafted in a manner such that a set of encumbrances can be discovered in an inherited encumbrance's series entailed by neurons that are positioned at different layers on the FANN network. FANN has a special property that pushes the data-processing flow from input at each layer to the output layer of the neural network. After processing of the learning process in a phased manner, FANN-based algorithm can be applied in classification and determination methodology. In a number of cases, an effective output is achieved because of flexible and common modeling approach that is non-linear data-driven. However, an interference is created because of the noise signal ratio generated from the input data so that proper modeling is not possible due to lack of an appropriate neural network architecture design. FANN-based algorithm provides an effective theoretical modeling pattern for PV technology-based electricity production [54]. Kumaravel et al. presented an adaptive multilayer FANN-based architecture for computation of maximum power output in case of an isolated complex solar/wind energy system [55]. Saberian et al. applied FANN for determination of output power obtained from PV panel beginning with maximum, minimum, and mean temperature values at the accessible irradiance values [56]. There are numerous regression methods beginning from classical approaches, which include polynomial, autoregressive moving averages (ARMAs) methods, autoregressive integrated moving averages (ARIMAs), several other methods based on AI, and many other soft-computing techniques.

Different factors affecting working of the PV systems are temperature, wind speed, and solar radiation amongst many others. A optimal number of solar cell panels to be conjoined in the architecture of a PV system, capacity of the charge storage of the battery, and capacity of wind generator for use in specific complex applications are the parameters to be considered in any PV system. Optimized PV systems reinforced with AI technologies are particularly suitable for pollution-free solar-empowered electrification of villages, water-pumping systems, heating systems, and for several other systems of direct public use.

FL finds applications mainly in control engineering and is closer in operational strategy to human cognition processes called approximate reasoning [9, 57]. The modus operandi of control methodology finds its roots in FL and fuzzy control characteristics endowed with capacities of conversion of problems and their fuzzy solutions into a typical "human language" application. Fuzzy controller rules can be expressed in natural linguistic terminology along with operational strategy of a typical "human behaviour"; they are easy to grasp and can be modified readily. ANN can be defined as an accumulation of tiny individual processing elements interconnected with each other for processing of information. An input and a corresponding encumbrance form the core part of an incoming connection. Output of the element forms a function of the total value. Typically, an ANN is composed of many layers of interconnected neurons, which in turn, are connected to other neurons in the next layers.

GA is an optimal search technique that focuses on problems of evolution and inheritance of the living organisms by selection of the best-suited humans for reproduction. The three fundamental basics of genetic operators are selection, crossover, and mutation.

This chapter deals with application of AI techniques in smart energy systems. It is an effort to provide a brief of all the information on optimization of solar energy systems for common-life applications. This is the latest technological approach that opens up a theoretical modeling-based search technique for the best optimally designed architecture and control of the solar empowered and other types of energy systems. An AI-based system utilizes the past data, performance of the real-time system, and selection of the best-suited empirical model that suits several plausible solutions. This model can be optimized further by testing it with the past datasets of the real-time systems. AI technologies find applications in fields of engineering, military, economics, marine, medicine, in several solar energy system applications, and many others.

REFERENCES

1. Cheng, Lefeng, Yu, Tao. "A New Generation of AI: A Review and Perspective on Machine Learning Technologies Applied to Smart Energy and Electric Power Systems." International Journal of Energy Research 43 no. 6 (2019): 1928–1973.
2. Kalogirou, A. Soteris, Şencan, Arzu. "Artificial Intelligence Techniques in Solar Energy Applications." Solar Collectors and Panels, Theory and Applications (2010). DOI: 10.5772/10343.
3. Pearson, J. G. Peter, Foxon, J. Timothy. "A Low Carbon Industrial Revolution? Insights and Challenges from Past Technological and Economic Transformations." Energy Policy 50 (2012): 117–127.
4. Li, Fan, Peng, Xiaoqi, Wang, Zuo, Zhou, Yi Wang, Wu, Yuxia, Jiang, Minlin, Xu, Min. "Machine Learning (ML) – Assisted Design and Fabrication For Solar Cells." Energy & Environmental Materials 9 (2019): 280–291.

5. Lachmann, Malin, Maldonado, Jaime, Wiebke, Bergmann, Jung, Francesca, Weber, Markus, Büskens, Christof. "Self-Learning Data-Based Models as Basis of A Universally Applicable Energy Management System." Energies 13 no. 2084 (2020): 1–42.
6. Ascencio-Vàsquez, Juliàn, Bevc, Jakob, Reba, Kristjan, Brecl, Kristijan, Jankovec, Marko, Topič, Marko. "Advanced PV Performance Modelling Based on Different Levels of Irradiance Data Accuracy." Energies 13 no. 2166 (2020): 1–12.
7. Bose, B. K. "Artificial Intelligence Applications in Renewable Energy Systems and Smart Grid – Some Novel Applications." Power Electronics in Renewable Energy Systems and Smart Grid: Technology and Applications (2019): 625–675. DOI: https://doi.org/10.1002/9781119515661.ch12.
8. Xu, Yueqiang, Ahokangas, Petri, Louis, Jean-Nicolas, Pongràcz, Eva. "Electricity Market Empowered by Artificial Intelligence: A Platform Approach." Energies 12 no. 4128 (2019): 1–21.
9. Jang, Roger Jyh-Shing. "ANFIS: Adaptive-Network based Fuzzy Inference System." IEEE Transactions on Systems, Man, and Cybernetics 23 no. 3 (1993): 665–685.
10. Bose, B. K. "Expert System, Fuzzy Logic, and Neural Network Applications in Power Electronics and Motion Control." 82 no. 8 (1994): 1303–1323.
11. Gündoğdu, Fatma Kutlu, Cengiz, Kahraman. "Spherical Fuzzy Sets and Spherical Fuzzy TOPSIS Method." Journal of Intelligent and Fuzzy Systems 36 no. 1 (2019): 337–352.
12. Duojie, Jia-hua, Haidong, Zhang, Yanping, He. "Possibility Pythagorean Fuzzy Soft Set and Its Application." Journal of Intelligent and Fuzzy Systems 36 no. 1 (2019): 413–421.
13. Kumanan, S., Jesuthanam, C. P., Ashok Kumar, R. "Application of Multiple Regression and Adaptive Neuro Fuzzy Inference System for the Prediction of Surface Roughness." International Journal of Advanced Manufacturing Technology 35 (2008): 778–788.
14. Wieprecht, Silke, Tolossa, Habtamu G., Yang, Chih Ted. "A Neuro-Fuzzy-Based Modelling Approach for Sediment Transport Computation." Hydrological Sciences Journal 58 no. 3 (2013): 587–599.
15. Savkovic, Borislav, Kovac, Pavel, Dudic, Branislav, Rodic, Dragan, Taric, Mirfad, Gregus, Michal. "Application of an Adaptive "Neuro-Fuzzy" Inference System in Modeling Cutting Temperature during Hard Turning." Applied Sciences 9 no. 3739 (2019): 1–13.
16. Takagi, T., Sugeno, M. "Fuzzy Identification of Systems and Its Applications to Modeling and Control." IEEE Transaction on Systems, Man, and Cybernetics 15 (1985): 116–132.
17. Toosi, Nadrajan Adel, Kahani, Mohsen. "A New Approach to Intrusion Detection Based on an Evolutionary Soft Computing Model Using Neuro-Fuzzy Classifiers." Computer Communications 30 (2007): 2201–2212.
18. Zadeh, L. A. "Calculus of Fuzzy Restrictions." Fuzzy Sets and Their Applications to Cognitive and Decision Process. Proceedings of The US-Japan Seminar on Fuzzy Sets and Their Applications held at The University of California, Berkeley, California, USA (1974, 1975): 1–39.
19. Huynh, V. N., Ho, T. B., Nakamori, Y. "A Parametric Representation of Linguistic Hedges in Zadeh's Fuzzy Logic." International Journal of Approximate Reasoning 30 (2002): 203–223.
20. Lo, S.-P. "An Adaptive-Network Based Fuzzy Inference System for Prediction of Workpiece Surface Roughness in End Milling." Journal of Materials Processing Technology 142 no. 10 (2003): 665–675.
21. Boyacioglu, M. A., Avci, D. "An Adaptive Network-based Fuzzy Inference System (ANFIS) for the Prediction of Stock Market Return: The Case of Istanbul Stock Exchange." Expert Systems with Applications 37 (2010): 7908–7912.

22. Cross, Valerie, Sudkamp, Thomas. "Patterns of Fuzzy Rule-Based Inference." International Journal of Approximate Reasoning 11 no. 3 (1994): 235–255.
23. Ishibuchi, H., Nakashima, T., Nii M. "Fuzzy If-Then Rules for Pattern Classification." In Ruan, D., Kerre E. E. (eds.) Fuzzy If-Then Rules in Computational Intelligence. (The Springer International Series in Engineering and Computer Science Springer) Springer, Berlin Heidelberg, p. 533 (2000): 267–295.
24. Mamdani, E. H., Assilian, S. "An Experiment in Linguistic Synthesis with a Fuzzy Logic Controller." International Journal of Man-Machine Studies 7 no. 1 (1975): 1–13.
25. Ying Cui, Qi Guo, Jacqueline P. Leighton, Man-Wai Chu. "Log Data Analysis with ANFIS: A Fuzzy Neural Network Approach." International Journal of Testing 20 no. 1 (2020): 78–96.
26. Zhang, Guoqiang, Patuwo, Eddy B., Hu, Y. Michael. "Forecasting with Artificial Neural Networks: The State of the Art." International Journal of Forecasting 14 (1998): 35–62.
27. Huang, Yiqun, Kangas, J. Lars, Rasco, A. Barbara. "Applications of Artificial Neural Networks (ANNs) in Food Science." Critical Reviews in Food Science and Nutrition 47 no. 2 (2007): 113–126.
28. Artificial Neural Networks. In Xingui, He, Shaohua, Xu, Zheng, Zibin, Lyu, R. Michael (eds.) Process Neural Networks. Advanced Topics in Science and Technology in China. Springer, Berlin Heidelberg (2009): 20–42.
29. Heyn, Jakob, Gümbel, Philip, Bobka, Paul, Dietrich, Franz, Dröder, Klaus. "Application of Artificial Neural Network in Force-Controlled Automated Assembly of Complex Shaped Deformable Components." Procedia CIRP 79 (2019): 131–136.
30. Srinivas, M., Patnaik, L. M. "Genetic Algorithms: A Survey." Computer 27 no. 6 (1994): 17–26.
31. McCall, John. "Genetic Algorithms for Modelling and Optimisation." Journal of Computational and Applied Mathematics 184 no. 1 (2005): 205–222.
32. Chaudhry, S.S., Luo, W. "Application of Genetic Algorithms in Production and Operations Management: A Review." International Journal of Production Research 43 no. 19 (2005): 4083–4101.
33. Jennings, P.C., Lysgaard, S., Hummelshøj, J.S. "Genetic Algorithms for Computational Materials Discovery Accelerated by Machine Learning. npj Computational Materials 5 no. 46 (2019): 1–6.
34. Doganaksoy, N., Hahn, J. G. "Data Mining: A Gateway to Better Data Gathering." Statistical Analysis and Data Mining: The ASA Data Science 1 no. 4 (2009): 280–283.
35. Goodman, Arnold. "Emerging Topics and Challenges for Statistical Analysis and Data Mining." Statistical Analysis and Data Mining: The ASA Data Science 4 no. 1 (2011): 3–8.
36. Hassani, H., Huang, X., Silva, S. E., Ghodsi, M. "A Review of Data Mining Applications in Crime." Statistical Analysis and Data Mining: The ASA Data Science 9 no. 3 (2016): 139–154.
37. Hoffmann, D., Apostolakis, J. "Crystal Structure Prediction by Data Mining". Journal of Molecular Structure 647 (2003): 17–39.
38. Salah, C. B., Chaabene, M., Ammar, M. B. "Multi-Criteria Fuzzy Algorithm for Energy Management of a Domestic Photovoltaic Panel." Renewable Energy 33 (2008): 993–1001.
39. Altas, I. H., Sharaf, A. M. "A Novel Maximum Power Fuzzy Logic Controller for Photovoltaic Solar Energy Systems". Renewable Energy 33 (2008): 388–399.
40. Gomez, V., Casanovas, A. "Fuzzy Logic and Meteorological Variables: A Case Study of Solar Irradiance." Fuzzy Sets and Systems 126 (2002): 121–128.
41. Gomez, V., Casanovas, A. "Fuzzy Modelling of Solar Irradiance on Inclined Surfaces." Solar Energy 75 (2003): 307–315.

42. Paulescu, M., Gravila, P., Tulcan-Paulescu, E. "Fuzzy Logic Algorithms for Atmospheric Transmittances of Use in Solar Energy Estimation." Energy Conversion and Management 49 (2008): 3691–3697.
43. Şen, Z. "Fuzzy Algorithm for Estimation of Solar Irradiation from Sunshine Duration." Solar Energy 63 (1998): 39–49.
44. Gouda, M. M., Danaher, S., Underwood C. P. "Quasi-Adaptive Fuzzy Heating Control of Solar Buildings." Building and Environment 41 (2006): 1881–1891.
45. Lygouras, J. N., Botsaris, P. N., Vourvoulakis, J., Kodogiannis, V. "Fuzzy Logic Controller Implementation for a Solar Air-Conditioning System." Applied Energy 84 (2007): 1305–1318.
46. Mellit, A., Kalogirou, S. A., Shaari, S., Salhi, H., Hadj Arab, A. "Methodology for Predicting Sequences of Mean Monthly Clearness Index and Daily Solar Radiation Data in Remote Areas: Application for Sizing A Stand-Alone PV System." Renewable Energy 33 (2008): 1570–1590.
47. Rehman, S., Mohandes, M. "Artificial Neural Network Estimation of Global Solar Radiation Using Air Temperature and Relative Humidity." Energy Policy 36 (2008): 571–576.
48. Mellit, A., Pavan A. M. "A 24-h Forecast of Solar Irradiance Using Artificial Neural Network: Application for Performance Prediction of a Grid-Connected PV Plant at Trieste, Italy. Solar Energy 84 no. 5 (2010): 807–821.
49. Benghanem, M., Mellit, A., Alamri S. N. "ANN-Based Modelling and Estimation of Daily Global Solar Radiation Data: A Case Study." Energy Conversion and Management 50 no. 7 (2009): 1644–1655.
50. Mubiru, J., Banda, E. J. K. B. "Estimation of Monthly Average Daily Global Solar Irradiation Using Artificial Neural Networks." Solar Energy 82 (2008): 181–187.
51. Zagrouba, M., Sellami, A., Bouaicha, M., Ksouri, M. "Identification of PV Solar Cells and Modules Parameters Using the Genetic Algorithms: Application to Maximum Power Extraction." Solar Energy 84 no. 5 (2010): 860–866.
52. Loomans, M., Visser, H. "Application of the Genetic Algorithm for Optimisation of Large Solar Hot Water Systems." Solar Energy 72 (2002): 427–439.
53. Varun, S. "Thermal Performance Optimization of a Flat Plate Solar Air Heater Using Genetic Algorithm." Applied Energy 87 (2010): 1793–1799.
54. Gligor, A., Dumitru, C.-D., Grif, H.-S. "Artificial Intelligence Solution for Managing a Photovoltaic Energy Production Unit." Procedia Manufacturing 22 (2018): 626–633.
55. Kumaravel, S., Ashok, S. "Adapted Multilayer Feedforward ANN Based Power Management Control of Solar Photovoltaic and Wind Integrated Power System. In Innovative Smart Grid Technologies-India (ISGT India), 2011 IEEE PES (2011): 223–228.
56. Saberian, A., Hizam, H., Radzi, M. A. M., Ab Kadir, M. Z. A., Mirzaei, M. "Modelling and Prediction of Photovoltaic Power Output Using Artificial Neural Networks." International journal of Photoenergy 2014 (2014): 1–10.
57. Takagi, H., Hayashi, I. "NN-Driven Fuzzy Reasoning." International Journal of Approximate Reasoning 5 no. 3 (1991): 191–212.

12 Energy Efficiency

Har Lal Singh[1], Sarita Khaturia[1], and Mamta Chahar[2]
[1]Department of Chemistry, School of Liberal Arts and Sciences, Mody University of Science and Technology, Sikar, Rajasthan, India
[2]Department of Chemistry, Nalanda College of Engineering Chandi, Nalanda, Bihar, India

CONTENTS

12.1 OVERVIEW OF ENERGY EFFICIENCY

12.1.1 What is Energy Efficiency?

The aim of energy efficiency is to minimize the amount of energy requirements. Energy consumption of heating and cooling can be minimized by insulating a building. Traditional bulbs can be replaced by installing LED lighting to minimize the energy required for illumination of home. We can adopt more efficient technologies, methods, or processes to improve the energy.

Consumers can save energy costs by minimizing energy consumption by using technologies which are energy efficient. Minimizing the energy consumption also helps to get rid of greenhouse gas emissions problem. International Energy Agency indicates that enhanced energy efficiency in various sectors could minimize 33% of the world's energy requirements up to 2050, and it can also help to control global carbon footprints and greenhouse gas effect [1].

Sustainable or renewable energy policy is based on two foundation pillars named energy efficiency and renewable energy [2]. Foreign imports of energy can be reduced by improving the efficiency of indigenous energy sources. In this way, less dependence on power imports will be proven beneficial to national security too. Furthermore, this strategy of energy efficiency can boost our national economy by reducing energy consumption.

The Vienna Climate Change Report mentions that energy efficiency could be proven to attain emission minimizations at very low cost [3]. It is also noticeable fact that two international standards (ISO17743 and ISO17742) also describe a proven methodology for calculation of energy efficiency for all countries and cities [4, 5].

12.1.2 Benefits of Energy Efficiency

We can save thousands of dollars by minimizing the cost of purchased energy with energy efficiency. There are enormous benefits which can be achieved by energy efficiency after minimization of energy consumption [6]. All these benefits of energy efficiency provide enhanced health, enhanced indoor conditions, minimized climate change impact, minimized air pollution, and security against the price risk for energy users. We have different methods for the calculation of money-saving benefits. There are huge economic benefits than the saved energy [7].

12.1.3 Energy-Efficient Appliances

Modern home appliances like freezers, clothes and dish washers, ovens, stoves, and electric dryers consume very less power in comparison to older appliances. Thus, they significantly minimize energy consumption. Current energy-efficient refrigerators are five-star-rated, and hence, they use 40% less energy than the old ones (Figure 12.1) [8]. Greenhouse gas emission and carbon footprints can be significantly reduced by replacing older appliances with modern energy-efficient appliances [9]. Modern smart power-saving appliances also minimize the energy usage by automatically turning them off when they are not needed to operate. Energy labeling is a proven method which is used on energy-efficient appliances [10]. The energy

FIGURE 12.1 Energy efficiency rating scale for smart homes.

efficiency impact is more on peak demand and less on off-peak demand because it depends on the time of appliance use.

12.1.4 Energy-Efficient Building Design

Buildings are major energy users, so they play an important role in energy efficiency enhancements. Their energy use in buildings depends on the indoor conditions. The energy is consumed in buildings for lighting, cooling, heating, and ventilation (Figure 12.2) [11]. If we divide consumed energy by built area of building, then we get energy efficiency of that building [12].

$$\text{Energy efficiency} = \text{Energy consumed/Built area}$$

The energy efficiency enhancement measures are taken to minimize energy requirement such as by insulating the building. Increased use of natural light is advised.

FIGURE 12.2 Energy-efficient buildings.

Surroundings of location greatly impact on the maintenance of its heating and light conditions. Greenery surrounding the building can provide shade to the building. In winters, designing the building according to the hemisphere and facing increases the time and amount of sunlight/solar heating, which in turns minimizes the energy consumption. Energy loss can be minimized by energy-efficient windows, insulated walls, and compact building design.

Dark shading roofs produce more heat than light-colored or white surfaces. Less dark surfaces reflect some heat inside the building. Less dark roofs of the buildings utilize 40% less energy for cooling in comparison to darker roofs buildings. More energy is saved by white roofs in summers [10].

Less artificial light is needed if we place doors and windows properly. LED lights and fluorescent lamps should be used, which consume 66% less energy than traditional light bulbs. Nowadays, commercial smart meters are in trend at homes and offices, which monitor energy consumption and help customers to cut their electricity bills. Wireless sensor networks are used by such smart meters for communication.

12.2 OVERVIEW OF ARTIFICIAL INTELLIGENCE

Artificial intelligence (AI) can be defined as intelligence of machines which is not natural as human brain, but it is comparatively more powerful in solving complex and ambiguous problems than human brain. AI algorithms learn and get trained to simulate human brain and provide better insights when executed on unknown information. The main goal of AI is to make machines think and perform tasks more intelligently than human brain [13]. Applications of AI technology are increasing dramatically because of its wide scope in many sectors, including energy, defense, agriculture, aviation, ecommerce, healthcare, finance, education, automobiles, etc.

12.2.1 Artificial Intelligence and Big data

AI and Big Data are interconnected technologies which are widely used in the management and processing of huge supervised and unsupervised data for the prediction of unknown patterns. Combinations of both technologies have applications in energy sector, healthcare, telecommunication, defense, weather forecasting, etc. [14, 15]. The approach of AI and Big Data is used to optimize energy efficiency by providing decision-making insights [16]. Big Data is massive data generated by sensors, satellites, social media websites, and financial and industrial offices (Figure 12.3) [17]. AI techniques are used to process this massive Big Data, which generates useful insights that are utilized in decision-making. Big Data acts as a raw material provider to AI for analysis. AI-based machine learning algorithms are used to extract valuable information from Big Data by many big companies such as IBM, Oracle, Microsoft, Intel, and NVIDIA [18].

12.2.2 Artificial Intelligence and Cognitive Computing

Cognitive computing is a technological approach based on AI framework, which simulates, replicates, or mimics the human brain by applying machine learning

FIGURE 12.3 Artificial intelligence (AI).

models for the solution of real-time complex problems in a better and faster way. IBM's Watson is a well-known cognitive platform [19]. Cognitive computing-based automated systems are being used in energy sector in improvement of uptime and accurate outage prediction, thus increasing energy efficiency by reducing energy consumption. It enables the consumers in monitoring and controlling the energy usage.

12.2.3 Artificial Intelligence and Machine Learning

Machine learning is a subset of AI in which machines are used to learn from available data from various supervised or unsupervised datasets. After learning the data, tasks are executed on new datasets to get useful insights to solve real-time problems. Machine learning helps customers in taking smart decisions by providing prediction models [20]. Machine learning algorithms are used to learn known and unknown data, which prepares a predictive model for the solution of various kinds of problems. Nowadays, machine learning is being widely used in energy sector for the solution of the problems related to minimization of energy consumption.

12.2.4 Artificial Intelligence and Deep Learning

Deep learning is a machine learning technique which learns the datasets in hierarchal manner. Algorithms of deep learning can automatically learn a large amount of unsupervised data from multi-dimensional datasets. Conceptually, deep learning is based on neural networks and it is implemented through multi-level neural networks [21]. Deep learning approach can be used to solve energy-related problems by mining sensors or satellite-generated images. It provides valuable insights to minimize energy usage. Deep learning-based frameworks are installed in energy industries for decision-making analytics in prediction and estimation of energy consumption.

12.3 HOW TO ACHIEVE ENERGY EFFICIENCY USING AI

There are numerous energy efficiency arrangements and blends of arrangements accessible today for a specific circumstance. The utilization of AI technologies permits us to plan, screen, and control energy consumption consistently. These arrangements are financially suitable also, and are relied upon to pay in a brief time frame (Figure 12.4) [22].

- **Step 1: Energy Audit**
  Energy audits help in every sector such as municipalities, industrial, and commercial companies, facilitate to realize proper energy use, and help to identify energy waste management and improvement. Energy-efficiency audits are performed in organizations and their operations to identify improvement potentials. By using AI and data mining, energy consumption is identified to minimize the energy losses [23].
- **Step 2: Implementation of Solutions**
  Renewable energy and AI-based solutions are implemented to monitor, control, and manage the energy usage to get optimum energy efficiency. Energy consumption is minimized by implementing AI-based application of energy-saving solutions.

FIGURE 12.4 Energy efficiency using AI.

12.4 RENEWABLE ENERGY AI TECHNOLOGIES

Two pillars of sustainable energy are supposed to be energy efficiency and renewable energy. These two methodologies must grow simultaneously so as to settle and limit CO_2 emission. Energy demand can grow by the efficient use of energy so that energy consumption as well as dependence on fossil fuels is minimized. If energy demand and consumption increases with time, then use of renewable power sources also increases to fulfill the power demand. Similarly, complete carbon emission can only be minimized by minimizing the power demand using clean energy sources such as renewable energy sources. So, both the pillars of sustainable energy strategy are equally important and require simultaneous attention (Figure 12.5) [24].

Renewable energy sources such as solar energy and wind energy are naturally available to be used for unlimited times. Conversely, non-renewable energy sources (fossil fuels) are viewed as limited. Maximum use of renewable sources is increasing for electricity generation. Still, there are boundaries to more extensive execution identified with strategy and innovation. Scientists and organizations are investigating the scope of improvement in efficiency and accessibility of renewable or sustainable energy technology.

Energy forecasting – AI algorithms are trained and used to make exact forecasts, assisting with recognizing supply and demand of energy.
Energy efficiency – AI is utilized to follow and improve the energy efficiency.
Energy accessibility – AI is utilized to demonstrate energy cost reserve funds and give proposals for investments in smart homes and buildings [25].

FIGURE 12.5 Renewable energy AI technologies.

12.4.1 Artificial Intelligence Technology for Energy Forecasting

12.4.1.1 Xcel

Renewable energy sources face big challenges because of the variable behavior of climate and weather. Power demand and supply fluctuate making renewable energy sources unstable. Xcel addresses such challenges by accessing accurate weather reports for smooth management of renewable energy sources. AI is used to mine the weather reports and information collected from satellites and renewable power stations. Insights generated from data mining of such information helps to manage energy usage and increase energy efficiency of renewable power systems during bad weather conditions [26]. Xcel in collaboration with other companies, is also using AI for the inspection of power transmission lines in the United States. Xcel energy is one of the power suppliers in eight states of the United States. The company offers many AI-enabled energy-saving solutions to customers and businesses.

12.4.1.2 Nnergix

Weather plays an important role in the renewable energy industry because variable behavior of weather could disrupt and destabilize the power demand and supply of renewable power generation systems. Solar and wind power generation systems are weather-dependent. Nnergix is the biggest weather analytics data provider to the energy industry. It provides weather forecast information to energy suppliers, which are incorporated with AI-based predictive analytics to help energy providers in making predictions related to demand and supply of renewable power. In this way, energy efficiency of renewable systems can be optimized by using accurate forecast of weather. Nnergix utilizes data mining and AI algorithms to analyze the combined information collected from the energy industries and satellites for the efficient management of power demand and supply [27].

12.4.2 Artificial Intelligence Technology for Energy Efficiency

12.4.2.1 Verdigris Technologies

Verdigris Technologies was founded by Mark Chung in 2011 at California, the United States. Verdigris Technologies uses AI in its energy-saving equipment, which monitors and forecasts the energy consumption in the building, and helps customers to find out where they can save energy and money. The AI systems are cloud-based that incorporate AI and send alerts to minimize or optimize peak hour demand of energy use. Their Internet of Things (IoT) hardware is 100% certified and approved and can be implemented in all kind of homes and buildings. Their AI-enabled hardware is easy to install and data can be tracked using laptop, mobile, or tablet. Forty-two smart sensors of their system use just 10 watt and there is no battery, and hence, no problem of charging their hardware [27]. Smart sensors send information to clouds using Wi-Fi and that data provides useful insights to the customer to reduce energy usage. Verizon Ventures is the biggest investor of Verdigris which has invested almost 18 million dollars in this AI technology firm. AI algorithm of equipment identifies and monitors the home appliances by their electrical footprint and generates comprehensive analysis of energy usage, which is sent to cloud-based servers of Verdigris.

12.4.2.2 Google DeepMind

DeepMind is an AI-based company involved in making AI systems used for learning the energy and other problems and then solving them by AI-enabled predictive analysis. This London-based company came into existence in year 2010. Later, it was taken over by Google in year 2014. DeepMind is an AI company which cooled down the server farms of Google by 40% [25]. Earlier, DeepMind started its business incorporating AI for the measurement and improvement of energy usage. Neural network systems were trained by utilizing data centers for the optimization to provide information of energy consumption and temperature data collected by thousands of sensors. Energy efficiency was taken as proportion of whole energy consumption and energy consumption in information technology tasks.

12.4.3 Artificial Intelligence Technology for Energy Accessibility

12.4.3.1 PowerScout

PowerScout is an online marketplace and data science company founded by Attila Toth from Sanfransisco, the United States. PowerScout is an AI-based startup funded and backed by Google and US Department of Energy. It provides industrial and home improvement projects to reduce energy consumption of smart buildings. PowerScout incorporates AI-based data analytics to analyze energy consumption information of homes and industries of customers on online marketplace. It uses AI-based algorithm to help meet customers with suppliers of energy products in an online marketplace. AI helps as an advisor to potential energy customers offering proposals of renewable energy products for their homes making improved smart homes [25].

12.4.3.2 Verv

Verv is an AI-based solution, which is used to help customers by supplying energy information on home appliances so that customers can manage or reduce energy consumption at their homes. It is actually an AI-based technology which was developed by a British company named Green Running Ltd. Verv is built on their AI-based platform smart hub, which generates information about energy usage of home appliances so that customers could know energy usage and energy cost of each and every home appliance. Whenever a customer turns on the home appliance, AI algorithm starts working and managing energy costs. Verv provides real-time information of electricity consumption by home appliances. It also provides notifications and tips to minimize energy usage and carbon footprints of home. Customers can access this AI-based application online using their laptop, cell phone, or tablet [25].

12.5 SECTOR-WISE AI-ENABLED ENERGY EFFICIENCY TECHNOLOGIES

Industrial, commercial, and municipal sectors are recognizing and identifying the scope of energy saving by applying energy-saving strategies by using AI. They are now seriously thinking to implement energy-saving techniques and technologies, which incorporate AI and renewable energy sources. AI applications have large-scale, long-term benefits for the better management, planning, monitoring and controlling, and reducing the energy consumption and carbon footprints (Figure 12.6) [28].

12.5.1 Energy Efficiency in Industry

Industries are the main energy consumer in which energy in huge quantity is utilized to run various industrial plants and processes. Electricity, gas, and fossil fuels are used in various industrial operations. Some industries manage their energy requirements by creating fuel from waste products and utilizing them in industries themselves [29]. Energy-efficient electric motors are used to avoid energy losses. AI-based applications are used in industries to monitor and improve the energy efficiency, thus reducing energy consumption in various industrial processes.

FIGURE 12.6 Artificial intelligence for smart cities.

AI-based applications monitor a large number of motors, pumps, and compressors, so that AI can help decision-makers to select energy-efficient devices and reject energy-consuming devices. Various factors affect the efficiency of such components; so, AI applications generate information to optimize process control utilizing good practices to save energy consumption in industries. AI helps to optimize air compressors and pumps by detecting and fixing the leaks, improving their productivity by 20–50% [30].

Energy cost is minimized by using AI-based applications for the optimization practices, monitoring, and controlling the industrial processes, thus reducing energy consumption (Figure 12.7) [31]. Many industries consume energy intensively, and most of them never try to become energy-efficient. AI-based industries are becoming successful in saving energy.

AI and data mining assist the industries in managing processes efficiently by improving energy efficiency.

Information is continuously generated by AI-based applications, which is used to identify energy wastage, and then energy efficiency practices are followed to minimize energy usage. Energy is saved by minimizing energy wastage [25].

- **Step 1: By Data mining**
  Various issues and bottlenecks are identified by data mining, and then the best energy-saving practice is applied in industry.
- **Step 2: By Algorithms**
  Best equipment of desired values is selected to minimize energy consumption by using AI algorithms.
- **Step 3: Combination of Data mining and Algorithms**
  AI algorithms and analysis of data mining is combined to minimize energy consumption improving energy efficiency [25].

FIGURE 12.7 AI is powering the energy industry.

12.5.1.1 How to Minimize Industrial Energy Wastage

AI technology is used to minimize energy wastage in various industrial processes by the following ways:

1. Forecasting the demand and supply of energy incorporating renewable energy sources to reduce fuel consumption and energy consumption.
2. Limiting industrial breakdowns – Sensor data and operational information generated by AI systems are used to minimize or limit the breakdowns.
3. Managing power grids autonomously – Power grids are managed by AI applications to efficiently manage various renewable and non-renewable energy sources such as solar, wind, oil, and gas.
4. Identifying optimum process set-points to maximize the plant outputs.
5. Reducing energy wastage in heating and cooling processes of industry.
6. Making improved and energy-efficient combined heat and power (CHP) plants.
7. Optimizing performance and maintenance-related problems to minimize the utility cost of compressed gases and water [32].

12.5.2 Commercial Sector

AI is revolutionizing the way energy management in commercial sector offices and buildings such as private hospital, big shopping malls, institutes, industries, hotels, gymnasiums, and schools and colleges. Energy cost is reduced and energy consumption is managed using AI-enabled systems [33].

Smart offices and smart building are made with the application of AI systems which minimize the energy consumption. Such AI-enabled systems generate information which is used to monitor, manage, and improve the complex energy-consuming processes. AI-enabled solutions manage the energy usage automatically. Energy consumption pattern of commercial buildings is monitored and analyzed by such AI applications to reduce energy consumption. AI applications are used to manage heating and cooling systems, heating, ventilation, and air-conditioning (HVAC) systems, gas and water supplying systems, and temperature controlling sensors based on weather conditions and energy demands.

12.5.3 Municipalities Public Sector

There is a saying that you can save only measurable things, and here comes the AI which saves energy after measuring it. AI helps public sector organizations and municipalities in solving energy challenges by helping them to change cities into smart cities.

Many areas of municipalities consume huge energy that vary from street lights and traffic lighting to heating and cooling of public offices, airports, railway stations, schools, colleges, temples, churches, hospitals, shipyards, and factories.

AI-enabled systems are used to provide information of energy consumption information, and then that information is used to manage energy supply according to the demand. It also takes weather as an energy-consumption factor. AI-enabled systems are used to utilize an array of renewable energy sources to offer the required energy and they engage fossil fuel energy sources, when needed [34].

12.5.4 Transport Sector

The energy efficiency in transport can be calculated as distance travelled by goods and passengers divided by the total energy. The energy input is assumed as combustion fuels and electrical energy [36]. The energy efficiency can also be considered as energy intensity [36]. In transport, the inverse of the energy efficiency is the energy consumption, which in turn reciprocates fuel economy. Fuel consumption is used to describe energy efficiency in transport [37].

To save the fuel, advanced and developed tires with less friction and resistance are used. In this way, mileage is increased by 3.3% by maintaining proper air pressure in tires [36]. Fuel consumption can also be improved by 10% by changing the dirty and clogged air filters of older vehicles [38]. Fuel efficiency can be increased by using eco-friendly turbochargers, which allow a comparatively smaller displacement engine.

In transport sector, by increasing energy efficiency, we can double the energy-efficient vehicles. The cutting-edge design vehicles have accomplished fuel efficiency up to four times. This is the reason why electric vehicles are rising. Electric engines are double fuel-efficient in comparison to internal combustion engines [39].

12.6 CONCLUDING THOUGHTS AND FUTURE OUTLOOK

The sustainable power source part or renewable energy sector is a developing monetary power and a successful methodology for the optimization of environmental sustainability. The major industrial sectors are incorporating AI by expanding the capacity and limit of data analytics. Changing behavior of weather poses various challenges, which force providers to depend on traditional energy sources for the satisfaction of consumer needs. In this manner, AI-enabled energy forecasting systems give guarantee of information required to handle fluctuations that may adversely affect the planning, designing, and operations according to the requirements.

It is feasible to achieve higher energy efficiency by using modern AI and machine learning-based approaches and techniques. The year 2015 was a standard year for sustainable confirmation or renewable confirmation of G7 and G20 commitments for the growth, implementation, and improvement of compete energy efficiency. However, for the solution of problems and proliferated fast usage will need guarantee of financial benefits and political will. Stages or platforms will play a vital role for energy buyers and organizations in cost investment funds and energy efficiency.

REFERENCES

1. Diesendorf M. & Mark D. (2007) Greenhouse solutions with sustainable energy. University of New South Wales Press, Sydney, Australia.
2. Prindle B., Maggie E., Mike E. & Alyssa F. (2007). The twin pillars of sustainable energy: synergies between energy efficiency and renewable energy technology and policy. American Council for an Energy-Efficient Economy, Washington, DC.
3. Press Release. (2019). Vienna UN Conference Shows Consensus on Key Building Blocks for Effective International Response to Climate Change.

4. ISO 17742. (2015). Energy Efficiency and Savings Calculation for Countries, Regions and Cities. International Standards Association (ISO). Geneva, Switzerland. Retrieved 2016-11-11.
5. Weinsziehr T. & Skumatz L. (2016). Evidence for Multiple Benefits or NEBs: Review on Progress and Gaps from the IEA Data and Measurement Subcommittee. In Proceedings of the International Energy Policy & Programme Evaluation Conference, Amsterdam, the Netherlands, 7–9.
6. International Energy Agency. (2014). Capturing the Multiple Benefits of Energy Efficiency. OECD, Paris.
7. International Energy Agency. (2007). Capturing the multiple benefits of energy efficiency. OECD, Paris.
8. https://cdn.pixabay.com/photo/2013/07/12/18/55/energy-efficiency-154006-960-720.png.
9. McKinsey Global Institute. (2009). Pathways to a Low-Carbon Economy: Version 2 of the Global Greenhouse Gas Abatement Cost Curve. Retrieved February 16, 2016.
10. Environmental and Energy Study Institute. (2010). Energy-Efficient Buildings: Using Whole Building Design to Reduce Energy Consumption in Homes and Offices. Eesi.org. Retrieved 2010-07-16.
11. https://cdn.pixabay.com/photo/2019/04/03/12/04/home-4100193_960_720.jpg.
12. ENERGY STAR Buildings and Plants. (2019). Energystar.gov. Retrieved 2019-3-26.
13. What is artificial intelligence? (2016). Progressive Digital Media Technology News Retrieved from https://search.proquest.com/docview/1789725622?accountid=27468.
14. Elgendy N. & Ahmed E. (2014). "Big data analytics: a literature review paper." In Industrial conference on data mining, Springer, Cham, 214–227.
15. Emmanuel I. & Stanier C. (2016). Defining Big Data. In Proceedings of the International Conference on Big Data and Advanced Wireless Technologies (BDAW '16), Blagoevgrad, Bulgaria, 10–11 November 2016; ACM: New York, NY, USA, 5:1–5:6.
16. Moreno M. V., Dufour L., Skarmeta A. F., Jara A. J., Genoud D., Ladevie B. & Bezian J. J. (2016). Big data: the key to energy efficiency in smart buildings. Soft Computing, 20(5), 1749–1762.
17. https://cdn.pixabay.com/photo/2020/06/05/06/57/artificial-intelligence-5261742_960_720.jpg.
18. O'Leary D. E. (2013). Artificial intelligence and big data. IEEE Intelligent Systems, 28(2), 96–99.
19. Zheng N. N., Zi-yi L., Peng-ju R., Yong-qiang M., Shi-tao C., Si-yu Y., Jian-ru X., Ba-dong C. & Fei-yue W. (2017). Hybrid-augmented intelligence: collaboration and cognition. Frontiers of Information Technology & Electronic Engineering, 18(2), 153–179.
20. Louridas P. & Ebert C. (2016). Machine learning. IEEE Software, 10–115.
21. Goodfellow, Ian, Yoshua Bengio, & Aaron Courville. (2016). Deep learning. MIT Press, Cambridge, MA.
22. https://cdn.pixabay.com/photo/2018/04/07/09/48/hand-3298095_960_720.png
23. https://www.maximpact.com/artificial-intelligence-in-energy-efficiency/
24. https://cdn.pixabay.com/photo/2018/04/23/12/11/windmill-3344052_960_720.jpg
25. https://emerj.com/ai-sector-overviews/artificial-intelligence-for-energy-efficiency-and-renewable-energy/
26. Proctor, Cathy. (2018). Xcel's Forecast Rosy for Wind and Solar. Denver Business Journal, 4 May 2018, www.bizjournals.com/denver/news/2018/05/04/cover-storyxcel-s-forecast-rosy-for-wind-and-solar.html
27. Verdigris. (2017). Fast Company. Retrieved 2017-3-14.
28. https://cdn.pixabay.com/photo/2019/05/03/11/40/smart-4175713_960_720.png

29. Environmental and Energy Study Institute. (2015). Industrial Energy Efficiency: Using new technologies to reduce energy use in industry and manufacturing (PDF). Retrieved 2015-01-11.
30. City of Edmonton (2019). Street Lighting. Edmonton.ca. Retrieved 2019-3-26.
31. https://cdn.pixabay.com/photo/2020/06/01/13/23/workers-5246634_1280.jpg
32. http://maximpactblog.com/ai-can-cut-energy-consumption/
33. Dorf R. C. (1981). Energy factbook. [includes glossary].
34. Höjer M. & Josefin W. (2015). "Smart sustainable cities: definition and challenges." In ICT innovations for sustainability, Springer, Cham, 333–349.
35. Sustainability & Energy Efficiency. (2011). Empire State Building, Esbnyc.com., 2011-06-16. Retrieved 2013-08-21.
36. Lovins A. B. (2012). A farewell to fossil fuels: answering the energy challenge. Foreign Affairs, 91, 134.
37. Tuominen P., Francesco R., Waled D., Bahaa E., Ghada E. & Abdelazim N. (2015). Economic appraisal of energy efficiency in buildings using cost-effectiveness assessment. Procedia Economics and Finance, 21, 422–430.
38. Heat Roadmap Europe. Heatroadmap.eu. Retrieved 2018-04-24.
39. Heinrich B. S. (2018). Energy atlas 2018. Figures and facts about renewables in Europe.

13 Renewable (Bio-Based) Energy from Natural Resources (Plant Biomass Matters)

Rajesh K. Srivastava
Department of Biotechnology, GITAM Institute of Technology, GITAM, (Deemed to be University), Visakhapatnam, India

CONTENTS

13.1 INTRODUCTION

The present and future generations will find high energy consumption of non-renewable energy sources due to exponential growth of population globally, and lots of development works are going on that need massive quantity of energy consumption. Energy consumption in any form increases with economic growth, and research and development tasks can enhance economic growth via stimulation of the importance of renewable and non-renewable energy sources [1]. The importance of research and development works or reports with trade openness tasks can positively influence economic growth. Heterogeneous causality analysis reports can reveal the feedback influences, which is shown in the form of bi-directional causal association among economic developments and it further impacts on patterns of renewable energy consumption and non-renewable energy fuel consumption [1].

People at worldwide are found to utilize traditional fossil fuel energy sources (i.e., petroleum oil or natural gases) at very quantity still in today periods. Around 80% of electricity output is obtained from conventional fuel sources, and the rest of the electricity is gained from renewable resources in the present period at fast rate and in rising trends. An economic case study for renewable energy sources reported that they are the backbone of the global energy system with an increasable clear and proven form. In this regard, more incredible bang-for-buck renewable form of energy is found quite simple and cheapest for energy generation in many countries [2].

In 2016, the global renewable capacity had grown by a record amount with its cost reduction in a considerable manner. This strategy has improved the generation of wind and solar energies cost-effectively. In 2016, 161 GW of renewable energy was found from installed generation capacity with a 10% rise in the preceding years. REN21 reported that the global renewable energy policy covers 155 nations and 96% of the world population [3]. Wood pellets and compressed briquettes come as by-products from wood processing industries and are needed to reduce fuel energy consumption. Compressed biomass fuels can produce a considerable quantity of energy than logs form of wood. It was reported that the burning of wood biomass (raw timber or processed wastes) released toxic gases with different particle sizes or by-products into the atmosphere. Pellet production cost saved around 0.25 M£ or 3.9% per year for 10,000-tonne plants. Further, a 2% reduction in CO_2 gas equivalent emission is reported from footprint of pellets sources (3.9 g/MJ of fuel) [4].

In this regard, renewable energy production and its promotion are recognized and encouraged by the involvement of governments all over the world. It is also necessary to solve some of the challenges such as lower energy or high-cost value. These are considered by policy or regulatory uncertainty, high investment risks, and system integration of wind, solar, or other resources. In the world, biomass is considered as the fourth energy resource, and only 10% of energy from this source is reported yet in the form of biofuels [3]. This biomass utilization for renewable energy generation can benefit our farmers with avoidance of its burning impacts on their farming lands. In the current period, renewable power capacity can expand by 50% between 2019 and 2024 by using solar photovoltaic (PV) cells. It has shown that 1200 GW fuel power in the United States are obtained from installed power capacity. Renewable power capacity can increase from solar PV (5% in 2020 to 9% in 2030) [5]. This chapter discusses renewable energy and its sources, including plant biomass sources for energy production. Next, it discusses the importance of renewable energy production and its utilization to maintain energy security and a healthy environment.

13.2 RENEWABLE ENERGY AND ITS SOURCES

Renewable energy obtained from such biomass sources has never gone for depletion of energy and can be replenished within a particular lifetime. Solar, wind, geothermal, hydropower, and biomass energies are reported as good examples of renewable energy power stock that can boom as innovative activity or performance with the delivery of clean energy in the future. These are reported as American solar or wind energy productions utilized via an integrated national electricity grid to maintain appropriate reliability. Further, renewable energy production and utilization are

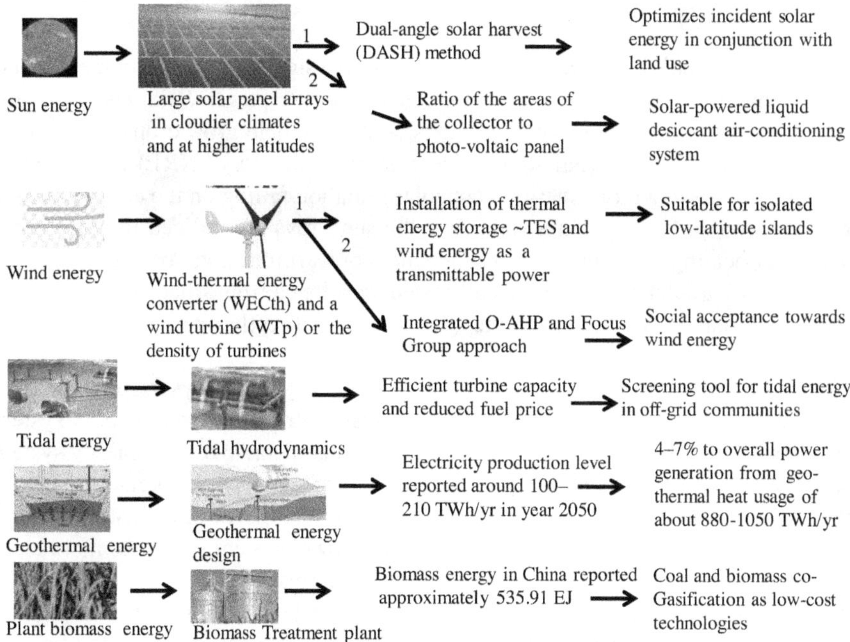

FIGURE 13.1 Different natural resource utilization for renewable energy, including green plants for their capability to store solar energy. All the sources of renewable energy are reported as clean energy sources with healthy environment and more energy security [7–13].

gradually increasing via the replacement of dirty fossil fuel sources in the power sectors that offered the benefits of reduced carbon emission or other fuel-generated pollutants, as shown in Figure 13.1 [6].

These renewable energy approaches help the entire rural communities with heating and lighting tasks. Also, the use of renewable energy has shown its goal of advancing the American electricity grids system that has made smart, more secure, or integrated forms across the region [14]. By 2050, renewable energy generation (around 95%) would be achieved, and Waterhouse Cooper has predicted 100% renewable energy utilization by 2050 in Africa. In the coming decades, the prices of solar PV panels would decrease around 99%, which might help in the generation of renewable energy in the United States with the creation of three times more jobs than that of fossil fuels. A lot of investments in the renewable energy generation sector has been reported surpassing the fossil fuel investment, and around $250 billion markets created from global renewable energy production [15].

The most common forms of renewable energy sources are solar, wind, hydroelectric, tidal, geothermal, and biomass energies. For solar energy generation, sunlight radiation on our planet reports an abundant and freely available energy resource. The amount of solar energy that reached to the earth's surface per hour is needed to capture the entire world's total energy requirement for the whole of the year. The amount of energy from sun varies depending on the time of day or season of the year, or geographical location [16].

13.2.1 Solar Energy

Solar energy in the United Kingdom has increased in significance or become popular to supplement the energy needs or usage related to solar power. Solar energy is utilized by humankind for thousands of years, and it can help grow crops, stay warm, and dry foods. The National Renewable Energy Laboratory (NREL) researchers report for harnessing more energy from solar radiation falling on the earth's surface per hour around the world in a year. And the sun's rays are utilized in many ways such as for heating water in home or in business of warming water or power devices. In this regard, a solar system or PV cell system developed from silicon or other semiconductive matters can transform the solar energy indirectly into an electric power system [17].

At small levels, the solar power system can harness the sun rays for powering the whole houses or villages. Their PV cell panel system or passive solar energy system for home design can utilize this solar energy. Further, a passive solar energy system for homes utilizes the solar energy via a south-facing window arrangement. Then, it is retained in warm condition via concrete matter, brick, tile, or other heat-absorbing materials storing the heat energy [11]. The solar power system generates more electric power for home needs, and it is possible to sell the excess quantity of power to the grids system. Nowadays, batteries are also considered as attractive ways of storing the extra solar energy that can be used at night, and scientists are working hard to find new advancements in blend forms or function arrangements such as skylight and roof shingles system [18, 19].

13.2.2 Wind Energy Source

Next, wind energy is also available as clean renewable energy sources, and wind farms are continuously increasing in the United Kingdom. Wind power has contributed to the National grid. This energy is harnessed for electricity generation with the making of wind power. In this regard, the turbine is used to drive the generator that feeds electricity into the National grid. Domestic or off-grid generation system is suitable for domestic wind turbine and generates wind energy from wind power. This energy source is like wind energy in a particular situation. It is a more viable commercial energy source, but it depends on property types for domestic uses via off-grid generation [20].

In this regard, wind energy can power the boats to sail the seas and can be used in a windmill to grind the grain. For many years, human has increased the turning of cheap dirty energy sources (i.e., coal or fracked gases sources). There are many reports on innovative and less expensive technologies that help capture wind or solar energy as a renewable energy form and become more critical power sources for the world, especially in the US energy (one-eighth of the total renewable) [21]. Many countries are found to expand renewable energy generation capacity at large or small scale by using rooftop solar panel for houses. And it can be sold to power banks or stations via applying back to the grid system to giant offshore wind farms locations [20, 21].

13.2.3 Hydro-Energy

Hydro-energy is also a renewable energy resource which is commercially developed. For this energy generation, a dam or barrier building and a large reservoir are used to create a controlled flow of water that can drive a turbine for electricity generation. This energy is more reliable than solar or wind power, and it can easily allow electricity for storage purposes that can be used at the demand peak [22].

Hydropower is utilized for electricity generation in the United States. This energy is dependent on fast speed moving water flow in a large river body or rapid rate descending water from a high point and converts the water forces into electric power that is generated by spinning of generators' turbine blades. Many national or international levels of large-sized hydroelectric plants or mega-sized dams are reported as non-renewable energy sources. Mega-sized dams can be diverted by reducing the natural water flows with restricting access for animal or human populations dependent on the river. Small-sized hydroelectric plants (capacity below 40 MW) are managed without environmental damaging that diverts a fraction of water flow [23].

13.2.4 Plant or Others Biomasses

Biomass is involved in the production of electricity via the burning of organic materials. This mode of wood-burning is not appropriate for clean energy generation. The generation of biomass energy is a cleaner and more energy-efficient process, which is carried out by converting agricultural or domestic waste into solid, liquid, or gas fuel, as shown in Figure 13.2. Biomass can generate power at a much lower commercial and environmental cost. In Turkey, the total agricultural residue is around 7.5×10^4 kilotonnes, and the theoretical energy potential of farm residues is reported around 9.9×10^5 TJ. Further, biomass energy is dependent on the spatio-temporal variation profiles and their characteristics. Their experts still assess biomass source quantification and its energy potentials via an analysis of 3.6×10^5 TJ for total residues [24].

Root and rhizomes of *Rubia yunnanensis* in the wild form are reported as Chinese medical parts as they exhibit anti-tumor properties and are found to cultivate at a large scale. This plant biomass is reported to contain photosynthetic pigment contents or characteristics with the chlorophyll fluorescence parameter, biomass productions, and secondary metabolites quantity [25]. Drought and rewatering conditions in *Rubia yunnanensis* have shown its growth condition in four water conditions such as well-watered conditions (C), moderate water conditions (MD), severe drought (SD), and rewatering after moderate drought (RMD) conditions [26].

In these conditions, plant growth patterns were studied via comparison to control one. Under moderate water conditions (MD) and severe drought (SD), photosynthetic pigments decreased photochemical efficiency. Due to a reduction in maximum potentials of PSII efficiency (Fv/Fm), it has proportionally affected to the vigorness, growth, and physiological functions of this plant. Further, the effective quantum yield of photochemical conversion (outcomes) and electron transport rate (ETR) were compared with their results. Next, limited photosynthetic capacity reduces the

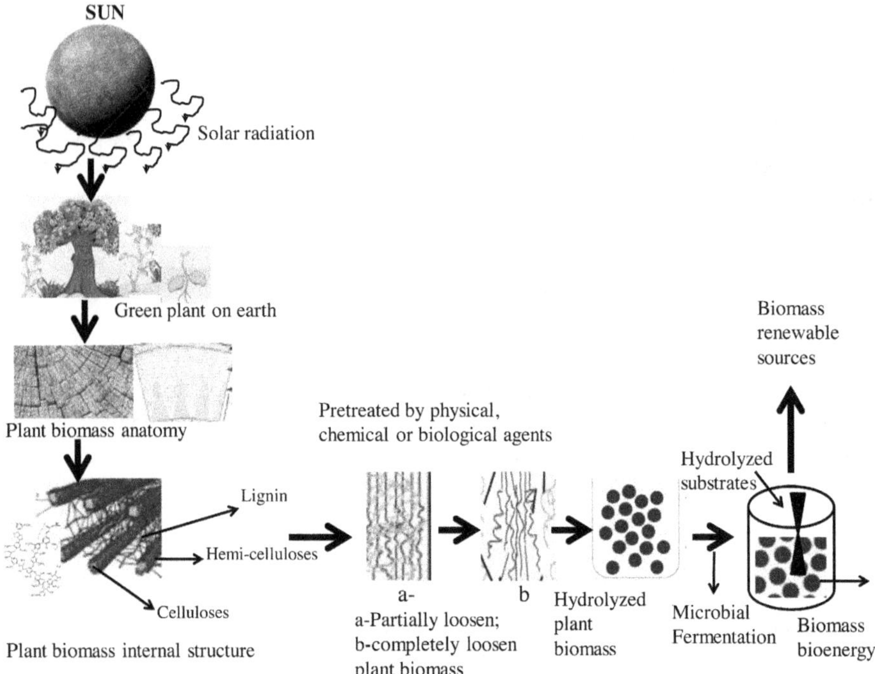

FIGURE 13.2 Plant biomass utilization for biomass energy as renewable energy source and green plant has capability to store solar energy in the form of chemical energy that can be transformed for biofuel energy via microbial fermentation.

biomass accumulation rate and Rubiaceae-type cyclopeptides accumulation (RAs) contents with its yield in root and rhizomes [25, 26].

Further, long hydraulic retention time (HRT) value reports to low degradability or slow degradation of lignocelluloses. But serial anaerobic digestion (AD) plants under 15 days of HRT period as the single stage of AD exhibit high rate of methane conversion efficiency (65% COD), and it has enhanced the degradation of slow degradation fractions (prolonged solid retention time ~52.2 days) of complete systems [27]. The methane fuel recovery from wet AD plant alone in the serial AD is found to suppress with 39 days HRT in single-stage AD. It is reported using appropriate HRT periods for labile fractions of aquatic plants or another microbial recirculation system. The serial wet-AD and SS-AD plants were reported as a suitable technology for underwater plant biomass treatment system with recalcitrant cell walls and labile cytoplasm systems [27, 28].

13.3 PRETREATMENT AND SACCHARIFICATION APPROACHES FOR PLANT BIOMASSES

Different pretreatment methods applied for plant biomass hydrolysis are physical conditions, chemical reagents, physicochemical or biological conditions or agents, or combined forms. But chemical pretreatment is reported as expensive due to the use of more amounts for lignocellulosic biomasses hydrolysis. However, more efficient pretreatment is combined treatment as compared to a single approach pretreatment. There are different pretreatment methods applied for biomass hydrolysis that are quickly converted into biofuels with microbial fermentation. These fermentation processes combined with pretreatment are used for biohydrogen, biomethane, bioethanol, or biodiesel synthesis [29].

A further report finds various pretreatment methods. Alkaline pretreatment is gaining more attention due to work with efficient catalysts. Many researchers have discussed the Kraft pulping industry that generated a considerable quantity of white, black, or green liquor alkaline solution during paper processing or making processes. So this green liquor (GL) from Kraft pulping processes is applied for lignocelluloses biomasses pretreatment tasks. Further, green liquor dregs (GLD) as alkaline waste products from crafting pulp processes were extensively studied for their performance with lignocellulosic pretreatment tasks [30].

13.3.1 Utilization of Fermentative Sugars for Renewable Energy

There are many biomass resources. The potential of de-oiled algal biomasses (DAB) residue is discussed and utilized as an alternative source for the synthesis of bioethanol or polymers under a controlled bio-refinery approach. Also, hybrid pretreatment approach shows a high quantity of sugar solubilization (0.6 g/g DAB), individual physicochemical (0.481 g/g DAB), or enzymatic method (0.484 g/g DAB). Further, these fermentable sugars (obtained from hybrid pretreatment approach) are used for bioethanol synthesis using microbial strain *Saccharomyces cerevisiae.* This fermentation has achieved maximum bioethanol yield (at pH 5.5~ 0.145 g/g DAB) and is higher than lower pH (0.122 g/g DAB) or higher pH (0.102 g/g DAB) [27].

There are many crops with biomass residues. And these available resources (such as wheat straw, rice husk, or stalk; cotton straw, corn stover, or sugarcane bagasse) assess for the estimation of energy production potential or capacities in Pakistan (the year 2018 to 2035), and there are about 40 tonnes of crops residues that are found for power generation. In 2018, crop residues removal was around 50% [31] and can be seen in Table 13.1.

This removal of crop residues is found to translate around 11,000 MW electricity generations from crop residues in 2018. This electricity capacity is gradually increasing by up to 16,000 MW (by the year 2035), depending on crop residue trends in the year 2001. In Pakistan, suitable conditions for potential regions for installing 100 MW capacity biomass-fitted power plants were reported with the assessment of crop residues density and an equivalent collection radius (Re) of 50 km. Punjab province is fair in sustainable agriculture residues availability with better road infrastructure and 7000 MW cumulative capacity at various or different locations reported [41].

Most of the crop residues or other feedstock sources are used for bioconversion or fermentation process. In this process, plant biomass, microbial biomass, or industrial wastes can be utilized for biofuel production. In bioconversion processes, several microbial cells modify via engineering or genetic tool modification. It operates to produce many valuable chemicals, including biomass energy, via sustainable resources or agricultural residues. In this regard, engineered *E. coli* is made for fuel

TABLE 13.1
Biomasses Sources for Bioenergy via Utilizing Efficient Pretreatment and Microbial System

Biomass Energy	Energy Production Parameter	Reference
1. Wheat straw form thirty winter season and wheat variety reported 34–35% glucose (for cellulose), or 7–20% xylan (for hemicellulose) and 18–21% for lignin.	Cellulosic bioethanol synthesis is reported in the Pacific Northwest, the USA. Dilute acid and hot water as optimal pretreatment used for variety wheat straw hydrolysis with good sugar yield.	[32]
2. The hydrothermal-ethanol method applied with morphological changes of lignin as a wheat straw during separation.	Lignin removal efficiency with ethanol yield (5.3% v/v) on hydrothermal treatment with good ethanol extraction. Lignin removal is a two-step hydrothermal ethanol method.	[33, 34]
3. Wheat straw pretreatment via ammonium sulfite (AS) reported for efficient acetone, butanol-ethanol (ABE) production.	The microbial strain *Clostridium acetobutylicum* ATCC 824 strain in separate hydrolysis, or fermentation processes and simultaneous mode of saccharification or fermentation reported. 6% to 9% ABE products increased butanol (titer ~9.5–12.6 and yield ~1.2–1.73 g/kg raw WS).	[35]
4. Liquid fuel yield with high rate of liquefaction, pyrolysis and gasification processes reported. Thermochemical conversion rate for 10 tonnes per four sugarcane bagasse that are converted into liquid crude biofuels.	Crude biofuels from bagasse reported by utilization of liquefaction, pyrolysis, or gasification. Thermochemical conversion ratio and fining conversion ratios affected the profitability. Next conversion ratios can sharply affect the minimum selling prices of products.	[36]
5. Effect of MgO and CaO in pyrolysis utilized for biofuel production and Cao as a catalyst with improved kinetic of SCB pyrolysis.	Catalytic and non-catalytic pyrolysis reported a fixed-bed reactor at temperature range (i.e., 200–480°C). It showed rapid decomposition zone (260–340°C) and slow decomposition zone ((340–480°C).	[37]
6. Two different strategies used for generation of sugarcane bagasse hydrolysates at productivity (Qp ~0.5 $g.L^{-1}h^{-1}$) or yield (Yp/s~0.4 g/g) as a solvents.	Butanol (6.4 $g.L^{-1}$), acetone (4.5 $g.L^{-1}$), and ethanol (0.6 $g.L^{-1}$) showed in *C. acetobutylicum* DSM 6228 fermentation. This bacterial species-mediated fermentation process reported hydrolysate of sugar bagasse.	[38]
7. Second step amylase use for glucose yield comes from broken rice hydrolysis. Amylase activity in the crude extract (358 U/g substrates) reported and partially purified enzyme at 4–5.5 pH, reported 54% yield for rice residue into reducing sugars in 10 h.	Three-step bioprocess used in ethanol production from broken rice. First amylase enzyme from *Rhizopus* microspores reported. And the third step is fermentation for glucose substrate via *S. cerevisiae* with high ethanol (95%) of theoretical value.	[39]

Biomass Energy	Energy Production Parameter	Reference
8. Phenolic acid exhibited to inhibitory effects on ethanol fermentation for rice straw substrates. Enzyme cellulase produced different ratios of free phenolic acids to soluble conjugated phenolic acid via fermentation efficiencies.	Removal of free phenolic acids from the straw hydrolysates has increased (twofold) the ethanol productivity. Phenolic acid derived from rice straw is the major factor and affects ethanol production via fermentation by *Pichia stipitis*.	[40]
9. Cornstalk separation from flower, stem, cob, husk or leaf used for ethanol synthesis.	The highest yield (92% and 86%) of glucose and production or concentration (24 and 17 $g.L^{-1}$) reported from cob and flower, respectively.	[29]

or other chemicals from different sustainable resources or agricultural residues. This modified *E. coli* can utilize alternative renewable feedstock (C1-gases or methanol). Further, these microbes can enhance fuel production yield and productivity via metabolic engineering effort [42].

13.3.2 Technical Analysis of Biomasses Resources on Environment

All the equipment involved in the pyrolysis process/technology evaluated environmental influence with a study of financial viability for coffee pulp. Biomass with the lowest impact on the environment includes Pinus sawdust, castor husk, or coffee pulp. And for this, biomass under the global warming potential (GWP) has shown more value than 700 kg CO_{2eq} for all the biomasses studies [42].

GWP has done categories of biomasses and it was reported due to emission of high amounts of methane and carbon dioxide gases. These toxic gases are released during biomass pyrolysis and all the equipment is shown with high impacts on types of equipment of separator, pyrolyzer, or the cyclone. The least favorable nature of biomasses shown with the environment and biomass energy is provided viable options from financial points of view of consideration or environmental performance for three Mexican biomass volarization approaches (castor husk, coffee pulp, or pinus sawdust) via using pyrolysis technology [42, 43].

13.3.3 Benefits of Bioenergy Production

Wood biomass is used for bioenergy biosynthesis by exhibiting the main and important benefits such as high production and storage of energy security, lesser or no dependency on the fossil fuel resource, mitigation, climatic impact revitalization of economic values or levels for rural people with the creation of new jobs. Bioenergy biosynthesis is concerned with environmental and socio-economic influences. There is a need for ecological, economic, and social sustainability, reported during the bioenergy biosynthesis and accessed via a set of multicriteria indicators. For bioenergy production and utilization study, life cycle assessment (LCA) is applied for exploring environmental performance in the Alpine area, North Italy [29].

Further, the impact of the environment on wood-based bioenergy plants is created by the utilization of local crops or other origin plant resources (from wood industry or forestry operation). These have investigated the amount of CO_{2eq} emission quantity (0.25 kg CO_{2eq} kWh^{-1}), and fossil fuel (0.09 CO_{2eq} kWh^{-1}) and these values have been analyzed or calculated for bioenergy generating plants. There are reports on lower value than fossil fuel-based power plants, but environmental performance for biomass energy plants is influenced by emission or methane gas consumption [44].

The next reports on plant biomass energy reported that the carbon emission trading of a market-based tool controls the carbon emission level. Evaluation of the best scenarios for deployment for clean energy solutions is done by including technology, supply chain, carbon emission level, or investment analyses. Further, the feasible and sustainable nature of bioenergy is found for different markets. Adaption of LCA and techno-economic analysis (TEA) tool for bioenergy production is compared for the cost-effectiveness of four bio-based energy production pathways with diesel-based power synthesis in Canadian remote and rural communities. These bioenergy synthesis pathways have exhibited the potential of carbon offsets and their credits with the baseline scenario. Additional tool (i.e., multicriteria decision-making or MCDM platform) has helped to make a complete decision for creating a platform on biomass energy [45–48].

13.4 CONCLUSION AND FUTURE PROSPECTS

This chapter discusses the different forms of renewable energy required to fulfill the world's need for strength. This energy utilization can make the world for more energy security conditions with a cheap energy production rate. From ancient times, people are using conventional energy or non-renewable energy for their development works as well as for the domestic or daily energy needs. And this non-renewable energy consumption has created big challenges for its future stocks or availability, increased fuel cost, and continuously growing demand due to more population growth in the world. Next, this energy consumption has created environmental issues such as toxic gases or by-products, greenhouse gas emissions, and climatic changes. This chapter focuses on the different renewable resources and biomass sources for renewable or bioenergy production. This energy can help in a surplus quantity of total energy due to less carbon emission. In this chapter, different biomasses are utilized as cheap and best feedstocks for biomass energy. An efficient pretreatment and microbial system is needed for its hydrolysis and its utilization for biofuel production. This biomass energy can maintain a healthy environment by fulfilling bioenergy for human kinds.

13.5 ABBREVIATIONS

$: US dollar; AA: Amylase activity; ABE: Acetone-butanol-ethanol; AD: Anaerobic digestion; AS: Ammonium sulfite; CaO: Calcium oxide; CO_2: Carbon dioxide; CO_{2eq}: Carbon dioxide emission quantity; COD: Chemical oxygen demand; CPG: CO_2-plume geothermal; DAB: Deoiled algal biomass; EGR: Enhancing gas recovery; ETR: Electron transport rate; g/MJ: gram/megajoule; GHGs: Greenhouse gases; GL: Green liquor; GLD: Green liquor dregs; GW: Gigawatt; GWP: Global warming potential; HPR: Hydrogen production rate; HRT: Hydraulic retention time; ISTGS: Integrated solar-tri-generation system; KJ/mol: Kilojoule/mole; L/kg: Liter per kilogram; kWh^{-1}: Kilowatt/hour; LCA: Life cycle assessment; LFR: Linear Fresnel reflector; M£: Million Euro; MCDM: Multicriteria decision-making; MD: Moderate water conditions; MgO: Magnesium oxide; MW: Megawatt; MWe: Megawatt electricity; NREL: National Renewable Energy Laboratory; PV: Photovoltaic; PSII: Photosystem; RAs: Rubiaceae-type cyclopeptides accumulation; REN: Renewable Energy; RMD: Re-watering after moderate drought; SCB: Sugarcane bagasse; SD: Severe drought; SS-AD: Solid state; TJ: Tera-joules; TWh: Terawatt hour; TEA: Techno-economic analysis; US: United States; U/g: International unit per gram; UK: United kingdom; USA: United States of America; v/v: Volume by volume; W-AD: serial wet; WS: Wheat straw.

REFERENCES

1. Zafar, M. W., Shahbaz, M., Hou, F., & Sinha, A. (2019). From nonrenewable to renewable energy and its impact on economic growth: The role of research & development expenditures in Asia-Pacific Economic Cooperation countries. Journal of Cleaner Production, 212, 1166–1178.
2. Sharif, A., Raza, S. A., Ozturk, I., & Afshan, S. (2019). The dynamic relationship of renewable and nonrenewable energy consumption with carbon emission: A global study with the application of heterogeneous panel estimations. Renewable Energy, 133, 685–691.
3. Wesseh, P. K., & Lin Jr., B. (2020). Energy substitution and technology costs in a transitional economy, Energy, 203, 117828.
4. Rudolfsson, M., Agar, D. A., Lestande, T. A., & Larsson, S. H. (2020). Energy savings through late-steam injection – A new technique for improving wood pellet production. Journal of Cleaner Production, 254, 120099.
5. Agyekum, E. B. (2020). Energy poverty in energy rich Ghana: A SWOT analytical approach for the development of Ghana's renewable energy. Sustainable Energy Technologies and Assessments, 40, 100760.
6. Atif, M., Hossain, M., Alam, M. S., & Goergen, M. (2021). Does board gender diversity affect renewable energy consumption? Journal of Corporate Finance, 66, 101665.
7. Kafka, J., & Miller, M. A. (2020). The dual angle solar harvest (DASH) method: An alternative method for organizing large solar panel arrays that optimizes incident solar energy in conjunction with land use. Renewable Energy, 155, 531–546.
8. Wang, Y., Fan, Y., Wang, D., Liu, Y., Qiu, Z., & Liu, J. (2020). Optimization of the areas of solar collectors and photovoltaic panels in liquid desiccant air-conditioning systems using solar energy in isolated low-latitude islands. Energy, 198, 117324.
9. Yamaki, A., Kanematsu, Y., & Kikuchi, Y. (2020). Lifecycle greenhouse gas emissions of thermal energy storage implemented in a paper mill for wind energy utilization. Energy (online 12 June 2020), 118056.
10. Caporale, D., Sangiorgio, V., Amodio, A., & Lucia, C.D. (2020). Multi-criteria and focus group analysis for social acceptance of wind energy. Energy Policy, 140, 111387.
11. Lafleur, C., Truelove, W. A., Cousine, J., Hiles, C. E., Buckham, B., & Crawford, C. (2020). A screening method to quantify the economic viability of off-grid in-stream tidal energy deployment. Renewable Energy (online 11 June 2020).
12. Longa, F. D., Nogueira, L. P., Limberger, J., van Wees, J. -D., & der Zwaan, B. (2020). Scenarios for geothermal energy deployment in Europe. Energy (online 10 June 2020), 118060.

13. Yan. P., Xiao, C., Xu, L., Yu, G., Li, A., Piao, S., & He, N. (2020). Biomass energy in China's terrestrial ecosystems: Insights into the nation's sustainable energy supply. Renewable and Sustainable Energy Reviews, 127, 109857.
14. Nassar, I. A., Hossam, K., & Abdella, M. M. (2019). Economic and environmental benefits of increasing the renewable energy sources in the power system. Energy Reports, 5, 1082–1088.
15. Heng, Y., Lu, C.-L., Yu, L., & Gao, Z. (2020). The heterogeneous preferences for solar energy policies among US households. Energy Policy, 137, 111187.
16. Karytsas, S., & Theodoropoulou, H. (2014). Socioeconomic and demographic factors that influence publics' awareness on the different forms of renewable energy sources. Renewable Energy, 71, 480–485.
17. Ma, Z., Bao, H., & Roskilly, A. P. (2019). Seasonal solar thermal energy storage using thermochemical sorption in domestic dwellings in the UK. Energy, 166, 213–222.
18. Fthenakis, V., Mason, J. E., & Zweibel, K. (2009). The technical, geographical, and economic feasibility for solar energy to supply the energy needs of the US. Energy Policy, 37(2), 387–399.
19. Almakrami, H., Wei, Z., Lin, G., Jin, X., Agar, E., & Liu, F. (2020). An integrated solar cell with built-in energy storage capability. Electrochimica Acta, 349, 136368.
20. Bahaja, A. B. S., Mahdy, M., Alghamdi, A. S., & Richards, D. J. (2020). New approach to determine the Importance Index for developing offshore wind energy potential sites: Supported by UK and Arabian Peninsula case studies. Renewable Energy, 152, 441–457.
21. Li, Y., Huang, X., Tee, K. F., Li, Q., & Wu, X.-P. (2020). Comparative study of onshore and offshore wind characteristics and wind energy potentials: A case study for southeast coastal region of China. Sustainable Energy Technologies and Assessments, 39, 100711.
22. Basheer, M., & Elagi, N. A. (2019). Temporal analysis of water-energy nexus indicators for hydropower generation and water pumping in the Lower Blue Nile Basin. Journal of Hydrology, 578, 124085.
23. Zhong, W., Guo, J., Chen, L., Zhou, J., & Zhang, J. (2020). Dangwei, future hydropower generation prediction of large-scale reservoirs in the upper Yangtze River basin under climate change. Journal of Hydrology, 588, 125013.
24. Avcıoğlu, A. O., Dayıoğlu, M. A., & Türker, U. (2019). Assessment of the energy potential of agricultural biomass residues in Turkey. Renewable Energy, 138, 610–619.
25. Zhang, Y., Qin, C., & Liu, Y. (2018). Effects of population density of a village and town system on the transportation cost for a biomass combined heat and power plant. Journal of Environmental Management, 223, 444–451.
26. Miao, Y., Bi, Q., Qin, H., Zhang, X., & Tan, N. (2020). Moderate drought followed by re-watering initiates beneficial changes in the photosynthesis, biomass production and Rubiaceae-type cyclopeptides (RAs) accumulation of *Rubia yunnanensis*. Industrial Crops and Products, 148, 112284.
27. Kumar, V., Kumar, P., Kumar, P., & Singh J., (2020). Anaerobic digestion of Azolla pinnata biomass grown in integrated industrial effluent for enhanced biogas production and COD reduction: Optimization and kinetics studies. Environmental Technology & Innovation, 17, 100627.
28. Iweh, N. S., Koyam, M., Akizuki, S., Ban, S., & Tod, T. (2020). Novel wet-solid states serial anaerobic digestion process for enhancing methane recovery of aquatic plant biomass. Science of the Total Environment, 730, 138993.
29. Kumari D., & Singh, R. (2018). Pretreatment of lignocellulosic wastes for biofuel production: A critical review. Renewable and Sustainable Energy Reviews, 90, 877–891.
30. Sewsynker-Sukai, Y., David, A. N., & Kan, E. B. G. (2020). Recent developments in the application of kraft pulping alkaline chemicals for lignocellulosic pretreatment: Potential beneficiation of green liquor dregs waste. Bioresource Technology, 306, 123225.

31. Javed, U., Ansari, A., Aman, A., & Qade, S. A. U. (2019). Fermentation and saccharification of agro-industrial wastes: A cost-effective approach for dual use of plant biomass wastes for xylose production. Biocatalysis and Agricultural Biotechnology, 21, 101341.
32. Fitria, Ruan, H., Fransen, S. C., Carter, A. H., Tao, H., & Yang, B. (2019). Selecting winter wheat straw for cellulosic ethanol production in the Pacific Northwest, U.S.A. Biomass and Bioenergy, 123, 59–69.
33. Li, J., Feng, P., Xiu, H., Li, J., Yang, X., Ma, F., Li, X., Zhang, X., Kozliak, E., & Ji, Y. (2019). Morphological changes of lignin during separation of wheat straw components by the hydrothermal-ethanol method. Bioresource Technology, 294, 122157.
34. Prasad, S., Malav, M. K., Kumar, S., Singh, A., Pant, D., & Radhakrishnand, S. (2018). Enhancement of bio-ethanol production potential of wheat straw by reducing furfural and 5-hydroxymethylfurfural (HMF). Bioresource Technology Reports, 4, 50–56.
35. Qi, G., Huang, D., Wang, J., Shen, Y., & Gao, X. (2019). Enhanced butanol production from ammonium sulfite pretreated wheat straw by separate hydrolysis and fermentation and simultaneous saccharification and fermentation. Sustainable Energy Technologies and Assessments, 36, 100549.
36. Ramirez, J. A., & Rainey, T. J. (2019). Comparative techno-economic analysis of biofuel production through gasification, thermal liquefaction and pyrolysis of sugarcane bagasse. Journal of Cleaner Production, 229, 513–527.
37. Pradana, Y. S., Daniyanto, Hartono, M., Prasakti, L., & Budiman, A. (2019). Effect of calcium and magnesium catalyst on pyrolysis kinetic of Indonesian sugarcane bagasse for biofuel production. Energy Procedia, 158, 431–439.
38. Gomes, A. C., Rodrigues, M. I., Passos, D. F., de Castro, A. M., Maria, L., Anna, M. S., & Pereira Jr., N., (2019). Acetone–butanol–ethanol fermentation from sugarcane bagasse hydrolysates: Utilization of C5 and C6 sugars. Electronic Journal of Biotechnology, 42, 16–22.
39. Ranke, F. F. B., Shinya, T. Y., Figueiredo, F. C., Núñez, E. G. F., Cabral, H., & Neto, P. O. (2020). Ethanol from rice byproduct using amylases secreted by *Rhizopus microsporus* var. oligosporus. Enzyme partial purification and characterization. Journal of Environmental Management, 266, 110591.
40. Li, P., Cai, D., Luo, Z., Qin, P., Chen, C., Wang, Y., Zhang, C., Wang, Z., & Tan, T. (2016). Effect of acid pretreatment on different parts of corn stalk for second generation ethanol production. Bioresource Technology, 206, 86–92.
41. Kashif, M., Awan, M. B., Nawaz, S., Amjad, M., Talib, B., Farooq, M. Nizami, A. S., & Rehan, M. (2020). Untapped renewable energy potential of crop residues in Pakistan: Challenges and future directions. Journal of Environmental Management, 256, 109924.
42. Zhao, C., Zhang, Y., & Li, Y. (2019). Production of fuels and chemicals from renewable resources using engineered *Escherichia coli*. Biotechnology Advances, 37(7), 107402.
43. Parascanu, M. M., Sánchez, P., Soreanu, G., Valverde, J. L., & Sanchez-Silva, L. (2019). Mexican biomasses valorization through pyrolysis process: Environmental and costs analysis. Waste Management, 95, 171–181.
44. Buonocore, E., Paletto, A., Russo, G. F., & Franzese, P. P. (2019).Indicators of environmental performance to assess wood-based bioenergy production: A case study in Northern Italy. Journal of Cleaner Production, 221, 242–248.
45. Ahmadi, L., Kannangar, M., & Bensebaa, F. (2020). Cost-effectiveness of small scale biomass supply chain and bioenergy production systems in carbon credit markets: A life cycle perspective. Sustainable Energy Technologies and Assessments, 37, 100627.
46. Zhang, L. Wang, Y.-Z., Zhao, T., & Xu, T. (2019). Hydrogen production from simultaneous saccharification and fermentation of lignocellulosic materials in a dual-chamber microbial electrolysis cell. International Journal of Hydrogen Energy, 44(57), 30024–30030.

47. Kumar, A. N., Chatterjee, S., Hemalatha, M., Althuri, A., Min, B., Kim, S.-H., & Mohan, S.V. (2020). Deoiled algal biomass derived renewable sugars for bioethanol and biopolymer production in biorefinery framework. Bioresource Technology, 296, 122315.
48. Fayiah, M., Dong, S., Li, Y., Xu, Y., Gao, X., Li, S., Shen, H., Xiao, J., Yang, Y., & Wesselld, K. (2019). The relationships between plant diversity, plant cover, plant biomass and soil fertility vary with grassland type on Qinghai-Tibetan Plateau. Agriculture, Ecosystems & Environment, 286, 106659.

14 Evolving Trends for Smart Grid Using Artificial Intelligent Techniques

Pooja Yadav[1], *Prakhar Chaudhary*[2], *and Hemant Yadav*[3]
[1]FET, MJP Rohilkhand University, Bareilly, India
[2]Accenture India, Pune, India
[3]Department of Computer Science, PGIE, Bareilly, India

CONTENTS

14.1 INTRODUCTION

Greenhouse gas (GHG) emission has increased since the 1700s and the main contributors to its global emissions are generation of heat and electricity. Carbon dioxide (CO_2) and methane (CH_4) are the gases which create disastrous effects on the atmosphere, including warming of the oceans, shrinking snow, and elevating global average sea level. These events brought into notice when humans became sensitive to climate mitigations. Global warming moderation would require limiting of GHG emissions by offsetting carbon activities. A recent study showed that the energy consumption is rapidly increasing and anticipated to rise from 663 to 736 quadrillion British thermal unit by 2040 [1]. The consumption will increase the annual carbon emission from 31.2 to 45.5 billion metric tons [2]. Selecting renewable energy (RE) sources can effectively reduce CO_2 emissions in the environment.

The involvement of RE in the world power system is growing rapidly. Renewable power sector was raised by 14.1%, which represents net electricity capacity of 165 GW in 2016. This was 34% less than the electricity generated by coal power. Moreover, the gap is expected to be reduced to just 17% by 2022 [3]. In the end of 2016, total power generated by renewables was 2017 GW, which was sufficient enough to supply ~24% world's electricity [4]). The global demand of electricity has been anticipated to rise by 76% (4,800 GW) by 2030. Remarkably, much of electricity is required to meet up the demand. Sustainable production of this much electricity must embrace renewables instead of conventional power plants of today. The application of renewables at grand scale possesses lots of challenges.

Over the past few years, energy demand and consumption has increased, but at the same time, infrastructures of transmission and distribution are neglected. Most of the countries are using the same electricity grid which was first installed. Furthermore, these grids are based on the idea of centralized power generation from fossil fuel. These grids have significant inefficiency in transmission and distribution. Conventional solution to upgrade this system is to develop a new plant, transmission lines, equipment, etc., which is very difficult and time-consuming. Due to this, the power grids are in pressure with rapid load, which leads to higher cost and compromised reliance [5]. The solution to this challenge is transformation of the current electrical systems to smart grid (SG).

SG is an advanced electric grid where electricity, generated from renewable resources, and information flow in a two-way communication by utilizing the digital or analog information technology. It conglomerates to an intelligent power system where the information about the behavior of power generators and consumers flows in an automated fashion, which ultimately improves the economics, sustainability, and distribution of the produced electricity. It could be explained as a completely automated power distribution network, which screens and controls customer branch point guaranteeing two-way electricity and information flow among the plant and all the points. The intelligence together with control and communication system facilitates market transactions in real time and continuous interfaces among consumers, buildings, the electrical network, and its generation facilities. It helps in reducing loss during transmission of power which is really an encouraging fact for implementing renewables on large scale. Perhaps, it is interesting that it has many challenges in terms of telecommunications, cyber-security, and system engineering. Artificial intelligence (AI) principally works on exchange of information, its intelligence distribution, and automation.

In this chapter, we discuss that the integration of renewables with SG comes up with some new challenges, which enhance the research in the field of AI. SG technologies need algorithms by AI that could solve many problems such as consumers with unlike demand or generators with different instabilities. Therefore, we elucidate such concerns occur within the vital gears of smart grid, i.e., managing demand side, self-restoring network, and the occurrence of prosumers by presenting which gears and interactions are required to be smart, for making the SG successful.

14.2 SMART GRIDS—AN OVERVIEW

SGs are defined as "an electrical network that can intelligently incorporate the actions for all users, i.e., generators and consumers connected to it for proficiently delivering sustainable, economic and secure electricity supply" [6]. SGs are also known as intelligent grid/intelligrid and future grid. The SG conceptually means a grid which is designed to flow information and electricity in a two-way communication. It might sound similar to electrical grids, but this is the point at which the operation of SG varies from electrical grid. Conventionally, grids are dependent on consumption of electricity, which means if demand for electricity increases the operators start producing more electricity at power plants and supply to the grids. Nowadays, the electricity usage has evolved. People have started installing solar energy grids on their properties by which they have become "prosumers," which means they are now both producers and consumers. In order to meet the needs, governments are also providing grants for demonstrating SGs across the world. Using SGs makes the society to move from centralized energy system to various scale sources. In order to make these SGs work, some advance metering system is required which allows the two-way flow between the consumer and the utility, and hence, smart meters are the fundamental technology that facilitates SGs functioning. It is so unfortunate to know that most of the power grids that are in use currently are quite old and out-of-date, and hence, unreliable and inefficient in providing protection for faults. The United States and some other countries have recently started using these tremendous SGs developed after a lot of R&D activities [7]).

The SGs have many benefits if they would be constructed and implemented well. SGs advance the performance of energy resources and their distribution. For example, SGs make sure that rooftop solar would be utilized effectively. As there is no centralized system, loss of electricity while transmission would be minimal. It also helps in balancing the load via better supply and reduces the cost of electricity for everyone. Also, the problems in SGs can be diagnosed easily, and therefore, resolved quickly. Besides all these advantages, SGs have a major benefit of integrating renewables. Application of SG in renewables makes it enormous and encouraging for deployment of non-conventional energy in the future.

14.3 RENEWABLE ENERGY: CONCEPT AND AVAILABILITY

Energy can neither be created nor be destroyed. It just transforms from one form to another. There are several forms of energy and their resources which can be majorly divided into two types of resources, renewable and non-renewable energy resources. Non-renewable energy sources are the mostly utilized ones. Fossil fuels, coal, oil, and uranium are a few non-renewable energy sources out of which fossil fuels and (uranium) nuclear energy are the prominent energy resources. The use of fossil fuels has been increasing every year. It has been approximately doubled since late 70s. Most of the power generating systems use fuel as their chief energy resources due to which the world has to face oil crisis. Due to the rise in carbon footprints and demand of clean green future, application of RE sources is booming. The world's energy paradigm is shifting toward sustainable lower emissions of carbon and other pollutants.

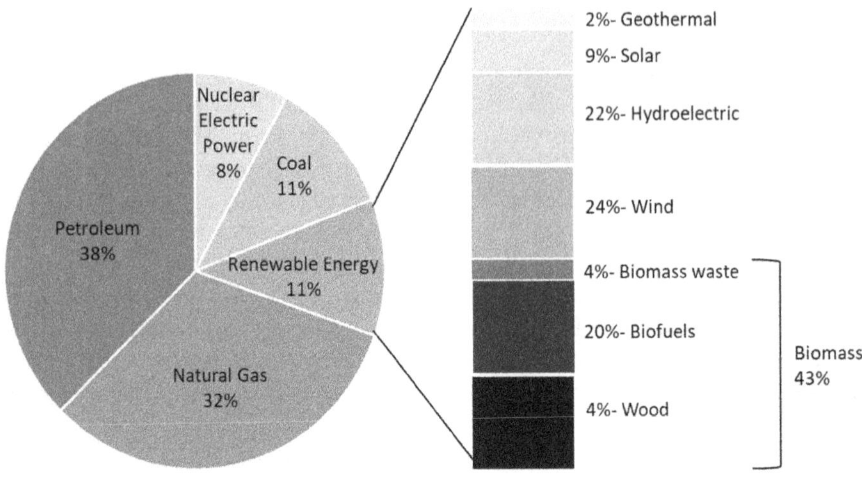

FIGURE 14.1 Share of energy sources in energy consumption.

RE, also known as clean energy, derives from natural resources/processes that are constantly replenished. For instance, sunlight keeps shining and wind keeps blowing, even if their accessibility is contingent on time and weather. It is energy from resources that refills naturally, but with limitations. This means, renewable resources are apparently inexhaustible in longer period of time, but with restricted availability in a given unit of time. RE has always been thought of as a technology for heating, transportation, lighting, and more harnessed from nature. Winds have power to sail the boats and windmills to generate energy. The solar energy, which is the major source of RE energy, can be used directly as bioenergy. According to the International Energy Agency, the total RE utilization was about 15,000 TWh in 2010 in which the hydro power was 3,438 TWh. In 2012, the wind energy utilization was 574 TWh. Though, the utilization of solar energy was lowest, about 100 TWh [8]. Figure 14.1 depicts the contribution of energy sources in energy consumption.

The key problems with the RE energy sources are their availability and cost. Contrasting to conventional sources, these resources cannot be dispatched when needed and this makes the energy consumption more expensive. Furthermore, over the past few centuries, humans are more prone to utilize cheaper energy sources such as coal and petroleum which is affecting the environment severely. Integration of SG with renewables facilitates the efficient utilization of non-conventional energy resources, which will be beneficial for environment as well as for the consumers.

14.4 INTEGRATING SMART GRID IN RENEWABLE ENERGY SYSTEM

One of the key performance indicators (KPIs) for smart grid stability is to keep balance between supply and demand by the use of smart meter and IT-OT integration, try to maintain the demand curve, reduce storage requirement, and provide dynamic tariff to the end consumers. Traditionally, it is used to be achieved by establishing demands at each grid point, optimum generation resources (mostly conventional energy) and by programming the electricity flow based on the physical model for economical and efficient supply of electricity. But, in today's time, we mostly talk about globalization of SG leading to have an extremely complex SG design and real time monitoring of operations.

Today's SG has huge segments of distributed RE system, alongside conventional generating power and energy storage systems. The exponential growth in electricity demand owing to the rising population and economic growth, increasing agitation of energy security combined with climate change and global warming concerns are some the key driving factors for RE energy integration. Easier said than done, there are many factors on which RE integration depends [9] such as RE share, size, and location of the network, energy conversion technology, the effect on system inertia, droop, power quality, system protection, etc.

RE sources provide great opportunities of de-carbonization, frequency regulation, voltage deviation, and are a good option to achieve a proper power generation mix. However, insecurity and intermittent generation of RE enforce stress on power system. To explicate the importance of RE, EU has fixed a goal to achieve 27% contribution of renewables in the total energy consumption by 2030 [10]. In Germany, transiting toward non-nuclear system, RE is the key alternative to achieve it [11]. In addition to these major decisions, there is a drastic increase in photovoltaic (PV) installation recently [12].

Therefore, with high dissemination of RE, the aim is to maintain the grid stability while meeting the demand. Another benefit is this integration would also reduce the carbon footprint and can act as an alternate ancillary service to be utilized for peak shaving and load management. The authors in ref. [12, 13] explained the reason behind why transit toward renewable energy as a stand-alone approach is not a good decision and coupling it with fossil fuels base power plant is a more viable option. Moreover, intermittent behavior of PV power generation poses instability in electricity supply. More precisely, this RE fails to guarantee stability and reliability of power supply [14]. Besides this, PV integration also has back-feeding issues, which impose operational challenges in power system [15].

Therefore, incorporation of these modern technologies into traditional power grid requires innovation and robust modeling approach, for co-existence of all non-renewable and RE sources. The increasing participation of renewables brings both positive and negative consequences. The obvious positive part is reliable power in remote locations, energy security, reduction in carbon footprint and global warming. However, issue arising due to RE integration is more challenging, and thus, needs further considerations.

14.5 ARTIFICIAL INTELLIGENCE IN SMART GRIDS

Broadly, artificial intelligence (AI) means the capability of a machine or any artifact to be able to accomplish any type of function that can illustrate human thoughts. Nilsson [16] defined AI as "An artifact having intelligent behavior which is able to do Reasoning, Perception, communication, learning and able to act intelligently in a complex environment, can be considered as an Artificial Intelligent ecosystem". In the context of computer systems, the term AI is considered as any system or program which can perform more complex tasks rather than simple programming. Herbert Simon and co-workers in the early 1950s led experiments in writing computer programs trying to imitate human through processes [17]. Experiments resulted in a theory which later on called Logic Theorist that consisted of previously proved axioms. Using heuristics, if a new logical expression is provided, it will try to search all possible operations to find out a proof of it. This was considered a major milestone in AI development. Around similar period, Shannon also came up with a research that possibly computers may play chess [18]. Furthermore, considering all the capabilities of AI, it can be segregated into sub-groups or branches such as solving problem, language processing, reasoning as well as general intelligence to solve complex problems.

In energy sector, mostly the research area is directly in experimenting various techniques for designing optimal system for operational management [19]. Numerous studies focus on predicting demand of energy in the market [20]. With the advancement of computational power and its decreasing cost, AI techniques have rapidly moved to another level coming up with methods such as deep learning (DL), reinforcement learning (RL), and their combinations, i.e., deep reinforcement learning (DRL), providing us with more matured methods of intelligence [21]. These application fields cover areas such as load consumption forecasting, demand response, anomaly detection, stability analysis, electric vehicle, and all related technical fields of SG. Most of these areas are covered in the subsequent sections.

14.5.1 Artificial Neural Network (ANN)

Neural network [22] typically emulates the biological nervous system and is the most generic form of AI. In a typical neural system, nerve cells receive signals from adjoining receptors via dendrites, and process that electrical pulse in the cell body, and then, transfer the signal through nerve fiber known as axon. Conceptualizing similarly in terms of electrical model, biological neurons have linear activator, which is followed by a non-linear inhibiting function. The amount of weighted input excitation is collected by linear activation function, whereas all the signals are captured by non-linear inhibiting function. This is how electrical neurons bond resulting in signals [23]. Therefore, neural network has the property of non-linear pattern recognition that basically mimics human brain associative memory property. The biggest advantage of ANN is its ability of mapping input and output relationship without making complex mathematical formulations among the inputs [24]. It then extracts the non-linear relationship between these variables using training process of network and with the help of pattern reorganization functions network will predict the future output behavior of load demand. ANN has a potential for being utilized in renewables and SG. Some of the applications mentioned in ref. [25] are as follows:

- Load forecasting of end consume
- Wind power generation forecasting
- Real-time anomaly detection and control of power system
- Feedback signals estimation
- Noise-free filtering of signals
- Accurate scheduling and forecasting of generation and storage
- Seamless control of system elements
- In electricity trading market, real-time pricing predictions of electricity

14.5.2 Genetic Algorithm

Genetic algorithm (GA) is an algorithm inspired by the process of evolution in biology [26]. Basically, GA has inspiration from the living organisms and the way they get adapted in different environments by continuous evolution and inheritance. These algorithms imitate how the growing population evolves by the selection of fittest individuals for reproduction. Hence, GAs are basically searching and optimization techniques which are based on the fundamentals of genetics and natural selection. The procedure is to maintain the population of knowledge structure representing candidate solution and let it evolves over time through competition. Recently, GA has come up with a wide range of applications in machine learning, intelligent search, and optimization problems. GA is basically controlled by three operators of genetics, i.e., mutation, crossover, and selection [23].

In the energy sector, the applications of GA are mostly found in different optimization use cases. One of the scenarios was developed for finding an optimal power generation mix with renewable energy in isolated islands [27]. Another scenario is about optimal distribution and sizing of connected grid system [28].

14.5.3 Fuzzy Logic

Fuzzy theory is the simplification of traditional set of theory, which was proposed by Zadeh [29]. Fuzzy logic (FL) handles fuzzy logical statements and sets for displaying real world intellectual problems. In contrast to conventional sets, fuzzy set comprises all elements of universal set with varying values in interval [0, 1]. The main reason behind the popularity of FL algorithms is that they have a good accuracy when it comes to address the uncertainty issues and lexical imprecision. Some of the essential properties of FL are as follows [30]:

- There is a limiting case of exact reasoning
- Degree is one of the key features to be observed
- Collection of variables which are elastic are considered as knowledge

FL exploits degree notation instead of direct true or false. It follows "Thumb rule" approach which usually human beings use for making their decision. In this way, FL method provides approximation rather than exact outcome. FL controllers work on production rules like "if-then" statements, which make it capable of making intellect decisions in vulnerable and variable situations. Particularly to energy sector, predicting solar radiations [31] and solar-tracking mechanism [32] is its application in renewables energy producing system.

14.5.4 Deep Learning

Deep learning (DL) is a more intriguing part of machine learning and can be considered as a subset of it. Originally, DL was a result of multi-layer ANN; we can consider it as large deep neural network which consists of many layers within layers, which eventually enhance the accuracy of the system, but increase its complexity and execution time. Therefore, DL exploits multiple layers of non-linear data for supervised or unsupervised feature extraction and transformation for pattern classification and analysis. DL overcomes the shortcomings of ANN for the issues in which it is not able to handle a significant amount of complex data, and its irregularity which is apparently found in many applications such as natural speech, images, and videos. There are different modules of DL which are used depending on the data set and the expected accuracy. The most popular once are convolutional neural network (CNN) and recurrent neural network (RNN). The basic difference between both of them is that CNN is mostly popular for spatial distribution kind of data, whereas RNN gives better result in time series data. CNN is basically a standard neural network, which is designed to recognize images using weights which recognize an object based on its edges. Whereas RNN assigns edges with time and feeds it into the next time step instead of the next layer in the same time step. RNN is majorly used to recognize text or speech signal sequences. It is more like a hierarchical network where input is processed in a tree fashion and input sequence doesn't have any time aspect [33]. So, the underlying architecture within DL consists of popular ANN algorithms, where training models on graphics processing unit (GPU) and tensor processing unit (TPU) support distributed learning of DL applications. There are many open source frameworks available in the market like Tenser Flow, MXNet, Thenano, etc. [34].

14.5.5 Reinforcement Learning

Reinforcement learning (RL) is used to make a sequence of decision by training machine learning models. Typically, RL works as game-like situation. To get the desired action to be performed, intelligence system with try and error approach either gets reward or penalties for the actions it performs. The final goal is to maximize the reward count by performing series of actions in response to the changing environment. Therefore, RL has four basic components: agent, rewards, environment, and action. The advantage of RL's approach is that it can perform sequence of decision-making under dynamic environment. Though RL operates with limited knowledge of environment and quality of the decision mostly does not get updated with changing feedback.

The basic difference between the RL approach and the supervised learning is that the RL decisions are made sequentially, i.e., output depends on the current state of the input, and the next output will depend on the previous input; whereas in supervised learning, decision is made on the initial input. There are some popular algorithms for RL implementation such as deep deterministic policy gradients (DDPG), Q-learning, deep Q net (DQN), and state–action–reward–state–action (SARSA). The most wide implementation areas where RL is often used are robotics, gaming, and navigation [35].

14.5.6 Deep Reinforcement Learning

Deep reinforcement learning (DRL), as the name suggests itself, is a combination of RL and DL and among one of the most trending types of machine learning because of its capability to solve a broad range of multifaceted decision-making chores. DRL leverages the property of perception from DL and accurate decision-making from RL. Therefore, DRL competency increases many folds to implement a variety of task with high dimensional raw inputs and policy control. It works on reward or penalized-based approach based on the actions, and those actions leading to the expected outcome are rewarded. With series of trial and error, knowledge base keeps on updating which makes it ideal for dynamic environments. This RL technique embedded with the deep learning approach has led to phenomenal results. Some of the areas where it is applied are robotics, neuro linguistic programming (NLP), transportation, video games, computer vision, finance, and healthcare. AlphoGo has been a fascinating research area which has shown the great success of DRLs. On the other hand, Uber is also looking forward to use DRL to handle real cars.

On the other hand, development of computation power has led the development of better algorithms. Most of these algorithms like DL & DRL require high GPU to train such models. Some of the applications of these use cases of AI in renewable energy systems [36] are explained below explaining AI's significant role in the engineering design process.

14.6 APPLICATION OF ARTIFICIAL INTELLIGENCE IN SMART GRID SYSTEM

Traditionally, power grid analysis, control and monitoring were primarily dependent on physical modeling and numerical-based calculations. Due to this, they lack in addressing uncertainty and incapable of handling huge data getting generated literally every second. With the advancement of SG and feeding RE such as solar and wind with high involvement of customers has led grid to be more volatile and complex environment. In spite of development of new technology platforms such as advanced metering infrastructure (AMI), wide area monitoring systems (WAMS) and energy management system are producing massive data and a huge database for AI algorithms to train their models. With this complex and uncertain environment, the challenge for power system is to ensure security and stable operations. Numerical calculations based on physical models, which used to happen traditionally, are non-practical or inaccurate because those were based on some assumptions and simplifications and with so much of data being poured, they are incapable to manage uncertainties and nonlinearities. The biggest contributor of this variability is the generation of solar and wind energies.

Nowadays, SG has many monitoring systems which are installed for equipment level monitoring and supervision. These advance systems generate equipment level data having many configured parameters and features. Like in case of smart meters, features like consumption pattern can be used for price forecasting and non-invasive load decomposition, etc. To come up with such insights, algorithms like DL and RL can be utilized and would be an effective measure for load shifting and demand response. Specifically, DL algorithm can also be used to extract such features so that anomalies can be detected, including equipment fault alerts, malicious attacks, and equipment tempering.

14.6.1 Load Forecasting

With high contribution of RE in SG, it has raised the challenges of uncertainty and possibility to forecast precise scheduling. Therefore, high forecast accuracy is the need of the hour for multiple time prospects related with regulations, scheduling, dispatching, and commitment of units from grid. Although there always exist many techniques for load forecasting, but with smart meter, smart sensors, and advance energy management system in place, they lack in handling a large amount of data. And this is where AI-based techniques play a vital role because of their capability to handle complex data, generate real-time relationship, and come out with optimized strategy with better insights [37].

As the size of real-time data increases day by day, with its complexity on the other hand, finding a single effective solution is bit tough. Many solutions have been found using any of the techniques, but always had some or other discrepancy. So, experts are exploring how we can join these different techniques together to achieve more accuracy in load forecasting. In a paper, the authors of ref. [38] have tried to join CNN and K-means algorithms together to forecast hourly electricity loads. To achieve this, 1.4 million of load records were taken in consideration and K-mean algorithm was used to cluster them into subsets. These subsets as a result were then trained by CNN to have satisfying experimental results. Another case study [39] was performed on one-year urban data set to forecast the hourly load pattern. This was performed using deep belief network (DBN) technique embedded with parametric copula models and was found more accurate than traditional prediction models such as ANN, support vector regression, and extreme learning machines.

Load forecasting at consumer level is another area where demand response is required to be efficient. With smart meters being installed at each consumer or building level, consumption data collected can be used for load forecasting enabling DL technique. To have better understanding of load pattern at 1 hour and at 1 min data set level, paper [40] did it by performing 2 long short-term memory algorithms based on neural network architectures.. But it was found that accuracy for one-minute load was not good and when same was implemented having sequence to sequence (S2) architecture, results were excellent in both kinds of data sets. In another paper, the authors of ref. [41] tried to do building level load forecast using CNN and got satisfactory results. Therefore, from these references, it can be inferred that DL algorithms are one of the viable options to be used for load forecasting, because of their capability to handle a large volume of data from different sources. However, their performance may vary depending on the attributes present on different data sets.

14.6.2 Demand Response

The idea behind electricity availability across the year is achieved by having perfect balance between the supply and demand over the grid. But with lots of data getting pored every second and thus monitoring an increased number of factors for demand management has become a critical task. To implement demand response, it is crucial to identify and predict the energy flexibility at demand side. At consumer level, consumption monitoring can be done by nonintrusive load monitoring (NILM) technique, where it can produce consumption at appliance level in individuals' home. Factored Four Way Conditional Restricted Boltzmann method was mentioned by the authors of ref. [42] to predict energy flexibility in real time. Similarly [43], near-real time tariff can be proposed to consumer using RNN technique as per their consumption pattern. Dynamic pricing is an effective method to control demand response and encourage customer to participate in energy trading. However, lack of consumer side demand information and consumption pattern makes it a bit challenging to have dynamic pricing in place at wholescale market.

Additionally, with all IT-OT integration and remote access of household level equipment, it is essential to owners to comprehend the significance of automated actions and their consensus should be there to delegate control on intelligent devices by service provider. In order to achieve this, it is important to have flexible autonomy and the service provider should be allowed to control the system to some extent [44].

14.6.3 Anomaly Detection of Electrical Equipment

Reliability of power system operations hugely depends on how accurately we detect fault/defect at equipment level. Nowadays, in SG, even though equipment are very advance and have inbuilt capability for anomaly detection, due to complexity of recorded data and varying input data, fault detection at early stages is still a challenge. In the following literature, we can see how DL is utilized to monitor insulators, transformer, and transmission lines. The authors in ref. [45] have tried to extract feature of insulators and identify the faults using high-level discriminative CNNs' ability to extract those with achieving accuracy of 93%. The authors in ref. [46] have tried to identify the power transformer failures using oil chromatography online monitoring data by applying classification method based on DL. Traditionally, transformer failure detection was not possible because of unlabeled fault samples and it used to fail in judging fault types. The classification techniques used for anomaly detection gave a basis for power transformer fault diagnosis in which unlabeled data from oil chromatogram and a small number of labelled data from dissolved gas-in oil analysis were used for training process. This approach was found better (in testing data) than the other proposed methods, which are radio, back propagation (BP) neural network, and support vector machine (SVM) method.

Another area is to identify the power line fault, which was discussed in ref. [38], where one modeling method was proposed based on sparse self-encoding neural network. In this approach, sub-band energy of wavelet decomposition was used as a parameter for DL neural network and to construct the structure. It was observed that fault recognition rate exceeds 99% in an experiment based on IEEE 34. With the help of references, we can say that DL is one of the effective methods for equipment level fault detection, but some issues are still open such as inability to handle small sample size, real-time detection, and classifying small differences between before- and after-fault conditions.

14.7 CONCLUSION AND FUTURE PERSPECTIVES

There is a significant motivation globally to reduce the dependence on fossil fuels and transit to low carbon economy for a guaranteed energy security in a sustainable manner. The change needs a principle re-engineering of the electrical grid. Upgradation of grids to SG enables numerous prosumers to use electricity, whereas automated network controls and algorithms uphold the operations of the SG and facilitate self-recovery whenever something damages. Various challenges such as uncertainty, dynamism, instability, and heterogeneity are the major issues regarding SGs. AI deals effectively with these factors and hence makes it possible for humans to rely upon the renewable energy systems for future energy demand. AI addresses all the issues such as sustainability, cyber-security, ethics, and efficiency, which makes the whole system intelligent. In conclusion, the AI methods that can be either adapted or utilized directly to solve intricate electric power system problems, have played, and will continue to play, a significant role in revolutionizing the entire energy sector from generation to distribution, so as to enable large-scale integration of renewable sources into energy networks.

14.8 ABBREVIATIONS

RE: Renewable energy; SG: Smart grid; AI: Artificial intelligence; ANN: Artificial neural network; FL: Fuzzy logic; DL: Deep learning; GA: Genetic algorithm; RL: Reinforcement learning.

REFERENCES

1. *International Energy Outlook 2019.* (n.d.). Retrieved June 25, 2020, from https://www.eia.gov/outlooks/ieo/.
2. Ringel, M., & Knodt, M. (2018). The governance of the European Energy Union: Efficiency, effectiveness and acceptance of the Winter Package 2016. *Energy Policy, 112*, 209–220. https://doi.org/10.1016/j.enpol.2017.09.047.
3. Outlook, E. (2016). *About this review.*
4. *World Energy Outlook 2019 – Analysis – IEA.* Retrieved June 25, 2020, from https://www.iea.org/reports/world-energy-outlook-2019.
5. Benysek, G. (2007). *Improvement in the quality of delivery of electrical energy using power electronics systems.* Springer-Verlag. https://www.springer.com/gp/book/9781846286483.
6. ETIP SNET. (n.d.). *The European Technology and Innovation Platform Smart Networks for Energy Transition.* ETIP SNET. Retrieved June 25, 2020, from https://www.etip-snet.eu/.
7. *"GRID 2030" A National Vision for Electricity's Second 100 Years.* (n.d.). Energy.gov. Retrieved June 25, 2020, from https://www.energy.gov/oe/downloads/grid-2030-national-vision-electricity-s-second-100-years.
8. Elias, S. (2013). *Reference Module in Earth Systems and Environmental Sciences.*
9. Morren, J., de Haan, S. W. H., Kling, W. L., & Ferreira, J. A. (2006). Wind turbines emulating inertia and supporting primary frequency control. *IEEE Transactions on Power Systems, 21*(1), 433–434. https://doi.org/10.1109/TPWRS.2005.861956.
10. Knopf, B., Nahmmacher, P., & Schmid, E. (2015). The European renewable energy target for 2030 – An impact assessment of the electricity sector. *Energy Policy, 85*, 50–60. https://doi.org/10.1016/j.enpol.2015.05.010.

11. Wurster, S., & Hagemann, C. (2018). Two ways to success expansion of renewable energies in comparison between Germany's federal states. *Energy Policy*, *119*, 610–619. https://doi.org/10.1016/j.enpol.2018.04.059.
12. Akbari, H., Browne, M. C., Ortega, A., Huang, M. J., Hewitt, N. J., Norton, B., & McCormack, S. J. (2019). Efficient energy storage technologies for photovoltaic systems. *Solar Energy*, *192*, 144–168. https://doi.org/10.1016/j.solener.2018.03.052.
13. Benedek, J., Sebestyén, T.-T., & Bartók, B. (2018). Evaluation of renewable energy sources in peripheral areas and renewable energy-based rural development. *Renewable and Sustainable Energy Reviews*, *90*, 516–535. https://doi.org/10.1016/j.rser.2018.03.020.
14. Vega-Garita, V., Harsarapama, A. P., Ramirez-Elizondo, L., & Bauer, P. (2016). Physical integration of PV-battery system: Advantages, challenges, and thermal model. *2016 IEEE International Energy Conference (ENERGYCON)*, 1–6. https://doi.org/10.1109/ENERGYCON.2016.7514038.
15. Bouhouras, A. S., Sgouras, K. I., Gkaidatzis, P. A., & Labridis, D. P. (2016). Optimal active and reactive nodal power requirements towards loss minimization under reverse power flow constraint defining DG type. *International Journal of Electrical Power & Energy Systems*, *78*, 445–454. https://doi.org/10.1016/j.ijepes.2015.12.014.
16. Nilsson, N. J. (1998). *Artificial intelligence: A new synthesis*. Morgan Kaufmann.
17. Krishnamoorthy, C. S., Rajeev, S., & Rajeev, S. (2018). *Artificial intelligence and expert systems for engineers*. CRC Press. https://www.taylorfrancis.com/books/9781315137773.
18. Shannon, C. E. (1950). XXII. Programming a computer for playing chess. *The London, Edinburgh, and Dublin Philosophical Magazine and Journal of Science*, *41*(314), 256–275. https://doi.org/10.1080/14786445008521796.
19. Quan, X. I., & Sanderson, J. (2018). Understanding the artificial intelligence business ecosystem. *IEEE Engineering Management Review*, *46*(4), 22–25. https://doi.org/10.1109/EMR.2018.2882430.
20. Ahmad, T., & Chen, H. (2018). Utility companies strategy for short-term energy demand forecasting using machine learning based models. *Sustainable Cities and Society*, *39*, 401–417. https://doi.org/10.1016/j.scs.2018.03.002.
21. Zhang, D., Han, X., & Deng, C. (2018). Review on the research and practice of deep learning and reinforcement learning in smart grids. *CSEE Journal of Power and Energy Systems*, *4*(3), 362–370. https://doi.org/10.17775/CSEEJPES.2018.00520.
22. Bose, B. K. (1986). *Power electronics and ac drives* (p. 416). Englewood Cliffs, NJ: Prentice-Hall. http://adsabs.harvard.edu/abs/1986ph...book....B.
23. Konar, A. (2018). *Artificial intelligence and soft computing: Behavioral and cognitive modeling of the human brain*. CRC Press.
24. Fausett, L. (1994). *Fundamentals of neural networks: architectures, algorithms, and applications*. Prentice-Hall.
25. Ciabattoni, L., Ippoliti, G., Longhi, S., & Cavalletti, M. (2013). Online tuned neural networks for fuzzy supervisory control of PV-battery systems. *2013 IEEE PES Innovative Smart Grid Technologies Conference (ISGT)*, 1–6. https://doi.org/10.1109/ISGT.2013.6497901.
26. Rich, E., & Knight, K. (1990). *Artificial intelligence* (Subsequent edition). McGraw-Hill College.
27. Senjyu, T., Hayashi, D., Yona, A., Urasaki, N., & Funabashi, T. (2007). Optimal configuration of power generating systems in isolated island with renewable energy. *Renewable Energy*, *32*(11), 1917–1933. https://doi.org/10.1016/j.renene.2006.09.003.
28. Hernández, J. C., Medina, A., & Jurado, F. (2007). Optimal allocation and sizing for profitability and voltage enhancement of PV systems on feeders. *Renewable Energy*, *32*(10), 1768–1789. https://doi.org/10.1016/j.renene.2006.11.003.

29. Zadeh, L. A. (1965). Fuzzy sets. *Information and Control*, *8*(3), 338–353. https://doi.org/10.1016/S0019-9958(65)90241-X.
30. Fullér, R. (1995). Neural fuzzy systems. *In Advances in soft computing series.* Springer-Verlag, Berlin/Heidelberg.
31. Şen, Z. (1998). Fuzzy algorithm for estimation of solar irradiation from sunshine duration. *Solar Energy*, *63*(1), 39–49. https://doi.org/10.1016/S0038-092X(98)00043-7.
32. Kalogirou, S. A. (2002). *Design of a fuzzy single-axis sun tracking controller.* https://ktisis.cut.ac.cy/handle/10488/17893.
33. Du, X., Cai, Y., Wang, S., & Zhang, L. (2016). Overview of deep learning. *2016 31st Youth Academic Annual Conference of Chinese Association of Automation (YAC)*, 159–164. https://doi.org/10.1109/YAC.2016.7804882.
34. Goodfellow, I., Bengio, Y., & Courville, A. (2016). *Deep learning.* MIT Press.
35. Sutton, R. S., & Barto, A. G. (n.d.). Reinforcement *l*earning: An *i*ntroduction (p. 352), MIT Press.
36. Kalogirou, S. A. (2001). Artificial neural networks in renewable energy systems applications: A review. *Renewable and Sustainable Energy Reviews*, *5*(4), 373–401. https://doi.org/10.1016/S1364-0321(01)00006-5.
37. Raza, M. Q., & Khosravi, A. (2015). A review on artificial intelligence based load demand forecasting techniques for smart grid and buildings. *Renewable and sustainable energy reviews*, *50*, 1352–1372. https://doi.org/10.1016/j.rser.2015.04.065.
38. Dong, X., Qian, L. & Huang, L. (2017). Short-term load forecasting in smart grid: A combined CNN and K-means clustering approach. *2017 IEEE International Conference on Big Data and Smart Computing (BigComp)*, 119–125. https://doi.org/10.1109/BIGCOMP.2017.7881726.
39. He, Y., Deng, J., & Li, H. (2017). Short-term power load forecasting with deep belief network and copula models. *2017 9th International Conference on Intelligent Human-Machine Systems and Cybernetics (IHMSC)*, *1*, 191–194. https://doi.org/10.1109/IHMSC.2017.50.
40. Marino, D. L., Amarasinghe, K., & Manic, M. (2016). Building energy load forecasting using Deep Neural Networks. *IECON 2016 – 42nd Annual Conference of the IEEE Industrial Electronics Society*, 7046–7051. https://doi.org/10.1109/IECON.2016.7793413
41. Amarasinghe, K., Marino, D. L., & Manic, M. (2017). Deep neural networks for energy load forecasting. *2017 IEEE 26th International Symposium on Industrial Electronics (ISIE)*, (pp. 1483–1488). https://doi.org/10.1109/ISIE.2017.8001465.
42. Mocanu, D. C., Mocanu, E., Nguyen, P. H., Gibescu, M., & Liotta, A. (2016). Big IoT data mining for real-time energy disaggregation in buildings. *2016 IEEE International Conference on Systems, Man, and Cybernetics (SMC)*, 003765–003769. https://doi.org/10.1109/SMC.2016.7844820.
43. Tornai, K., Oláh, A., Drenyovszki, R., Kovács, L., Pinté, I., & Levendovszky, J. (2017). Recurrent neural network based user classification for smart grids. *2017 IEEE Power Energy Society Innovative Smart Grid Technologies Conference (ISGT)*, 1–5. https://doi.org/10.1109/ISGT.2017.8086043.
44. Scerri, P., Pynadath, D. V., & Tambe, M. (2002). Towards adjustable autonomy for the real world. *Journal of Artificial Intelligence Research*, *17*, 171–228. https://doi.org/10.1613/jair.1037.
45. Zhao, Z., Xu, G., Qi, Y., Liu, N., & Zhang, T. (2016). Multi-patch deep features for power line insulator status classification from aerial images. *2016 International Joint Conference on Neural Networks (IJCNN)*, 3187–3194. https://doi.org/10.1109/IJCNN.2016.7727606.
46. Reddy, P. P., & Veloso, M. M. (2011, June 28). Strategy learning for autonomous agents in smart grid markets. *Twenty-Second International Joint Conference on Artificial Intelligence.* https://www.aaai.org/ocs/index.php/IJCAI/IJCAI11/paper/view/3346.

15 Introduction to AI Techniques for Photovoltaic Energy Conversion System

Siddharth Joshi and Nirav Karelia
Department of Electrical Engineering, School of Technology, Pandit Deendayal Energy, University, Gandhinagar, Gujarat, India

CONTENTS

15.1 INTRODUCTION

The energy engineering is the basic science among all sciences of mankind because energy acts as a critical factor for human development. Energy usually comes from various sources. Energy is used everywhere and whenever any work is done. We can harness the energy in the form of commercial and non-commercial way to perform our day to day tasks. We are using this energy from conventional sources. The estimation of conventional sources of the energy may vary, but there are limited reserves of such sources, particularly oil and gas. There is a probability that the sources may come to an end within few decades. Similarly, it is also applicable to coal reserves and crude due to their rapid use as a conventional fuel, which could vanish within the near future. Moreover, there is also an issue of price hike in traditional sources.

The energy consumption from any renewable or non-renewable sources is used to develop "human development index (HDI)" of any country. The HDI depends on the current scenario of per capita energy consumption of any country. The per capita energy consumption gives a rough idea about the standard of living and prosperity of the country. The average electrical energy per capita in terms of kilowatt hour per person per year is 1,181 kWh, and the average power per capita is 140 W per person. In case of the United States, it states that the average electrical energy per capita in terms of kilowatt hour per person per year is 12,071 kWh, and the average power per capita is 1,377 W per person. It means that the country comprises 7% of the world population that consumes around 32% of the total energy consumption of the world, whereas the other nation has 20% of the world population that consumes about 1% of the total energy consumption [1]. This creates disparity in the system. Still, power demand is increasing day by day. The similar cases can be compared for other developed and underdeveloped countries. Talking about industrial and technology revolution, the burning of fossil fuel and other sediments from the industries pollute the environment and release greenhouse gases in the atmosphere, which leads to increase annual Earth's temperature globally. Carbon dioxide is also produced by burning of fossil fuels; methane is produced during transportation of coal, oil, and natural gas; greenhouse gases like hydro-fluorocarbons (HFCs), sulfur hexafluoride (SF_6), etc., are generated by various industrial processes. Such issues lead to ozone layer depletion, loss of biodiversity, etc.. Tones of documentaries and articles are available on such problems due to rapid use of energy via fossil fuels and traditional sources. This issue has been defined thoroughly by Mr. Algore from the United States in his movie, "The Inconvenient Truth". He focused on increment in annual average temperature, ocean current patterns, global warming precipitation, upcoming novel virus (COVID 19 is one of them), loss of species, etc. [2]

As far as the power sector of our country is concerned, India comprises three power sectors at a glance trifurcated as central sector, state sector, and private sector. The sector-wise power generation is shown in Figure 15.1.

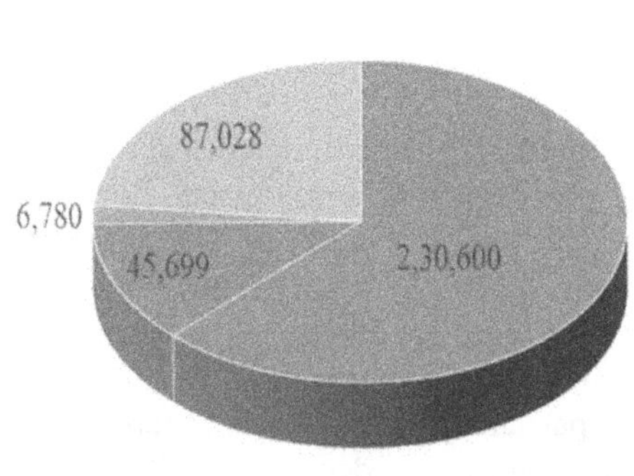

FIGURE 15.1 Present-day contribution of energy sources in the Indian grid [3].

TABLE 15.1
Bifurcation of Sources in Indian Grid in Present Day [3]

Sr. No.	Power Plant	Percentage Contribution
1.	Total thermal	62.8%
2.	Coal-based thermal	54.2%
3.	Lignite-based thermal	1.7%
4.	Gas	6.9%
5.	Diesel	0.1%
6.	Hydro (renewable)	12.4%
7.	Nuclear	1.9%
8.	RES	23.5%
Total generation at present		3,70,106 MW

The majority of the energy is still coming from conventional thermal power station and other conventional sources. The global energy outlook [4] is released; the new target of the renewable power sources is proposed; and the set target is achieved. These alternative energy sources are mainly obtained from the Sun. The Sun is the gigantic source for energy and provides ample of energy through which energy is harnessed. The sunlight striking to the Earth's surface in one and half hours is enough to satisfy the energy demand of world's energy consumption for a year. The answer of this is hidden in the increment in the penetration of renewable energy sources. Currently, several such renewable power sources like wind and solar systems have been installed, commissioned, and proposed. Table 15.1 typifies the present-day contribution of renewable power in the Indian grid. The Indian government sets a massive target in the years to come.

This chapter comprises mathematical modeling of small-scale photovoltaic energy conversion system (PVES), the interfacing of PVES with DC loads and its characteristics, followed by the concept of maximum power point tracking (MPPT) and its simulations analysis with DC load. The incorporation of the artificial intelligence (AI) technique in MPPT technique is also explained.

15.2 MATHEMATICAL MODELING OF SOLAR PHOTOVOLTAIC SYSTEM

The photovoltaic (PV) cell or a module works on the PV effect. The effect states that when the light falls on to the bulk material, it has characteristics to generate electricity using photons. The effect of recombination hall and electrons is indicated in plenty of studies [5, 6]. The PV cell or a module is modeled as a current source. The photocurrent then drives the load and the other assembly. The conversion efficiency is around 15–17% depending upon the type of bulk materials. The PV cell is a preliminary element of a PV module. The cells are usually connected in series. A number of series-connected cells form a module. The PV modules are rated with wattage capacity. The PV module produces DC power so that wattage power can be

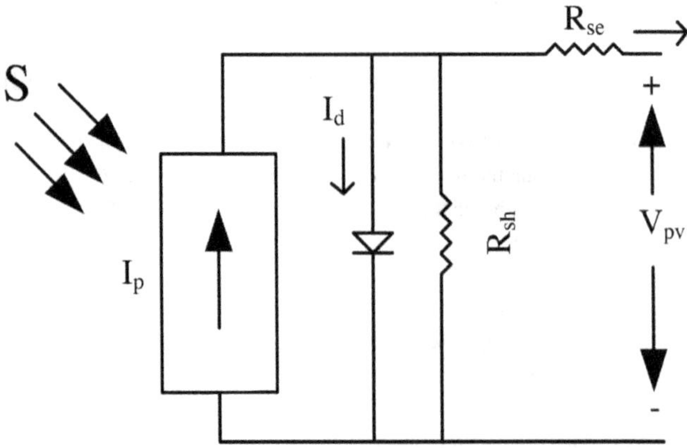

FIGURE 15.2 Mathematical model of solar cell [5, 6].

measured in peak rating, abbreviated as W_P. The parameters for PV module or cell comprise open circuit voltage (V_{OC}), short circuit current (I_{SC}), rated voltage or maximum power voltage (V_R), and rated current (I_R). One can get this parameter from the data sheet of PV module. Figure 15.2 shows single diode model and its equivalent circuit of single solar PV cell. The cell comprises single diode, shunt, and series resistance of the bulk material. The formation of the circuit is as under.

The PV cell or module can be demonstrated, as shown in Figure 15.3. A simplest equivalent circuit of a PV cell comprises a current source, which is driven by sunlight and with diode model shown in Figure 15.2. The first two parameters to be computed are open-circuit voltage and short-circuit current. The open circuit voltage is measured by connecting a DC voltmeter across open terminals of PV cell; likewise, the short circuit current is measured by connecting a DC ammeter by short circuiting the PV terminals. These measurements are shown in Figure 15.3. The simplest equivalent circuit connected with DC load is shown in Figure 15.4. Figure 15.4 demonstrates the photocurrent supplied to the DC load and equivalent

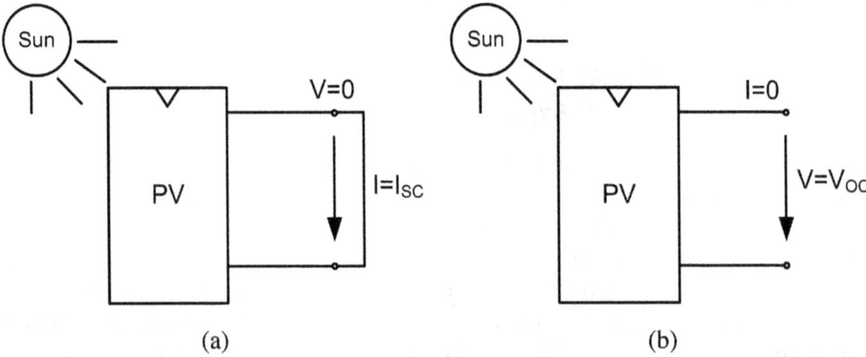

FIGURE 15.3 Measuring (a) short-circuit current (I_{SC}) and (b) open circuit voltage (V_{OC}) from photovoltaic (PV) cell/module [5, 6].

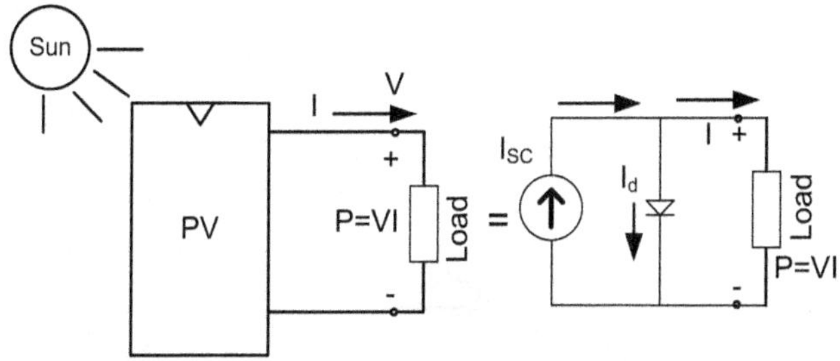

FIGURE 15.4 Simple model of PV cell or module [5].

circuit of PV connected with load. In Figure 15.4, one can apply Kirchhoff's current law (KCL) at common node point to develop the mathematical model of PV cell and is demonstrated in eq. (15.1)–(15.3).

In case of open-circuit voltage, following equations are modeled; the relationships between open-circuit voltage and short-circuit current are demonstrated in eq. (15.4) and (15.5).

A simple model of PV cell or module is demonstrated in Figure 15.4. The quantity I denotes the load current I_{SC} denotes current from PV, and I_d denotes the diode current. After considering all the modeling equations, eq. (15.6) is modeled, which drives a photocurrent to the load, and is also applicable for PV module, string, or array.

$$I = I_{SC} - I_d \tag{15.1}$$

$$I_d = I_0\left(e^{\frac{qV_d}{kT}} - 1\right) \tag{15.2}$$

$$I = I_{sc} - I_0\left(e^{\frac{qV_d}{kT}} - 1\right) \tag{15.3}$$

$$V_{OC} = \frac{kT}{q}\ln\left(\frac{I_{sc}}{I_0} + 1\right) \tag{15.4}$$

$$V_{OC} = 0.0257\ln\left(\frac{I_{SC}}{I_0} + 1\right) \tag{15.5}$$

$$I_{PV} = I_{ph} - I_{rs}\left(e^{\frac{q(V_{pv}+I_{pv}R_s)}{AkT}} - 1\right) - \frac{V_{pv} + I_{pv}R_s}{R_{sh}} \tag{15.6}$$

where I_{PV} is the load current in A, I_{ph} is the photocurrent in A, I_{rs} is the diode reverse saturation current in A, V_{pv} is the output voltage in V, q is the electron charge ($=1.609 \times 10^{-19}$) in Coulomb, A is the diode ideality constant, k is the Boltzmann's

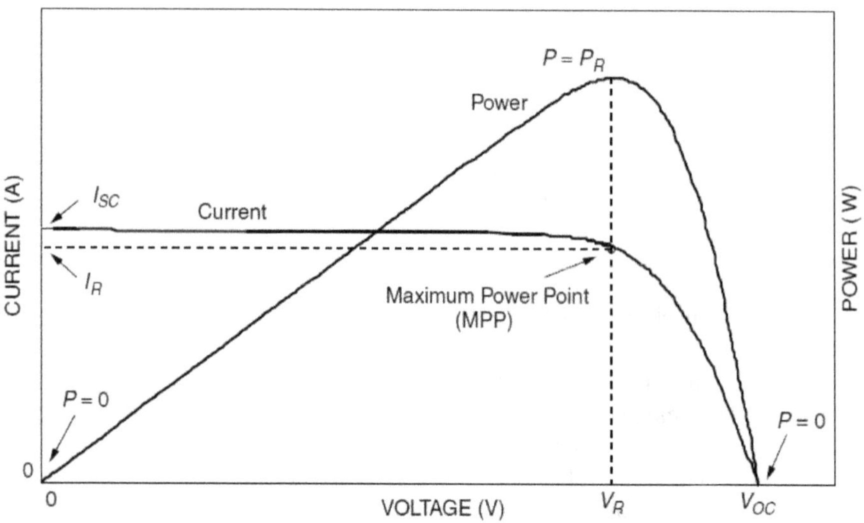

FIGURE 15.5 IV and PV curve of PV module [5].

constant (=1.38 × 10^{-23}) in Joule per Kelvin, T is the cell absolute temperature in Kelvin, R_s is the series resistance of PV cell in ohm, R_{sh} is the shunt resistance of PV cell in ohm.

The equations can be modeled and plotted with the various values of voltage and current with the change in radiation and temperature as well. The curve can be plotted in terms of voltage and current, i.e., IV curve, and power and voltage curve, i.e., PV curve. These curves are shown in Figure 15.5. The biggest rectangle as shown in Figure 15.5 intersects IV curve at maximum power point (MPP). This rectangle demonstrates the overall efficiency of the system also known as fill factor (FF) of the PV module. FF can be defined as the ratio of multiplication of rated voltage and rated current of PV module at particular climatic conditions to the multiplication of open-circuit voltage and short-circuit current. Mathematically, FF is shown in eq. (15.7).

$$FF = \frac{V_R I_R}{V_{OC} I_{SC}} \tag{15.7}$$

15.3 INTEGRATING PV MODULE WITH THE LOAD

The PVES plays an important role when the electrical load is connected to it. The load is either in the form of DC, AC, or a battery charging, and in case of grid-connected system, the grid acts as a load. In this section, PV connected with DC loads is discussed. The discussion is based on three types of load. The first one is the resistive load, followed by the DC motor, and the battery.

The PVES connected with the load and the load operate according to the line of intersecting of IV curve of the PV and IV curve of the load. This point of intersection is known as operating point of the load. If one can draw a line on x- and y-axes

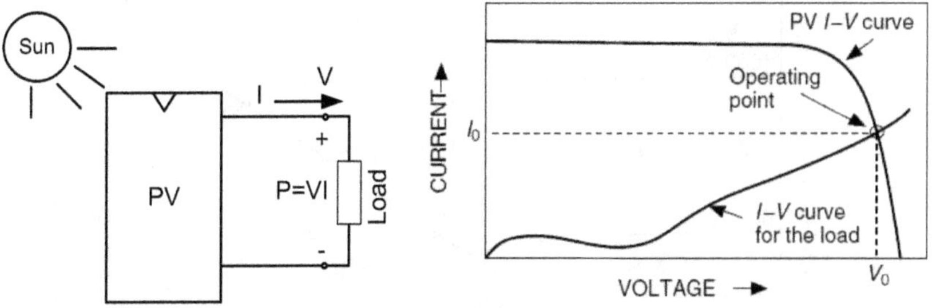

FIGURE 15.6 Concept of operating point for load and PV module [5].

corresponding to this point, the values of the voltage and current are obtained, and accordingly, the power consumed by the load is computed. Power consumption can also be computed from PV curve of solar PV module. The concept of operating point is shown in Figure 15.6. As shown in Figure 15.6, the load operates in a randomized manner and intersects the IV curve of PV at a point known as operating point. The operating point is a point at which both the PVES and the load operate satisfactorily.

Taking the case of resistive load, the operating point changes as the curves are shifted either from left to right or from right to left on IV characteristics of PV. The characteristic equation of the resistive type of the load is the ratio of voltage to current. The load characteristics decide the operating point of solar PV module. The operating point is the point of intersection where the characteristics of load intersect the IV curve of PV. With the change in resistance, the operating point is changed, as shown in Figure 15.7. According to the characteristics, if the value of resistance is increased, then the curve approaches to open-circuit voltage point and operates in constant voltage region; and if the value of the resistance decreases, then the curve approaches to short-circuit current point in the characteristics and operates in constant current region. There might be an impact of MPP with the change in climatic conditions. There can be multiple MPPs if change in radiation is considered

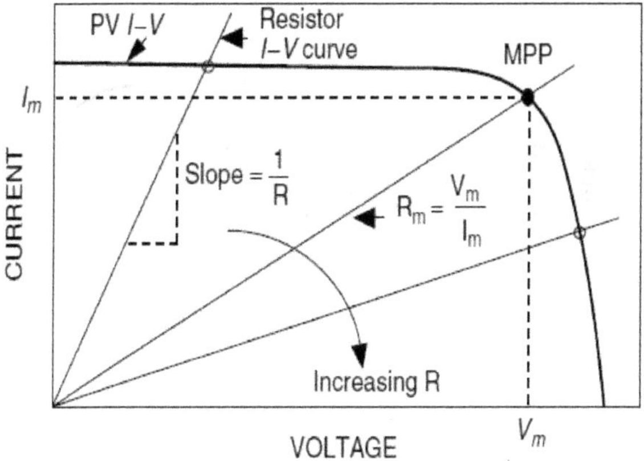

FIGURE 15.7 Interfacing of resistive load with PV module [5].

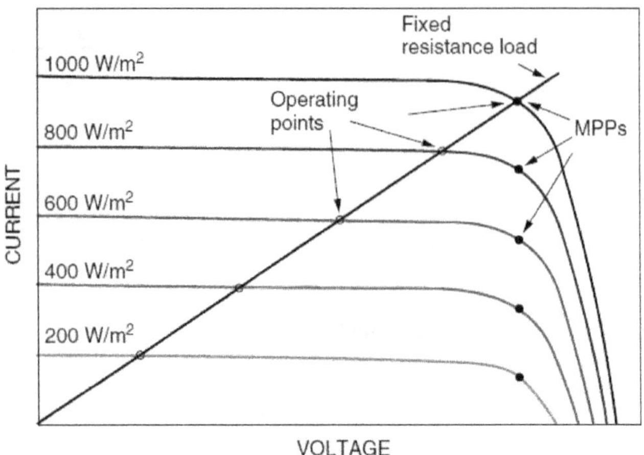

FIGURE 15.8 Multiple MPPs and operating points at fixed load and change in insolation [5].

in the discussion. This phenomenon is shown in Figure 15.8. In this condition, the load resistance must be kept at a fixed value.

Taking the case of DC motor, DC shunt motor is used particularly for pumping application. The interaction of IV curve of DC motor and IV curve of PV is shown in Figure 15.9. The mismatch of the curve is discussed in [5]. In the third case, i.e., the battery, the battery is used as backup for PVES during no sunshine hours. When the load drawn from the PV system is less, the battery operates in charging mode; and during no sunshine hours, it operates on discharging mode. The charging and discharging curve of the battery is shown in Figure 15.10. To operate these loads, there can be an interfacing unit such as buck or boost converter according to the application.

For example, if application seeks higher voltages, the boost converter is accommodated, and for lower voltages, buck converter is accommodated. For bidirectional

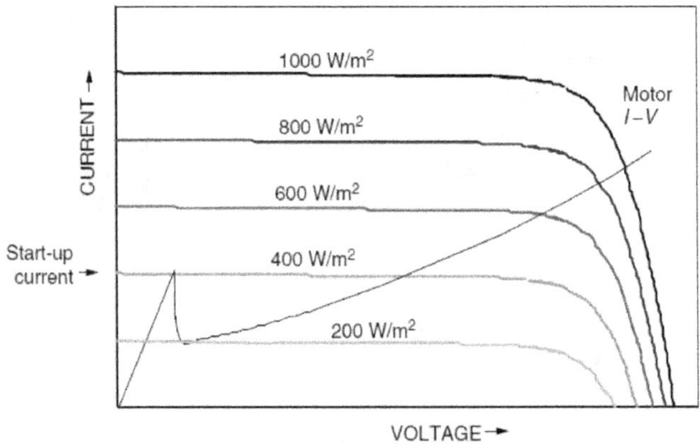

FIGURE 15.9 Interfacing of DC motor with PV module [5].

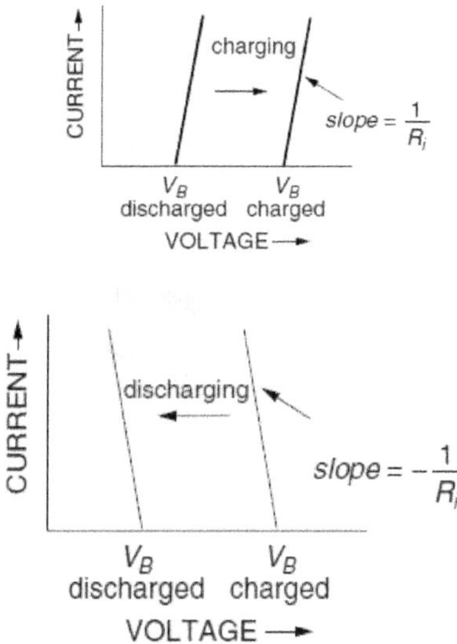

FIGURE 15.10 Charging and discharging of the battery [5].

power flow such as applications like battery energy storage system, the bidirectional converters are used. The buck converter is usually used for battery charging application for standalone battery bank system. Moreover, using bidirectional converter, these battery banks are charging and discharging as per the requirement. So, to operate these loads at the maximum efficiency or peak efficiency, the interfacing of MPPT algorithm has been accommodated with the system. The concept of MPPT is discussed in the next section.

15.4 CONCEPT OF MPPT AND TRACKING METHODS AND INTERFACING MPPT WITH PVES

The power from PV varies with the change in climatic conditions such as insolation of the Sun, temperature, etc. This is the reason it requires continuous process to track maximum power, otherwise the system tracks suboptimal points. So, the MPP is to be tracked with the change in atmospheric conditions. Moreover, if we assume for some time these conditions remain same, then also the load changes continuously. The concept of mismatching of such load as compared to IV curve is demonstrated in Figure 15.11.

To address the above-stated issues, the MPP trackers are installed in the system. The interfacing of MPPT algorithm block with the PVES is shown in Figure 15.12. The system shown in Figure 15.12 comprises the block diagram with control and power circuit. The system comprises DC-DC boost converter and DC load. The DC-DC boost converter is interfaced between the PV and a load. To control the switch modulator signal generates the duty cycle using the MPPT algorithm.

A number of MPPT methods have been proposed and reported till date for solar PVES. But in majority, perturb and observe (P & O) and incremental conductance

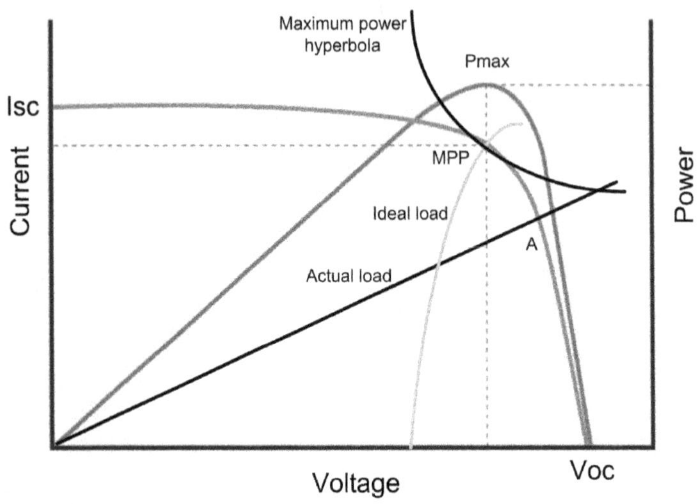

FIGURE 15.11 The concept of load mismatch and MPP tracking [7].

(INC) are used to track MPP. In this work, P & O-based MPPT method is simulated for solar PVES powered to DC load. The P & O algorithm works on both the variable voltage and variable current-based P & O method. In this work, we have simulated reference current or short-circuit current-based P & O method for simulation of PVES proposed in [8, 9] to track MPP. The work is validated through simulation analysis under sudden change in atmospheric conditions.

Authors in ref. [10] focused on the size of the perturbation in insolation of MPPT algorithm. If the perturbation increases, the power changes with the change in voltage and current depending upon perturbation in insolation of the Sun and temperature of the PV. The algorithm measures instantaneous values of voltage and current as a sample variable. Some of the algorithms are voltage reference-based, and some of them are current reference-based. Basically, current-controlled MPPT algorithms are used to take accurate sample of the current as this variable has smaller change

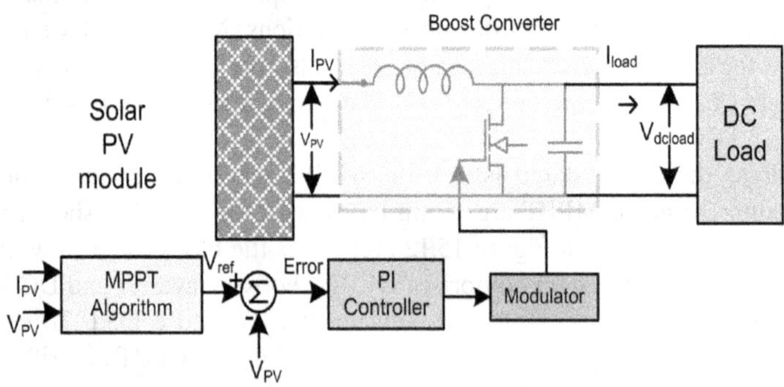

FIGURE 15.12 Interfacing of MPPT algorithm with the load.

compared to voltage, and the current of the PV module depends on the radiation. However, voltage sample is also important in the effectiveness of algorithm, but with the greater change in solar insolation and hence current, there is small change in the voltage. In this work both the cases, i.e., voltage reference and current reference, have been discussed through simulation analysis with P & O method.

The concept of P & O method depends on perturbation size of the chosen variable, i.e., voltage or current, which depends on solar radiation and temperature with its impact on power of the solar PV. The voltage-based P & O algorithm starts with the measurement of PV power and particular voltage. Then, with fixed value of perturbation, the reference voltage changes either in positive or negative direction by fixed increment or decrement of the voltage value. The power is computed with the new value, and it is compared with the old value of the power. Depending on the value of two consecutive powers with the signature, the next perturbation is decided. In this case, if the present power is more than the sample of power taken in past, then voltage further increases; this procedure continues till the operating point reaches up to MPP. After this point, the value of power starts decreasing. The generalized equation for the generation of reference voltage and current is stated in eq. (15.8), where, V_{ref} is the reference value of the voltage, V_{ref} $(k - 1)$ is the previous or past value of the voltage before change in radiation, and C is the perturbation size. The same phenomenon will happen in case of current perturbation algorithm using IV curve and PV curve of the solar PV module.

$$V_{ref}(k) = V_{ref}(k-1) \pm C \tag{15.8}$$

Other MPPT methods have been proposed till date. A brief about other MPPT techniques is mentioned here. The variable perturbation for sudden change in insolation is proposed in ref. [8]. The proposed algorithm combines three adaptive control algorithms by introducing two stage perturbations. Authors proposed new algorithm with conventional algorithm and proved adaptive algorithm better as compared to conventional algorithm. In addition to that in ref. [9], authors derived the equation of short-circuit current of photovoltaic with upper and lower bounds of current. The proposed algorithm is tested for various permutations and combinations of climatic conditions. The MPPT algorithm is proposed by the author in context of parallel connection of PV modules with MPPT [11]. The proposed MPPT controller is verified by connecting buck and boost converters; they modeled the converters and implemented the converters in the system. They have also incorporated the battery charging and discharging system with various modes of MPPT [12]. The comparison of various MPPT techniques for solar PV system is available in ref. [13]. They compared the various MPPT techniques in context of change in characteristics during operation with the change in climatic conditions. Author compared various MPPT techniques in ref. [14]. They mentioned and compared various MPPT techniques are used such as hill climbing technique, incremental conductance technique, fractional open circuit voltage technique, fractional short-circuit current technique, fuzzy logic control, implementation, neural network (NN)-based MPPT, and dP/dV or dP/dI with the feedback control method. The direct and indirect MPPT control methods are compared in [14]. These methods are compared in context of PVES interfacing with battery unit.

Figure 15.12 shows the layout of PVES connected with DC load with MPPT technique; the system consists of PV module, DC-DC converter, MPPT-based controller, and the DC load. The PVES generates DC power, which is in terms of voltage and current. The voltage and the current sample is taken from the system and these samples act as an input of the MPPT controller; the algorithm inside MPPT block generates reference voltage and/or current depending on the perturbation variable taken. As discussed earlier, various MPPT algorithms are proposed till date comprising various perturbation methods such as current perturbation that generates the reference current, voltage perturbation that generates the reference voltage, etc. Figure 15.12 shows the incorporation of voltage perturbation MPPT algorithm in PVES. In this work, the voltage perturbation and current perturbation are simulated for PVES.

The most rising and distinctive idea is to incorporate artificial neural network (ANN) in PVES in case of MPPT controllers. The authors in ref. [15] depicted the scope of NN and performed a thorough study of ANN-based PV MPPT techniques. They described that ANN MPPTs have their various effects. Authors in ref. [16] proposed a novel NN MPPT controller for PVES. Through MATLAB simulation results, they demonstrated the results and found better tracking correctness, response time, and overshoot of the suggested NN controller under fast changing isolation.

15.5 SIMULATION AND ANALYSIS FOR MPPT-BASED PVES

This section presents simulation analysis of MPPT-based PV conversion system. The simulation analysis has been performed using the P & O MPPT technique. The analysis is done with four cases: (1) voltage perturbation-based MPPT algorithm, (2) current perturbation-based MPPT algorithm, (3) impact of change in load in voltage and current-based perturbation algorithm, and (4) AI-based PI (proportional integral) controller used for PVES.

The first two cases comprise the simulation analysis using voltage perturbation algorithm MPPT and the variable current perturbation MPPT for generating the current reference and the voltage reference depending upon the type of variable used. The above-stated two algorithms are tested for variable irradiation and temperature, i.e., varying climatic conditions and change in load. The conventional PI controller is used for this test. The third case comprises the impact of change in load in the system; the system is tested for various permutations and combinations of load with change in irradiation and temperature of the PVES as well. The last case comprises incorporation of AI-based NN supported PI controller in the PVES. The applications of AI techniques in the field of renewable engineering are plenty. Nowadays, AI and NN-based applications took an emerging place in the current situation of the power system era. In addition to PI controller, the ANN-based proportional integral controller is used in case four

15.5.1 Case 1: Voltage perturbation-based MPPT algorithm

The voltage perturbation-based MPPT algorithm is simulated in PSIM ® 9.0 software for a 240 W PV panel feeding power to the DC load. The MPPT algorithm is implemented using a simplified C block in the software using digital coding of the

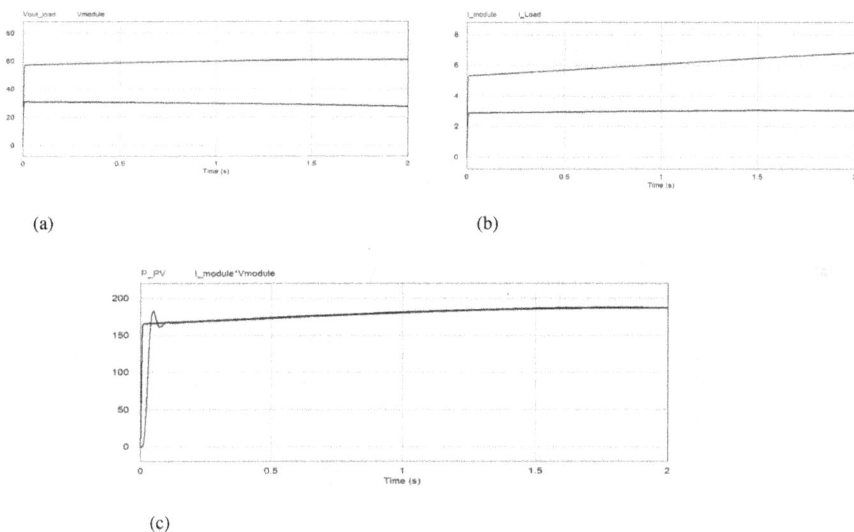

FIGURE 15.13 Response of PVES with MPPT at 800 W/m² insolation and 30°C. (a) Voutload_load – Voltage across load & V_module – Voltage across PV module, (b) I_module – Current drawn from PV module before DC-DC boost converter & I_load – Load current, and (c) P_PV = Power drawn from PV module & I_module*Vmodule = Power generation from module.

MPPT algorithm. The algorithm is available in [7]. The system has been tested for permutations and combinations for climatic conditions and load as well. The MPPT block generates reference voltage, which is then compared to actual voltage of the PV that leads the duty cycle generation of the DC-DC power converter. The block diagram of the system is shown in Figure 15.12.

Figure 15.13 shows the response of PV system with constant insolation and temperature with a constant load of 15 Ω and also the response of PVES with MPPT technique. The switching frequency of 15 kHz is used for this work; the value of L and C is calculated with the method mentioned in [17]. Figure 15.13(a) corresponds to voltage of module and voltage across load, and the voltage level changes by boosting the input of PV; Figure 15.13(b) typifies the module and load current; and Figure 15.13(c) provides power drawn from PV and multiplication of input and output voltage of PV module. Superimposing of these two graphs gives the efficacy of MPPT algorithm which is around 99.2% with fixed insolation and temperature level. The dynamic response of the same system with the change in insolation and temperature is shown in Figure 15.14. Figure 15.14(a) shows the input pattern for insolation and temperature given as an input to the PV module with time on the x-axis. Figure 15.14(b) and (c) shows the input and output voltages and current with PVES converter assembly. The voltage regulation with the change in climatic conditions is also observed from the responses. Figure 15.14(d) shows the efficacy of the MPPT with the change in climatic conditions for the PVES. Table 15.2 shows the dynamic response of the system.

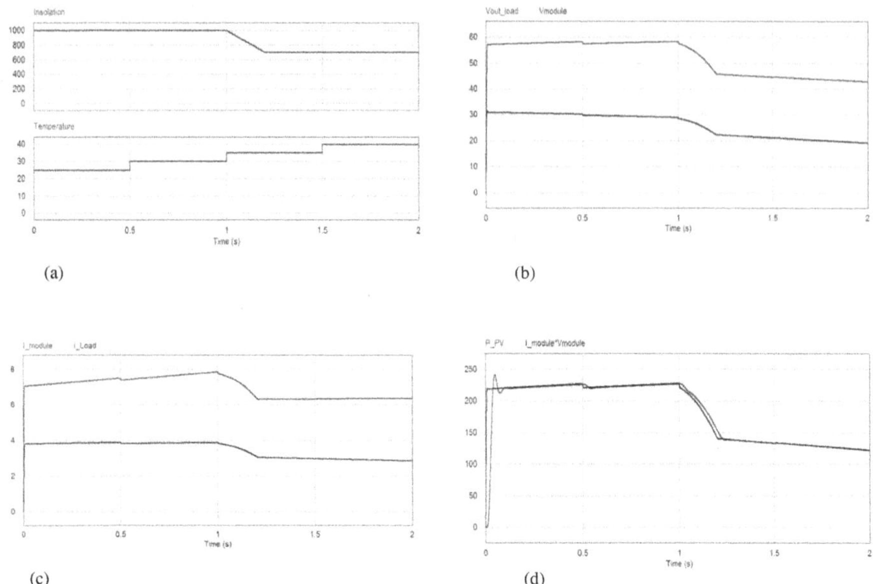

FIGURE 15.14 Response of PVES with MPPT with change in insolation and radiation (a) Change in insolation and radiation, (b) Voutload_load – Voltage across load & V_module – Voltage across PV module, (c) I_module – Current drawn from PV module before DC-DC boost converter & I_load – Load current, (d) P_PV = Power drawn from PV module & I_module*Vmodule = Power generation from module.

TABLE 15.2
Response of Variable Perturbation PVES with Change in Insolation and Tempature at Constant Load of 20 Ω

Sr. No.	Radiation (W/m²)	Temperature (°C)	Current drawn from PV I_module (A)	Load current I_Load (A)	Voltage across PV Module (V_module) (V)	Load Voltage Vout_load (V)
1.	1000	25	7.9	3.9	29.3	59.0
2.	900	30	7.6	3.7	28.0	56.4
3.	800	30	7.0	3.4	25.9	52.2
4.	700	25	6.2	3.0	23.0	46.3
5.	600	25	5.3	2.6	19.7	39.8
6.	500	35	4.49	2.21	22.06	44.39
7.	400	40	3.66	1.80	17.97	36.16

15.5.2 Case 2: Current perturbation-based MPPT algorithm

This case provides the simulation analysis of the single solar PV module connected with a DC-DC converter and with an incorporation current perturbation type of MPPT algorithm; for a single PV, panel of 240 W is simulated in this work. The parameter incorporated in solar PV module is from [17]. The code for MPPT is developed in the DLL (Dynamic Link Library) block or C block available in PSIM software. The closed-loop DC-DC converter is used for voltage regulation. Compared to voltage algorithm, the reference current is generated from MPPT algorithm by using the same control circuit shown in Figure 15.12. The control and power interfacing of current perturbation-based MPPT techniques is shown in Figure 15.15. In Figure 15.15, the reference current is generated from P & O algorithm, which is compared with the current drawn from PV module. The error signal is given to the controller to generate the duty cycle for the converter. The switching frequency of 15 kHz is used and the calculation of converter is done by the method available in [18]. The effectiveness and overall conversion efficiency of MPPT algorithm is presented in [19].

Figure 15.16 shows response for the single PV module of 250 W peak capacity connected with resistive load via DC-DC converter. The response is also tested for more number of modules in series to develop high power. The comparative analysis is done with and without MPPT algorithm for 30 such modules, 60 *Wp* rating of each connected in series. The modules are connected in series for development of high power through PV string. The block diagram and circuit diagram remain same as per Figure 15.15. The reference current is generated from MPPT algorithm,

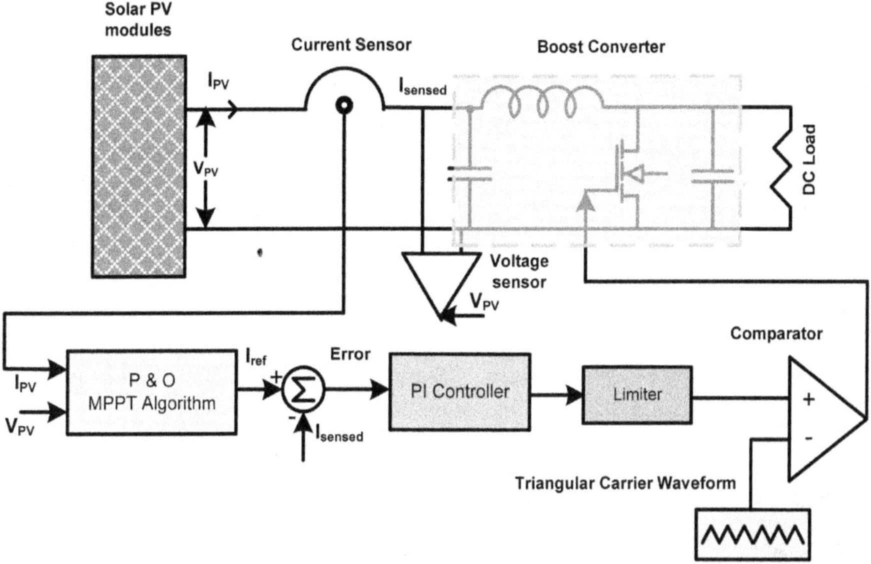

FIGURE 15.15 Current perturbation MPPT algorithm-based PVES.

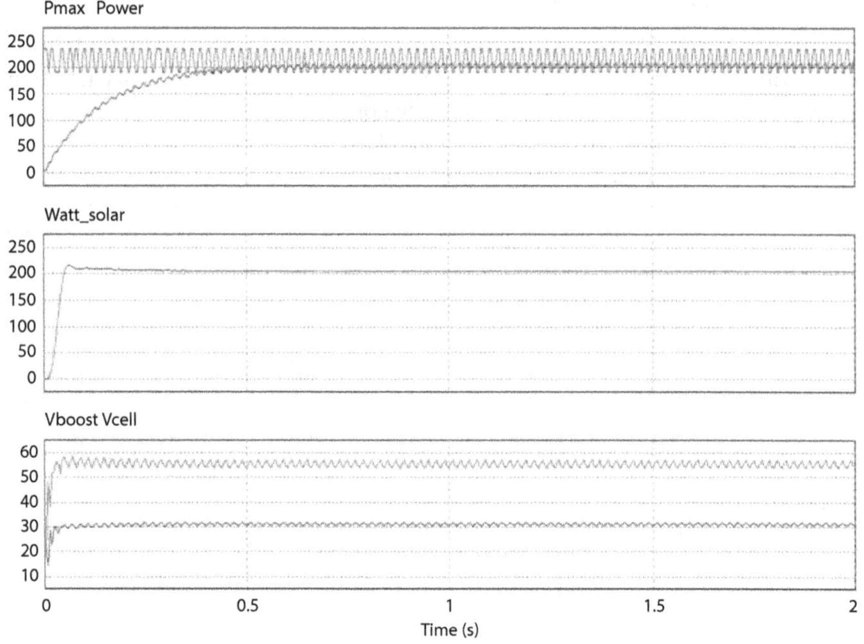

FIGURE 15.16 Result with maximum power point tracking (MPPT) algorithm. Power from PV (P_{max}), tracked power (Power), power consumed by load (Watt_Solar), output voltage of PV module (V_{cell}), and response of boost converter (V_{boost}) at a given perturbation in radiation.

which is compared with the actual PV current. Depending on the sensitivity of the current sensor, the suitable gain is selected in practical applications. The simulation analysis is done by incorporating suitable rating of DC-DC converter with varying atmospheric conditions. The perturbation is applied in the input between 1000 W/m^2 and 800 W/m^2 with constant temperature of PV module. The input is modeled in PSIM software by applying square wave as a continuous input. This is done because in the previous cases of the simulation, the insolation varied with sudden changes, and now in day to day life, it may vary in gradual manner except cloudy days. So, in this model, the simulation insolation varies between specific bounds and the simulation is performed. Figures 15.17 and 15.18 show a comparative analysis with and without MPPT algorithm under varying insolation.

15.5.3 Case 3: Impact of Change in Load in the System

This case comprises simulation analysis with change in the DC load. To observe the impact of load on the system, the range of load between 10 Ω and 40 Ω is considered. The results for 15 Ω load are shown in Figure 15.19. Like in cases 1 and 2,

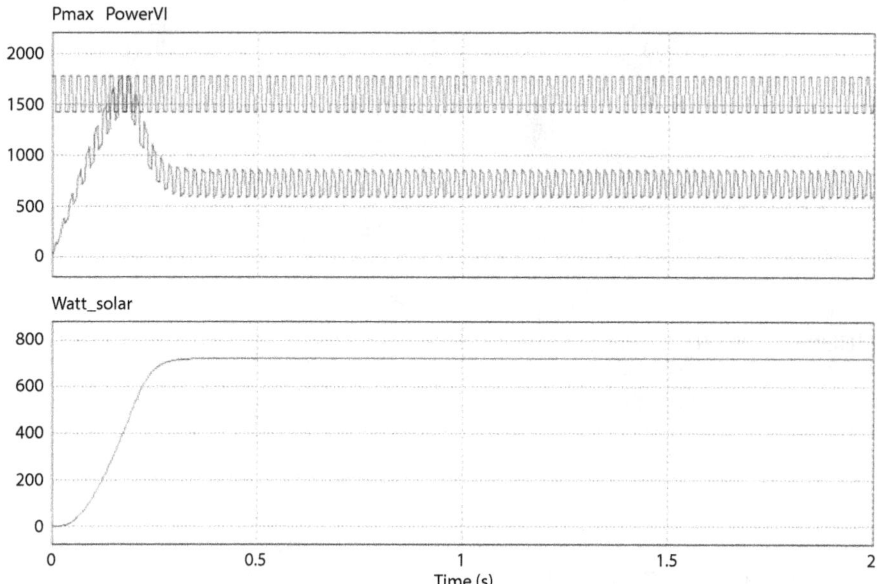

FIGURE 15.17 Response of P_{max} and $P_{tracked}$ without MPPT algorithm for series-connected 30 modules.

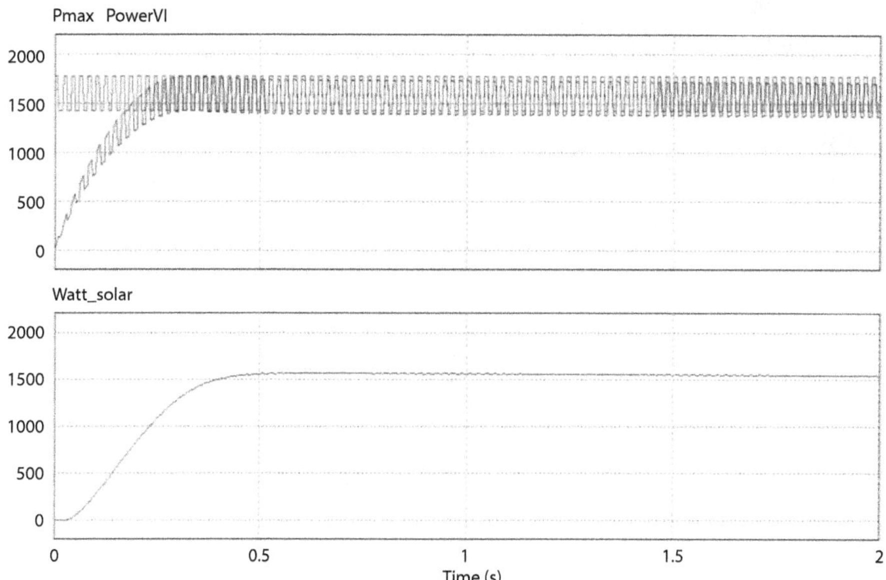

FIGURE 15.18 Response of P_{max} and $P_{tracked}$ with MPPT algorithm for series-connected 30 modules. Where P_{max} = maximum available power from PVES, Power VI = tracked power, Watt_solar = power consumed by load.

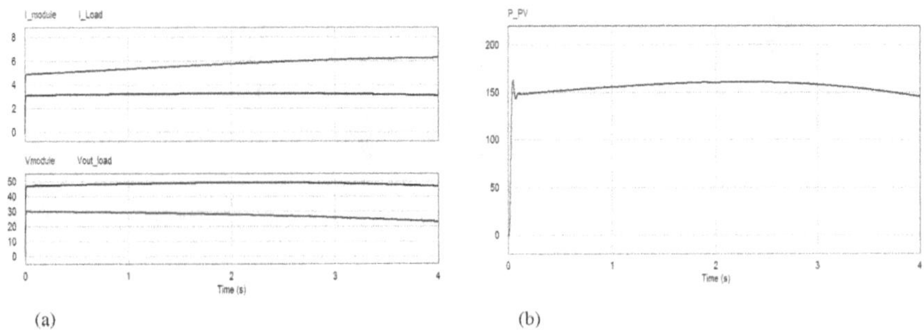

FIGURE 15.19 Results for constant insolation 700 W/m², 35°C, and 15 Ω resistive load. (a) Module current and load current, module voltage and load voltage. (b) Power drawn from PV to load at particular climatic condition and load, where, I_module = current drawn from PV module before DC-DC boost converter and I_load = load current, V_module = module voltage and Vout_load = voltage across load.

the results are plotted with voltage perturbation method. The values of various quantities such as current, voltage, and power drawn by load from PVES are shown in Table 15.3.

15.5.4 Case 4: Impact of AI-based PI controller in PVES

This case comprises incorporation of AI system-based PI controller and its comparison with fixed PI controller is demonstrated. The block diagram of the AI-based PI controller interfaced with P & O MPPT technique with PVES is shown

TABLE 15.3
Response of Variable Voltage Perturbation for PVES with Change in Insolation and Tempature for Variable Load

Sr. No.	Radiation (W/m²)	Temperature (°C)	Load Resistance (Ω)	Current Drawn from PV I_module (A)	Load Current I_Load (A)	Voltage across PV Module (V_module) (V)	Load Voltage Vout_load (V)	Power Drawn by Load (W)
1.			10	6.33	3.26	16.99	32.69	107.53
2.			15	6.24	3.07	22.98	46.19	141.71
3.			20	5.69	2.80	27.81	56.11	156.22
4.	700	35	25	4.93	2.42	29.96	60.57	145.57
5.			30	4.26	2.09	31.07	62.88	130.70
6.			35	3.74	1.83	31.77	64.33	117.26
7.			40	3.32	1.63	32.26	65.35	105.87
8.			15	8.58	3.94	27.26	59.14	233.79
9.	1000	25	25	6.31	2.83	31.94	70.93	200.91

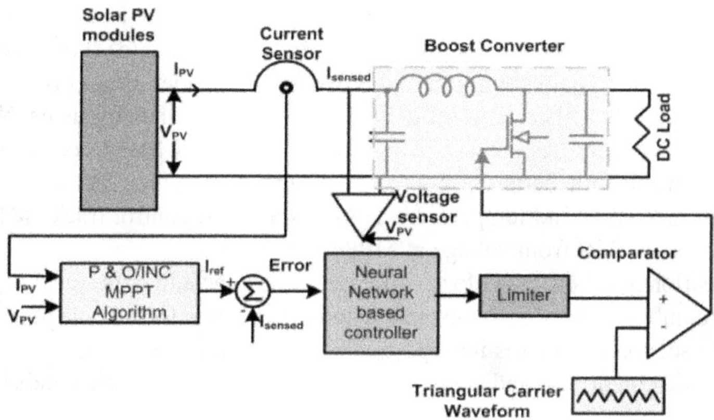

FIGURE 15.20 Block diagram of standalone PVES with AI-based NN controller.

in Figure 15.20. The conventional fixed gain PI controller is replaced by NN-based controller in Figure 15.20.

Generally, fixed gain PI controllers are employed to regulate the output voltage across load. The fixed gain controller might be a viable option for a small system. But for a large system, there might be some other option. To maintain voltage regulation, these controllers need to be tuned optimally. The other issues related to fixed gain PI controller have been discussed earlier. The NN-based controller is used to overcome the issue of voltage drift, which is presented over here. The comparative analysis is done for the NN-based system over fixed gain controllers. The simulation analysis is performed in MATLAB platform. The simulation is done for 1.5 kW solar PVES. A comparative analysis has been done for drift analysis between fixed gain controller and NN-based control for P & O MPPT method. The results obtained for change in climatic conditions are shown in Figure 15.21.

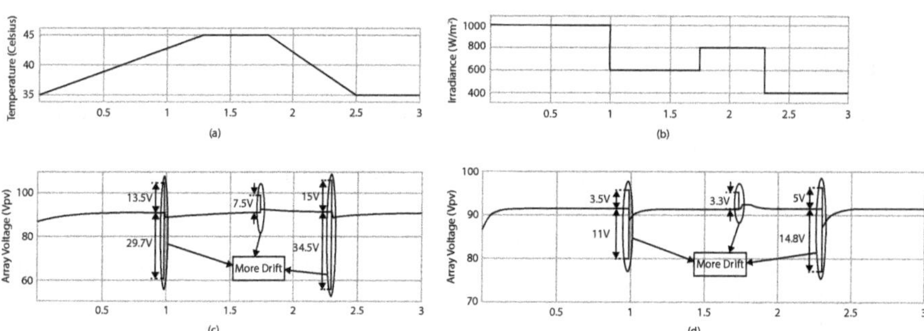

FIGURE 15.21 Comparative analysis for PI controller and AI-based NN controller (time in seconds). (a) Temperature plot, (b) Irradiance plot, (c) Vpv~time graph for PI based control, (d) Vpv~time graph for NN based control for P & O method

Figure 15.21 shows the variable insolation and temperature is applied to the PV array, as shown in Figure 15.21(a) and (b). Figure 15.21(c) and (d) shows a comparative analysis of fixed gain controller and NN-based controller. From Figure 15.21(d), one can observe that there is a reduction in the voltage drift by using NN-based AI-enabled controller. As shown in Figure 15.21, the NN-based control beats the PI-based control with better response to voltage drift effect. In case of sudden changes in insolation and temperature graph, NN-based control tracks MPP better with less voltage shift from voltage at MPP.

This section is a show case of various case studies, including the various permutations and combinations of solar insolation and temperature for a single module to PV array. The section comprises incorporation of the results with change in load with fixed climatic conditions, and change in load with variable climatic conditions.

15.6 CONCLUSION

This chapter represents a short review of IV and PV curve of PV module, the interconnection of module with various types of DC loads, and MPPT algorithms. This chapter presents MPPT-based PVES fed to DC load. The incorporation of MPPT system increases the system efficacy to ensure maximum power delivery to the load. The incorporation of PI controller tracks operating point around MPP in IV characteristics of PV. The system gives good results with change in climatic conditions and load as well. The issue of the drift in voltage waveform is reduced by incorporating an AI-based NN technique. This technique proves better as compared to conventional PI controller. The system has less transient drift compared to conventional controller. In order to maintain MPP and precise output, adaptive control such as NN is considered better as compared to fixed gain controller in varying climatic conditions. However, for a small system like single module, the fixed gain controller gives fair results with slowly varying climatic conditions. Apart from that, the P & O method is incorporated under slow variation in the climatic conditions, which provides better results. This system is best-suited for the incorporation of DC microgrid in standalone mode to increase the reliability of the system. This system finds a suitable place where grid connection is not possible, like hilly areas and remote locations.

REFERENCES

1. List of countries by electricity consumption https://en.wikipedia.org/wiki/List_of_countries_by_electricity_consumption
2. Algore Documentary Inconvenient Truth, Documentary by Mr. Algore.
3. World Energy Outlook 2019, IEA, Paris https://www.iea.org
4. Masters G. (2013) Renewable and Efficient Electric Power Systems. WI Publication.
5. Patel M. R., (2006) Wind and Solar Power Systems: Design, Analysis, and Operation, Second Edition.
6. Mukerjee, A. K., and Nivedita Thakur. (2011). Photovoltaic Systems: Analysis and Design. PHI Learning Pvt. Ltd.
7. Kollimalla, S. K. & Mishra, M. K. (2013). Novel adaptive P&O MPPT algorithm for photovoltaic system considering sudden changes in weather condition. In 2013 International Conference on Clean Electrical Power (ICCEP), IEEE, pp. 653–658.

8. Kollimalla, S. K. & Mishra, M. K. (2014). A novel adaptive P&O MPPT algorithm considering sudden changes in the irradiance. IEEE Transactions on Energy conversion, 29(3), pp. 602–610.
9. Gules, R., Pacheco, J. D. P., Hey, H. L. & Imhoff, J. (2008). A maximum power point tracking system with parallel connection for PV stand-alone applications. IEEE Transactions on Industrial Electronics, 55(7), pp. 2674–2683.
10. Kollimalla, S. K. & Mishra, M. K. (2014). Variable perturbation size adaptive P&O MPPT algorithm for sudden changes in irradiance. IEEE Transactions on Sustainable Energy, 5(3), pp. 718–728.
11. Koutroulis, E. & Blaabjerg, F. (2015). Overview of maximum power point tracking techniques for photovoltaic energy production systems. Electric Power Components and Systems, 43(12), pp.1329–1351.
12. Subudhi, B. & Pradhan, R. (2012). A comparative study on maximum power point tracking techniques for photovoltaic power systems. IEEE Transactions on Sustainable Energy, 4(1), pp. 89–98.
13. Esram, T. & Chapman, P.L. (2007). Comparison of photovoltaic array maximum power point tracking techniques. IEEE Transactions on Energy Conversion, 22(2), pp. 439–449.
14. Salas, V., Olias, E., Barrado, A. & Lazaro, A. (2006). Review of the maximum power point tracking algorithms for stand-alone photovoltaic systems. Solar Energy Materials and Solar Cells, 90(11), pp. 1555–1578.
15. Elobaid, L. M., Abdelsalam, A. K. & Zakzouk, E. E. (2015). Artificial neural network-based photovoltaic maximum power point tracking techniques: A Survey. IET Renewable Power Generation, 9(8), pp.1043–1063.
16. Messalti, S. (2015). A new neural networks MPPT controller for PV systems. In IREC2015 the Sixth International Renewable Energy Congress, IEEE, pp. 1–6.
17. http://powersimtech.com/wp-content/uploads/2013/04/Tutorial-Solar-Module-physical-model.pdf
18. Michael G. (2012). Design Calculations for Buck-Boost Converters. Texas Instruments Inc.
19. Joshi, S., Pandya, V. & Dhandhia, A. (2017). Simulation on MPPT based solar PV standalone system. PDPU Journal of Energy & Management, November 2017, pp. 35–42.

16 Deep Learning-Based Fault Identification of Microgrid Transformers

S. Poornima
Anna University, CEG Campus, Chennai, Tamil Nadu, India

CONTENTS

16.1 INTRODUCTION

Transformers form a vital part of AC as well as DC Power Systems (PSs). The microgrid (MG) structure has a group of interconnected industrial or commercial loads and distributed sources (renewable energy resources such as microturbines, photovoltaics, solar, and wind) with storage devices such as batteries and flywheels. The MG can operate in grid-connected or island mode to maintain continuity of supply. Also, the replacement of conventional generation by renewable sources has added economical, environmental and technical enhancements. The MG operation combined with information, telecommunication, measurement, and automation domains uplifted the system into a smarter, autonomous, and self-healing structure. On the other hand, MG suffers from poor short-circuit capacities as the conventional protection equipment still needs to be updated according to requirements. The fault states of

MG-integrated transmission or distribution systems get reflected as power flow fluctuations gradually in either direction, and thus the requirement of early identification of faults and appropriate protection of systems has become an emerging research area.

16.1.1 Artificial Intelligence in Power System

Artificial intelligence (AI) is a subset of computer science engineering that enables the programmed machine to mimic human learning. Any issues related to learning, forecasting, pattern classification and recognition, problem-solving and decision-making handled by human can be programmed in machine. This has led to evolution of expert systems (ESs), fuzzy logic (FL) systems, genetic algorithms (GAs), artificial neural networks (ANNs), etc. The efforts of these ESs eased the many constraints of optimization of generation and schedule, real and reactive power demands of generator, bus and transformers, forecasting of load and fault, economic load dispatch, protection of all PS equipment, etc.

16.1.1.1 Role of Artificial Intelligence in PS Operation

The complexity of integrated power networks has increased hugely because of unstable system expansion, reliability, dynamics and insufficient short-circuit current capacity. Irrespective of acquired data from conventional techniques, there is a necessity of remote monitoring of devices, real-time prediction of load and generation, identification and clearing of faults, and study of smart power grid integration. AI is being used to research on the solution of these problems. ES, ANN, FL and GA have individual applications on every phase of PS operation [1]. The application of AI techniques in the study of PS analysis is given in Table 16.1.

A plenty of research is being done using AI methods for optimizing generation cost, power flow, monitoring, and control, integration of nonconventional resources, economical investment, reliability, and security of PS.

16.1.2 Applications of Deep Learning in Power System

Deep learning (DL) is a subfield of machine learning (ML) to learn intuitive problems such as recognition of faces and recollection of spoken words or songs. The hierarchy of making a machine to learn complex concepts built as graph of experiences requires deep (many) layers in a neural network. Recognition of a speaker's accent can be done in a machine using DL. The 2D data initially applied as input pixels got featured into edges in the first hidden layer, corners in the second hidden layer, objects in the next layer and so on. The inherent feature extraction and classification of DL structures found many applications in medical images, PSs, communication systems, information sciences, etc.

The literature [7] reveals convolutional neural network (CNN), combined with K-means algorithms and Deep Belief Networks, does hourly load forecasting, while sequence-to-sequence-based long short-term memory (LSTM) architecture does minute and hourly demand forecasting based on smart meter data or web source. DL also predicts energy consumption enriched with weather data for a prototype system. DL has been used to estimate the states of insulators, transmission lines, transformers and generators. DL-based online fault diagnosis approach has been proposed using unlabeled online oil monitoring data and labeled dissolved gas analysis results.

TABLE 16.1
Application of AI Techniques

Technique	PS Applications
Expert systems [2]	• AC/DC network design, placement of capacitor, siting of power plant – *planning phase*. • Fault alarm and diagnosis, substation switching, daily and monthly load forecasting, dynamic security assessment, operational Energy Management System (EMS), maintenance scheduling, network restoration, management of real and reactive power on load and generation side, voltage control, load dispatch – *operation phase*. • Control system design, system protection coordination and control, text retrieval, condition monitoring and diagnosis, online safety measures – *analysis phase*.
Fuzzy logic [3]	• Generation expansion, system reliability – *planning phase.* • Security, load forecasting, voltage stability and management of reactive power control, dynamic generator rescheduling and maintenance, state estimation, system restoration and security assessment, fault diagnosis – *operation phase*. • Convertor control, control system designs, equipment condition monitoring – *analysis phase*
Artificial neural network [4, 5]	• Load forecasting – *planning phase.* • Optimal unit commitment and power flow, generator shutting, state estimation, security assessment, contingency analysis, fault detection and diagnosis, maintenance of substation equipment, system voltage stability assessment – *operation phase.* • Dynamic stability assessment, analogue and digital control of generator speed and voltage, identification of coherent dynamic equivalents, harmonic analysis, detection of bad/false data, system monitoring and equipment protection – *analysis phase.*
Evolutionary algorithms [6]	• Wind turbine location, generation expansion, network feeder routing, allocation of capacitor and sectionalizer, power flow optimization) – *planning phase*. • Unit commitment, valve point loading, evaluation of emission standard, hybrid plant coordination, load management and forecasting, contingency ranking, minimization of system power loss, maintenance scheduling, security assessment, alarm processing, fault diagnosis, system restoration, component control of Flexible AC Transmission Systems (FACTS) devices – *operation phase*. • Optimization of load flow, harmonic distortion and generator parameters, design of filter and stabilizer, hydrogenerator governor control, load frequency control – *analysis phase.*

CNN-based DL architecture is preferred to classify prefault, during fault and post-fault data of transformers. DL-based algorithms are proposed to identify the future threat of false data injection and power theft using phasor measurement units (PMUs) in practical studies. The self-adapting property of LSTM outperforms wind forecasting for wind farm generation. It also finds application in power quality assessment, automatic generation control (AGC) of interconnected grids, feeder restoration, electrical vehicle charging prediction, optimal commitment in photovoltaic plant, etc. To summarize, DL-based approaches may be implemented in smart grid for detection, prediction, control, optimization, and decision-making.

16.2 ML AND DL ARCHITECTURES OF FAULT IDENTIFICATION

The ML- and DL-based approaches can be effectively used for identification of nature, type and location of the fault. Normally, support vector machine (SVM), backpropagation neural network, FL, decision trees, K-nearest neighbors and extreme learning machines (ELMs) are proposed as ML techniques for fault analysis. These approaches outperform the conventional wavelet transforms, GAs, PMU approach, etc., either individually or in hybrid. Some of the popular architectures listed in Table 16.2 are suggested in literature for real-time implementation of fault detection system.

TABLE 16.2
DL Techniques for Power System Fault

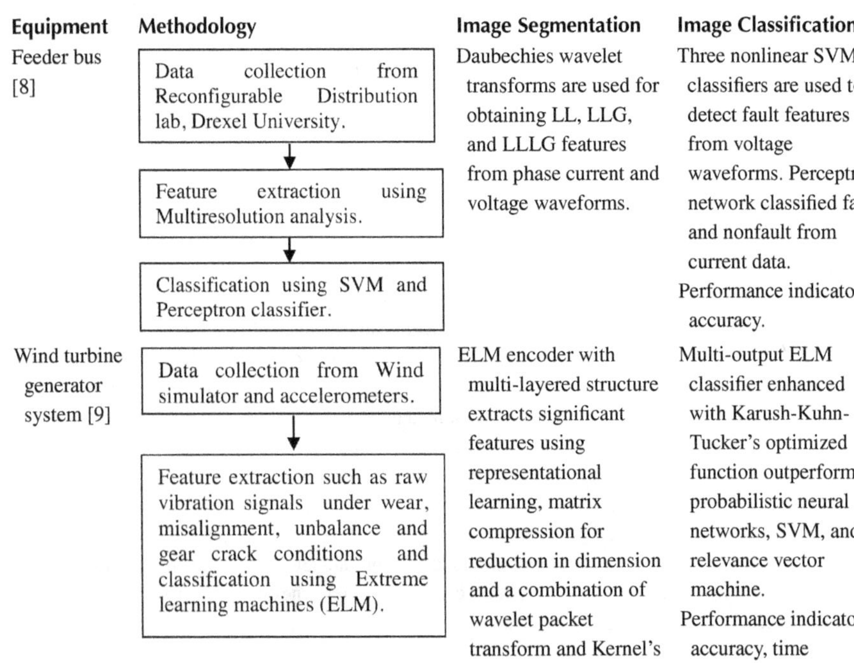

Equipment	Methodology	Image Segmentation	Image Classification
Feeder bus [8]	Data collection from Reconfigurable Distribution lab, Drexel University. → Feature extraction using Multiresolution analysis. → Classification using SVM and Perceptron classifier.	Daubechies wavelet transforms are used for obtaining LL, LLG, and LLLG features from phase current and voltage waveforms.	Three nonlinear SVM classifiers are used to detect fault features from voltage waveforms. Perceptron network classified fault and nonfault from current data. Performance indicator: accuracy.
Wind turbine generator system [9]	Data collection from Wind simulator and accelerometers. → Feature extraction such as raw vibration signals under wear, misalignment, unbalance and gear crack conditions and classification using Extreme learning machines (ELM).	ELM encoder with multi-layered structure extracts significant features using representational learning, matrix compression for reduction in dimension and a combination of wavelet packet transform and Kernel's principal component analysis (PCA).	Multi-output ELM classifier enhanced with Karush-Kuhn-Tucker's optimized function outperforms probabilistic neural networks, SVM, and relevance vector machine. Performance indicator: accuracy, time consumption.

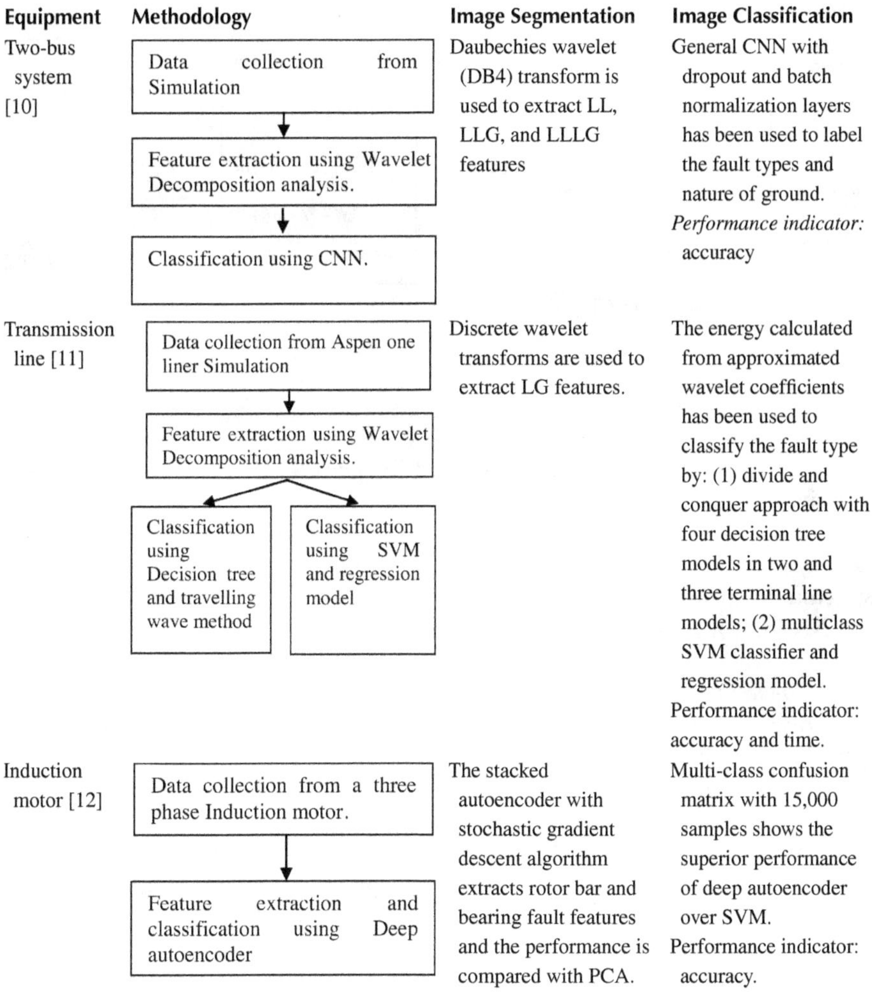

Equipment	Methodology	Image Segmentation	Image Classification
Two-bus system [10]	Data collection from Simulation → Feature extraction using Wavelet Decomposition analysis. → Classification using CNN.	Daubechies wavelet (DB4) transform is used to extract LL, LLG, and LLLG features	General CNN with dropout and batch normalization layers has been used to label the fault types and nature of ground. *Performance indicator:* accuracy
Transmission line [11]	Data collection from Aspen one liner Simulation → Feature extraction using Wavelet Decomposition analysis. → Classification using Decision tree and travelling wave method / Classification using SVM and regression model	Discrete wavelet transforms are used to extract LG features.	The energy calculated from approximated wavelet coefficients has been used to classify the fault type by: (1) divide and conquer approach with four decision tree models in two and three terminal line models; (2) multiclass SVM classifier and regression model. Performance indicator: accuracy and time.
Induction motor [12]	Data collection from a three phase Induction motor. → Feature extraction and classification using Deep autoencoder	The stacked autoencoder with stochastic gradient descent algorithm extracts rotor bar and bearing fault features and the performance is compared with PCA.	Multi-class confusion matrix with 15,000 samples shows the superior performance of deep autoencoder over SVM. Performance indicator: accuracy.

16.3 CNN AND CLASSIFIERS

CNN is an enhancement of multilayer perceptron developed based on a cat's visual cortex cell structure. The cells have sensitive receptive field to filter the edge and small and large locations of the visual pattern. The small filters connected as a subset of large ones generate strong nonlinear responses and hence encode the features of locally spatial inputs. This motivates the development of CNN, which followed initially Neocognitron and then LeNet designs.

16.3.1 CNN Architecture

Each layer of CNN is spatially three-dimensional (3D) and encodes image features inherently as depth using convolutional and subsampling layers. CNN is sequentially arranged by several convolutional and subsampling layers (3D) in feed-forward

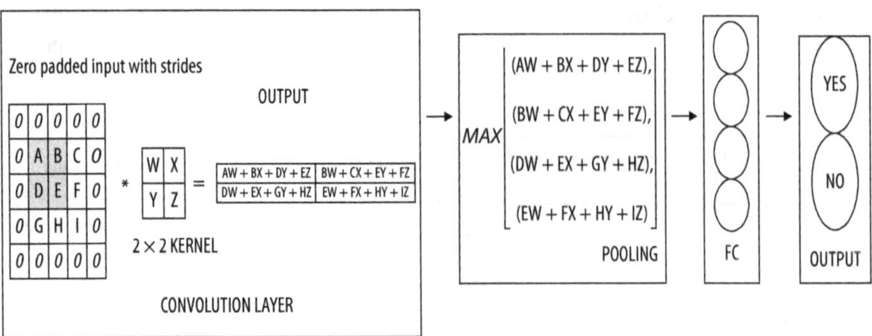

FIGURE 16.1 CNN architecture.

manner, followed by a final fully connected (FC) layer, which converts the 2D features into 1D data for the purpose of classification. The architecture accepts images of size 256 × 256, extracts features as segment using many hidden layers (>50) and forwards the output to the FC layer. A small modification of FC layer as classifier structure improves the performance further. The functions of basic building blocks of CNN (shown in Figure 16.1) are discussed as follows:

1. *Convolution layer*: The convolution operation is a discrete multiplication of pixel matrix, as shown in convolution block of Figure 16.1. The convolution between the weights of filter matrix and any spatial region of same size in nth layer gives activated hidden states of $(n + 1)$th layer. The special features of convolution layer [13] are as follows:
   - *Sparse interactions by choosing kernel size several times smaller than input image so that it requires less memory and improves efficiency.*
   - *The parameter-sharing feature enables the single set learning, as the weight of kernel is shared at all input values and becomes more efficient than product of a dense matrix in terms of storage.*
   - *Equivariant representation of output for a shift in input of an image vector eases edge detection in the neighbor layers too.*
2. *Strides*: The value by which the local receptive field (LCR) got shifted between the subregions (slowly at the center and faster at the edges) of the input image is called stride. Normally, strides of 1, 2, or 4 size are chosen for DL architectures. Larger the number of strides, better the receptive field feature and lesser the overfitting as the dimension of spatial region reduces.
3. *Padding*: The zero padding of convolution matrix chosen as valid or full or same convolution case makes the dimension of output vector equal to that of input. The information of edge pixels and overlapping border are not lost by using small kernels.
4. *Pooling layer:* The max-pooling is a nonlinear downsampling process that outputs a maximum value from the subregions of convoluted matrix. This reduces nonmaximal values computation in succeeding layers and provides a shifted invariance with increased robustness.

5. *Training*: The weights and biases of FC are equal to the number of LCR units. Training time of CNN is small as it has small number of weights. A 5-layer CNN needs 45 weights and 5 biases if LCR of size 3×3 has been chosen [14]. For larger images, training of each feature map is independent such that parallel training is possible to enable quick learning. The error between FC output and image label is backpropagated and repeated until the network completes the training.
6. *FC layer*: The FC has densely connected layers with one or two output neurons and uses linear or nonlinear activation function. It has the majority parameters than the convolutional layer. The hidden states of FC are connected to the previous spatial layer. The output layer of CNN depends on the requirement of application (regression or classification).

16.3.2 CNN Architecture with Pretrained Models

The layers of CNN extract edges, initially lines and then complex structures. The last convolution layers transferred the features to FC as segments with properly tuned weights according to specific application. This flexibility of CNN enables pretrained models in FC unit, which introduced transfer learning (TL). The FC unit of CNN with TL inclusions has been used for classification of 2D data. Different CNN structures derived for ImageNet Large Scale Visual Recognition Challenge which used for classification are briefed in Table 16.3.

TABLE 16.3
Popular CNN Architecture Used for 2D Classification

Pretrained Model	Architecture		Special Features
LeNet [15]	Convolutional layers	7	• Fixed input image size
	Pooling	Average	• Grouping and repetition of convolutional and pooling layers
	FC	2	• Increased number of filters with respect to network depth
	Activation function	Tanh (convolution) Softmax (output)	• Feature extraction and classification within architecture
AlexNet [16]	Convolutional layers	5	• Input image of size 227×227
	Pooling	Maximum	• First user of ReLU function
	FC	3	• Regularization using dropout between the FC layers
	Activation function	ReLU (convolution) and Softmax (output)	• Trained 650,000 neurons with 30 million parameters using two GPUs
	% Accuracy	84.75	• Direct connection between convolutional layers with two strides
	% Error	15.2	• Data augmentation

(Continued)

TABLE 16.3 (Continued)

Pretrained Model	Architecture		Special Features
GoogleNet [17]	Convolutional layers	22	• Input image of size 224 × 224 • Development and repetition of the inception module. • Used many 1 × 1 convolution kernels to reduce the number of channels. • 6.4 million parameters to train • Multiple error feedbacks in the network
	Pooling	Maximum	
	FC	3	
	Activation function	3 (2 ReLU, 1 Softmax)	
	% Accuracy	93.30%	
	% Error	6.67	
ResNet [18]	Convolutional layers	152	• Input image of size 224 × 224 • Insertion of skip links to handle gradient loss • Repeated use of residual blocks • 60.3 million parameters to train
	Pooling	Average	
	FC	5	
	Activation function	Softmax	
	% Accuracy	95.51%	
	% Error	3.6	

16.4 CNN FRAMEWORK FOR FAULT IDENTIFICATION

The deep layers tuned properly to extract features inherently enhance classification performance of 2D data. The pretrained models connected to the FC layer of CNN improve the system accuracy. AlexNet (AN) and GoogleNet (GN) are chosen in this work as classifier as only 120 voltage and current images are chosen as input. The input image size is initially modified according to classifier requirement. The features extracted in the convolution and pooling layers of classifiers are fed as input to the hidden layers of FC. The FC output after activation is classified into fault and nonfault classes. The performance is compared in terms of accuracy, sensitivity and precision among classifiers.

16.4.1 Dataset

The conversion of traditional distribution generation into an active network and its integration with grid develop more technical difficulties. Those can be operation, stability, estimation, protection or restoration phases of integrated systems. The modern research of PSs helps us to develop an automatic and smarter network that does not depend on human decision. Focus is being given to protection and stability of MG-integrated utility system using automation and intelligent techniques. The objective is to identify the fault condition of a wind farm transformer from the three-phase voltages and currents. The collection and processing of data has been explained to understand the fault classification using DL technique.

16.4.1.1 Microgrid Operation

Wind turbine systems organized as a wind farm are the backbone of renewable generation in India. It generated 37.699 GW with its installed capacity nationwide, as the state grids encourage this through proper guidelines to the developers. The operation on wind farm has influencing factors such as location, nodal voltages, real and reactive power limits, point of common coupling, and power losses.

The back-to-back converter enables power flow from grid to rotor during low wind speeds and vice versa at high wind speeds. Rotational speed of doubly fed induction generator (DFIG) determines the direction of power flow between grid and rotor through slip rings. Thus, DFIG configuration shown in Figure 16.2 enables power flow during sub-synchronous, synchronous, and super-synchronous modes, making the generation unique among other conventional system. The rotor absorbs and delivers active power through converters at sub-synchronous and super-synchronous speeds, respectively. The AC energy converted and filtered by the converter system is fed to transformer to step up the voltage to the utility level.

16.4.1.2 Preprocessing

As wind energy is an uncontrollable source, the study on voltage stability and transient stability of integrated grid system is necessary. A utility connected wind farm is considered for generation of fault signals in this work using Simulink [19]. The wind farm consists of six wind turbines rated to generate 1.5 MW each at 575 V. The Simulink model shown in Figure 16.3 has DFIG-configured wind energy conversion system (WECS) integrated with 60-Hz grid. The system is designed to deliver maximum energy by optimizing mechanical losses, power quality and turbine speed, which depends on wind speed. In the simulation, the wind speed is maintained at 15 m/s to have 1.2 pu turbine speed and 0 MVAr power using torque controller system. The wind farm is connected to a 575 V/25 kV transformer, which in turn is connected to 30-km feeder. The other end of the feeder is connected to a 25 kV/120 kV transformer, which integrates the WECS to 120-kV grid. The short-circuit fault is simulated for 0.2 s between the phase and neutral terminals of both the transformer terminals. The voltage and current of all the phases are stored for line-to-ground (LG), line-to-line (LL), and three-phase-to-ground (LLLG) faults. The fault is created at 0.017 s and cleared at 0.083 s of simulation duration.

All the amplitudes of voltage and current waveforms are plotted in per unit (PU) for all types of fault. The dataset consists of 160 fault images obtained under all abnormal conditions and 8 nonfault images. Table 16.4 shows all the types of short circuit voltages and currents at 25 kV busbar. The voltage and current images are normalized and

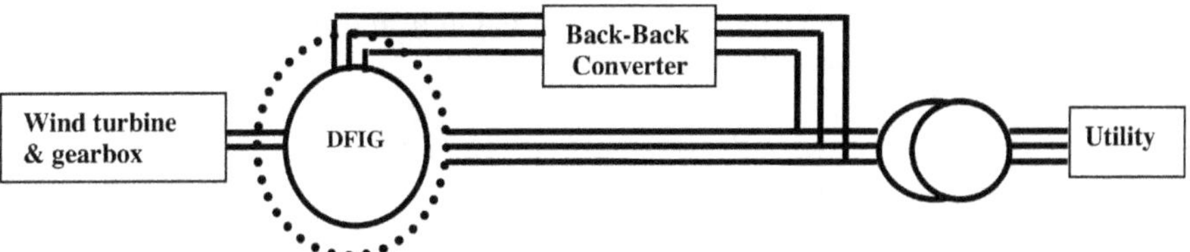

FIGURE 16.2 DFIG configuration of wind farm.

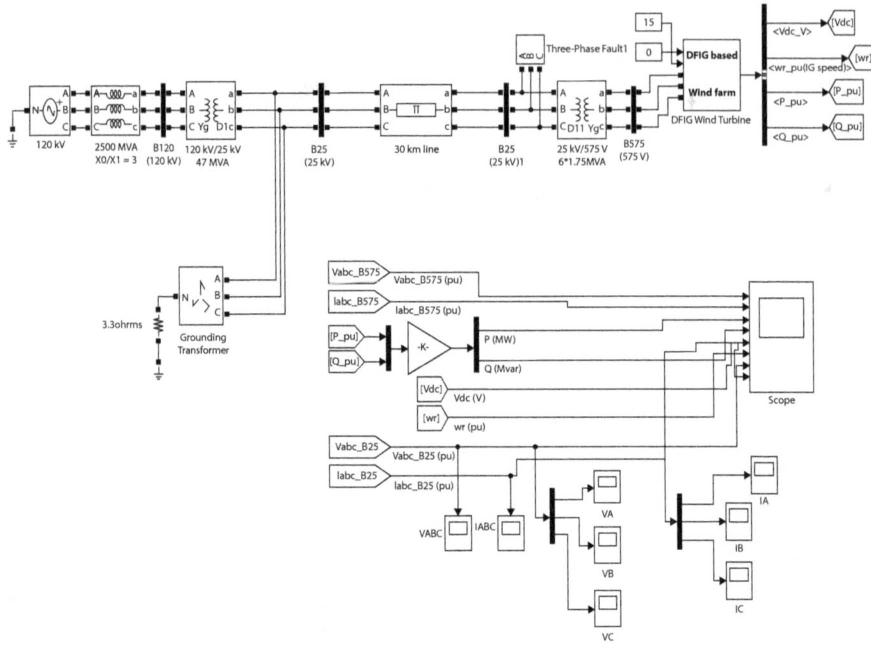

FIGURE 16.3 Simulink model of microgrid.

resized according to CNN structures. The architecture of AN accepts input image of size 227 × 227 × 3, while GN requires input image of size 224 × 224 × 3.

16.4.2 MATLAB Implementation

The following algorithm is implemented in MATLAB to evaluate the performance of CNN network with and without classifier nets. The dataset developed using wind farm configuration of Simulink model is fed to CNN input layer. The training parameters are set such that the test features are validated with good accuracy with minimum time consumption. The test features extracted at the FC layer are fed to AN and GN architecture to obtain optimum performance.

- *Step 1: Load the fault image dataset obtained using Simulink model of MG.*
- *Step 2: Count each label to count the number of categorized images.*
- *Step 3: Balance the split image folders to be fit for training and validation.*
- *Step 4: Load the CNN layers with strides, padding and pooling information.*
- *Step 5: Set the training parameters such as learning rate, bias, gradient method and epoch.*
- *Step 6: Pass the extracted test features to the last FC layer of CNN.*
- *Step 7: Replace output layers of CNN with the pretrained model.*
- *Step 8: Train the network for the classification using training options.*
- *Step 9: Peformance of the network can be measured in terms of confusion matrix (CM) parameters.*

TABLE 16.4
MG Fault Voltages and Currents

Fault	Three-Phase Voltages (pu)	Three-Phase Currents (pu)
LLG		
LG		

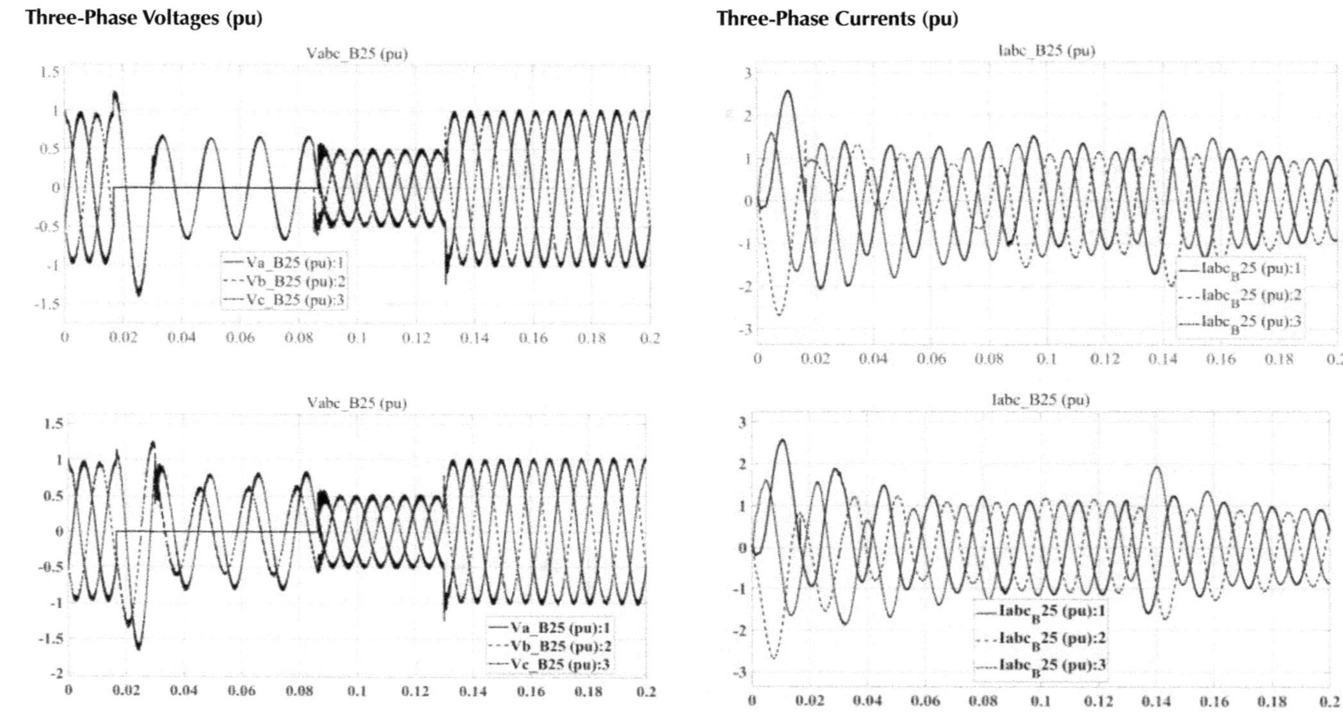

(Continued)

TABLE 16.4 (Continued)

Fault	Three-Phase Voltages (pu)	Three-Phase Currents (pu)
Three-phase		
LL		

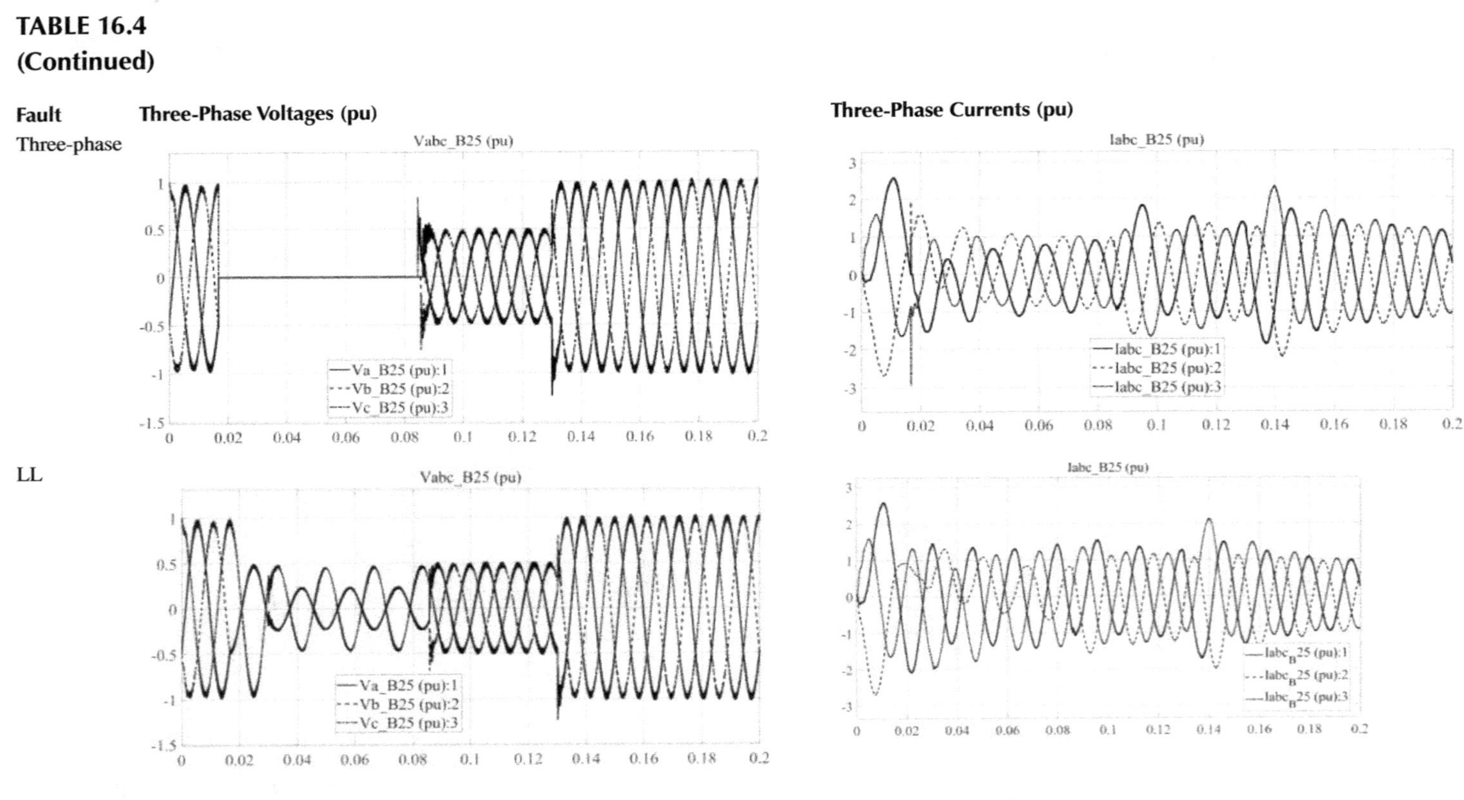

16.4.3 Performance Assessment

The CNN network has been trained with the preprocessed and normalized dataset. Two convolution layers with three kernels of size 3 × 3 and one stride have been considered. Batch normalization layers are located between convolution and maximum pooling layer. Stochastic gradient descent with momentum is chosen as solver, and learning rate is 1e-04. The activation functions ReLU and Softmax are used in the FC layers. The training of CNN layers achieved 94.12% accuracy with minimum loss. The inherent feature extraction and classification of chosen CNN structure has been successful after 30 epochs with 95.24% accuracy for the chosen data, as shown in Figure 16.4. For the AN and GN classifiers, the learning rate for bias and weight is chosen as 10 to achieve good learning characteristics. A mini-batch of size 8 and 55 image data has been chosen for AN and GN, respectively, for training the dataset. The dataset has been trained and validated on 80:20 ratio in both the classifiers. The basic AN structure with dropout and cross channel normalization features trained the input data to achieve 93.44% in 3 minutes. However, the training of GN classifier reached 100% in 301 s with same training options of CNN, as shown in Figure 16.5.

The classification performance of the CNN with and without classifiers has been compared in terms of CM parameters. The rows of CM indicate the output, while the targets are indicated by columns. The main diagonal gives the percentage of outputs matched with targets, whereas the off-diagonal elements give the mismatched values.

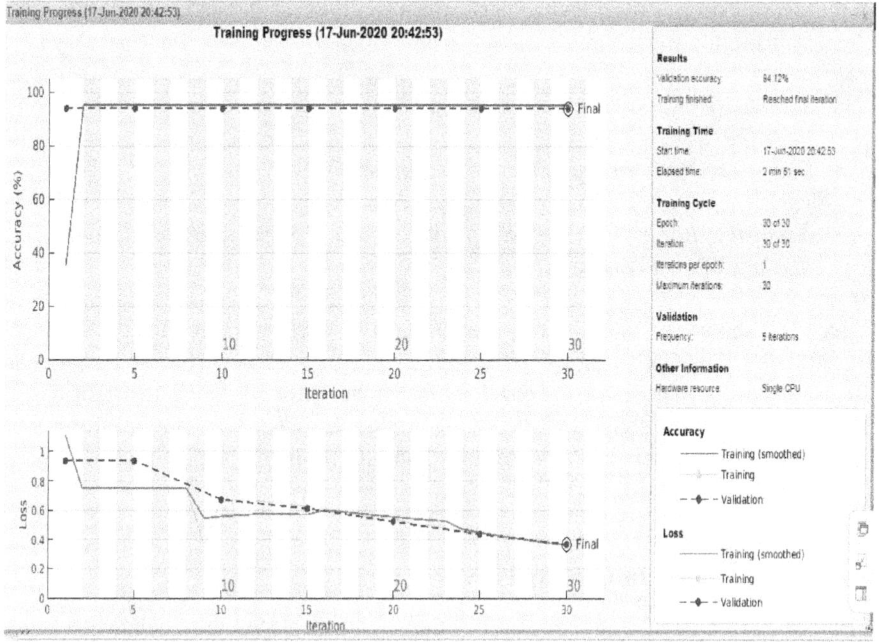

FIGURE 16.4 Training window of CNN.

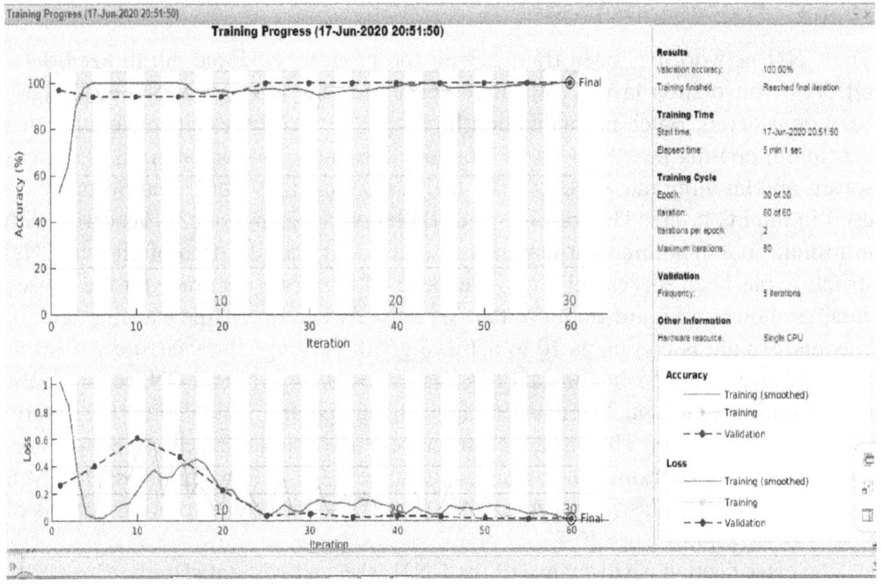

FIGURE 16.5 Training progress of GN.

The rightmost column gives the percentage of predicted values and the last row gives the true category percentage. The last element of the CM gives the overall accuracy of the classification. The following performance parameters are calculated using CM elements as follows:

- *Sensitivity: Ratio of true positive to the sum of true positive and false negative.*
- *Precision: Ratio of true positive to the sum of true positive and negative values.*
- *Accuracy: Ratio of sum of true positive and negative values to the sum of all the CM elements.*

Table 16.5 shows the classification comparison of fault and nonfault data of MG transformers. The CNN and AN have appreciable changes in performance between 10 and 30 epochs. There are no considerable changes after 30 epochs in all the cases. The obtained results show that the GN is precise and sensitive in its performance with maximum efficiency. The quantity of dataset limits the application of ResNet in this work.

16.5 REAL-TIME PROPOSAL

The smart grid employs automatic protection and control system to handle decision-making without human interventions. In the future, an AI technique-based intelligent protection relay will be employed in a substation for fast restoration of system operation.

TABLE 16.5
Fault Classification Performance

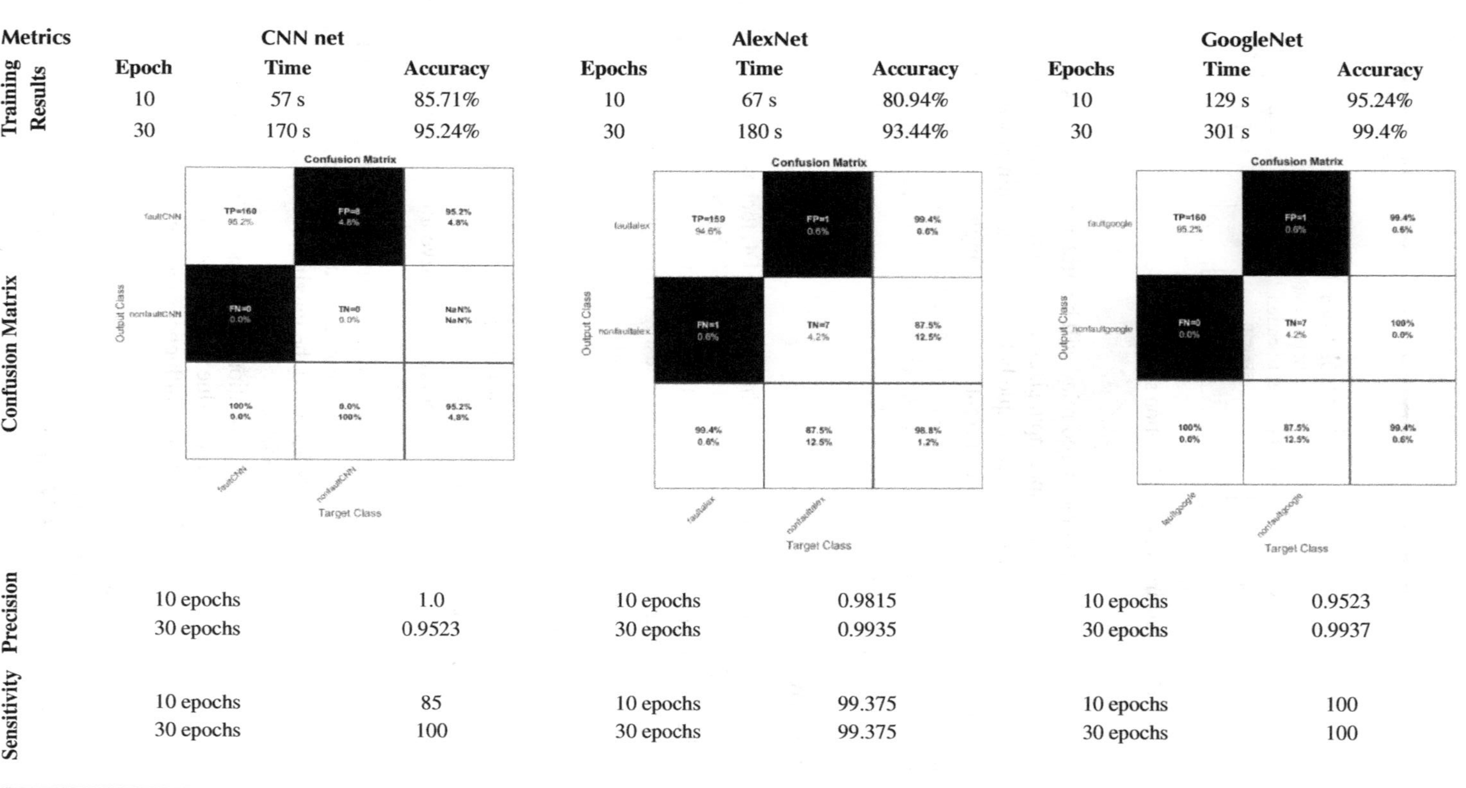

Metrics	CNN net			AlexNet			GoogleNet		
Training Results	**Epoch**	**Time**	**Accuracy**	**Epochs**	**Time**	**Accuracy**	**Epochs**	**Time**	**Accuracy**
	10	57 s	85.71%	10	67 s	80.94%	10	129 s	95.24%
	30	170 s	95.24%	30	180 s	93.44%	30	301 s	99.4%
Confusion Matrix	Confusion Matrix: TP=160 95.2%; FP=8 4.8%; 95.2% 4.8%; FN=0 0.0%; TN=0 0.0%; NaN% NaN%; 100% 0.0%; 0.0% 100%; 95.2% 4.8% (Output Class: faultCNN, nonfaultCNN; Target Class: faultCNN, nonfaultCNN)			Confusion Matrix: TP=159 94.6%; FP=1 0.6%; 99.4% 0.6%; FN=1 0.6%; TN=7 4.2%; 87.5% 12.5%; 99.4% 0.6%; 87.5% 12.5%; 98.8% 1.2% (Output Class: faultalex, nonfaultalex; Target Class: faultalex, nonfaultalex)			Confusion Matrix: TP=160 95.2%; FP=1 0.6%; 99.4% 0.6%; FN=0 0.0%; TN=7 4.2%; 100% 0.0%; 100% 0.0%; 87.5% 12.5%; 99.4% 0.6% (Output Class: faultgoogle, nonfaultgoogle; Target Class: faultgoogle, nonfaultgoogle)		
Precision	10 epochs		1.0	10 epochs		0.9815	10 epochs		0.9523
	30 epochs		0.9523	30 epochs		0.9935	30 epochs		0.9937
Sensitivity	10 epochs		85	10 epochs		99.375	10 epochs		100
	30 epochs		100	30 epochs		99.375	30 epochs		100

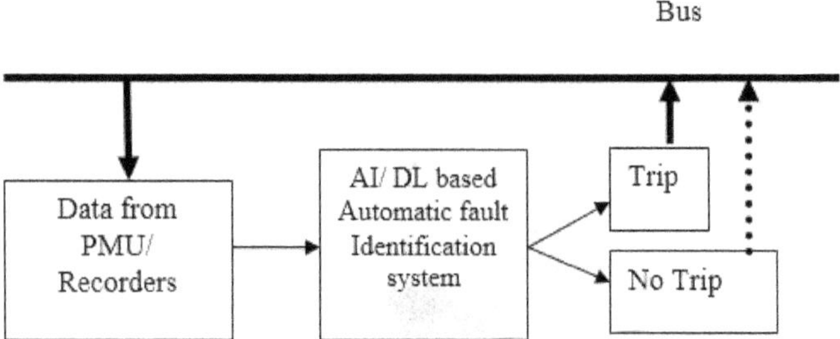

FIGURE 16.6 Smart fault identification system.

The advanced DL technique-based relay will be preferred due to its accuracy, adaptability, management of variable input sizes, and inherent features. A real-time implementation of a relay with DL technique helps us to maintain power quality, stability, and continuity. The voltage or current and phase angle data recorded in a PMU or digital recorders may feed the DL-based automatic fault detection system, as shown in Figure 16.6. The decision based on AI system output helps us to recover quickly the normal operation of system.

16.6 SUMMARY

The proposed work is done to highlight the use of DL techniques, which can be used for smart fault detection systems in PSs. A DFIG-based wind farm configuration has been simulated with all types of faults at its transformer's terminals. The fault voltages and current signals (recorded using PMU practically) under normal and abnormal conditions of MG have been processed into an image dataset. The CNN structure with and without TL extracted the features and their classification has the summary as follows:

- The GN confusion matrix identified the true positive and negative values appropriately than the others.
- The GN framework is more accurate, sensitive and precise than the others in the classification performance.
- The training of GN reached maximum efficiency at the cost of time. The CNN without TL achieved 3% more accuracy with extra 66 s than that of 10 epochs. However, GN achieved the maximum accuracy of AN and CNN (after 30 epochs) in 10 epochs.
- The time to perform classification increases with increase in number of iterations. The best accurate value is accomplished with time limitations as the depth of classifier network influences the both.

It has been concluded that GN gives enhanced accuracy in identification of fault data. It has been inferred that GN performed the classification more precisely. Although the CNN is also sensitive as GN, the latter becomes superior by its enhanced structure.

While this work explored CNN architecture and TL approaches for fault images, individual fault type classification of the same dataset can also be done further. The TL architecture for the classification of fault belongs to other MG components can be investigated in the future.

REFERENCES

1. Kevin Warwick, Arthur O. Ekwue & Raj Aggarwal. (1997). Artificial Intelligence Techniques in Power Systems. London: The Institution of Electrical Engineers.
2. G. Bretthauer, E. Handschin & W. Hoffmann. (1992). Expert systems application to power systems - State-of-the-art and future trends, IFAC Proceedings Volumes, 25(1), 463–468.
3. I.H. Altas & J. Neyens. (2006). A fuzzy logic load-frequency controller for power systems, in: International Symposium on Mathematical Methods in Engineering, Cankaya University, Turkey, 1–10.
4. Biswarup Bhattacharya & Abhishek Sinha. (2017). Intelligent fault analysis in electrical power grids, in: IEEE 29th International Conference on Tools with Artificial Intelligence (ICTAI), Boston, MA, 985–990, doi: 10.1109/ICTAI.2017.00151.
5. Anamika Yadav & Yajnaseni Dash. (2014). An overview of transmission line protection by artificial neural network: Fault detection, fault classification, fault location, and fault direction discrimination, Advances in Artificial Neural Systems, 2014(230382). http://dx.doi.org/10.1155/2014/230382
6. Nadil Amin, Aby Quaser Marowan & B. Chandra Ghosh. (2017). Application of genetic algorithm in power system optimization with multi-type FACTS, International Journal on Scientific and Research Publications, 5(5), 748–754.
7. Dongxia Zhang, Xiaoqing Han & Chunyu Deng. (2018). Review on the research and practice of deep learning and reinforcement learning in smart grids, CSEE Journal of Power and Energy Systems, 4(3), 362–369.
8. Nicholas S. Coleman, Christian Schegan & Karen N. Miu. (2015). A study of power distribution system fault classification with machine learning techniques, in: North American Power Symposium (NAPS), Charlotte, NC, 1–6, doi: 10.1109/NAPS.2015.7335264.
9. Zhi-Xin Yang, Xian-Bo Wang & Jian-Hua Zhong. (2016). Representational learning for fault diagnosis of wind turbine equipment: A multi-layered extreme learning machines approach, Energies, 9(379), 1–17. doi: 10.3390/en9060379.
10. Florian Rudin, Guo-Jie Li & Keyou Wang. (2017). An algorithm for power system fault analysis based on convolutional deep learning neural networks, International Journal of All Research Education and Scientific Methods, 5(9), 11–18.
11. Vidya Venkatesh. (2018). Fault classification and location identification on electrical transmission network based on machine learning method, MS thesis, Virginia Commonwealth University. https://scholarscompass.vcu.edu/etd/5582.
12. Shokoofeh Zare. (2018). Fault detection and diagnosis of electric drives using intelligent machine learning approaches, Electronic thesis and dissertations 7436, University of Windsor. https://scholar.uwindsor.ca/etd/7436.
13. Ian Goodfellow, Youshua Benigo & Aaron Courville. (2016). Deep Learning. Cambridge: MIT Press.
14. Sandro Skansi. (2018). Introduction to Deep Learning: From Logical Calculus to Artificial Intelligence. Cham: Springer International Publishing AG.
15. Y. Lecun, L. Bottou, Y. Bengio & P. Haffner. (1998). Gradient-based learning applied to document recognition, Proceedings of the IEEE, 86(11), 2278–2324.

16. Alex Krizhevsky, Ilya Sutskever & Geoffrey E. Hinton. (2012). ImageNet classification with deep convolutional neural networks, in: 25th International Conference on Neural Information Processing Systems – Volume 1 (NIPS'12), New York, NY, 1097–1105.
17. Christian Szegedy, Wei Liu, Yangqing Jia, Pierre Sermanet, Scott Reed, Dragomir Anguelov, Dumitru Erhan, Vincent Vannhoucke & Andrew Rabinovich. (2015). Going deeper with convolutions, in: IEEE Conference on Computer Vision and Pattern Recognition (CVPR), Boston, MA, 1–9, doi: 10.1109/CVPR.2015.7298594.
18. K. He, X. Zhang, S. Ren & Sun J. (2016). Deep residual learning for image recognition, in: IEEE Conference on Computer Vision and Pattern Recognition (CVPR), Las Vegas, NV, 770–778, doi: 10.1109/CVPR.2016.90.
19. https://www.mathworks.com/company/user_stories/hydro-qubec-models-wind-power-plant-performance.html

17 Power Quality Improvement for Grid-Integrated Renewable Energy Sources

A Comparative Analysis of UPQC Topologies

Nirav Karelia, Amit Sant, and Vivek Pandya
Department of Electrical Engineering, School of Technology, Pandit Deendayal Energy University, Gandhinagar, Gujarat, India

CONTENTS

17.1 INTRODUCTION

Due to development in semiconductor integration technologies, tremendous growth has been observed in the use of power electronic converters at distribution and utilization levels [1]. It is almost impossible to imagine any gadget, appliance, equipment, or machine that does not use power electronic converters in one or the other form. The frequent use of power electronic converters has introduced a huge amount of pollution in terms of voltage and current waveform distortion, which has led to huge amount of harmonics in the power system [2]. Many types of power electronic converters use high-frequency switching, which also injects a considerable amount of switching harmonics into the system [3]. In addition to this, nonlinearity of electrical machines and equipment also introduces harmonic pollution into the power system [4, 5]. This harmonic pollution results in substantial increment in the losses and hence in poor efficiency of the energy conversion [6]. Again, higher losses translate into higher operating temperature than the design, which reduces the life expectancy of the insulating material and ultimately results in equipment failure, thereby incurring huge economical losses [7].

Due to issues of environmental pollution and greenhouse gas emission, immense shift has been observed in energy generation through renewable sources [8]. Renewable energy sources are linked to the grid via power electronic interface in most of the cases, which contributes to harmonic pollution of the grid at source [9, 10]. The harmonic pollution at grid level results in higher losses, derating of the power transmission capabilities and harm to the equipment to which the source supplies electric power [11, 12]. These growing concerns over harmonic pollution in the power system result in more losses, resulting in poor operating efficiency and lesser life expectancy of equipment and machines, thereby leading to net financial losses and maloperation of digital instrumentation that ends up in faulty measurements or false tripping in worst cases [13].

As already mentioned, due to tremendous development in the large-scale integration of semiconductor chip, the data, and signal processing speeds have exponentially increased in the last couple of decades. The power electronic converters that produce harmonics can similarly mitigate the same with the help of high-speed signal processors, known as active power filters [14]. Active power filters have become a mature technology for harmonic and reactive power compensation of AC power networks with high penetration of nonlinear loads [15]. At distribution and utilization level, active power filters are implemented when the harmonic spectra are not fixed and are varying due to many parameters [16]. Again, there is a variety of classifications according to the configuration and application. Mainly, there are three variations in topological configuration, namely, series, shunt, and series-shunt, where series-shunt is a combination of series and shunt active filter [17]. One variety of series active filter is known as dynamic voltage restorer, which mainly deals with abnormalities such as voltage sag and swell, flickers, and harmonics in the source voltage [18]. Shunt active filter, however, deals with harmonic mitigation in load current as well as reactive power compensation and voltage stability [19]. Active power filters are also classified on the bases of topology, supply system, power rating, power to be compensated, etc. [20]. Although active power filters have many merits such as comprehensive power quality solutions, they also have demerits such as high power rating, huge cost, and sluggish response in some cases.

In unified power quality conditioner (UPQC), series active filter and shunt active filter are connected through DC bus with shunt capacitor, which is capable of handling almost all power quality-related issues [21]. There are many ways to classify the UPQC configuration, which will be dealt well in detail in the chapter. This chapter mainly focuses on UPQC-MC (multiconverter), UPQC-DG (distributed generation), and UPQC-I (interline), and compares as well as analyzes them for their operation when multiple sources are connected to enhance reliability of supply to the critical load in detail with respect to various power quality parameters, control strategy, operating complexities, power rating, and economical aspects.

17.2 INTRODUCTION TO UPQC

As already discussed in Introduction, UPQC is a combination of series active filter and shunt active filter connected through a common DC link capacitor [22, 23]. UPQC looks after majority of power quality issues, ranging from abnormalities of voltage supply, such as voltage sag, swell, flicker, and harmonic distortions, to current harmonic mitigation, reactive power compensation, and voltage stability on the load end. The schematic arrangement of UPQC is depicted in Figure 17.1. UPQC follows

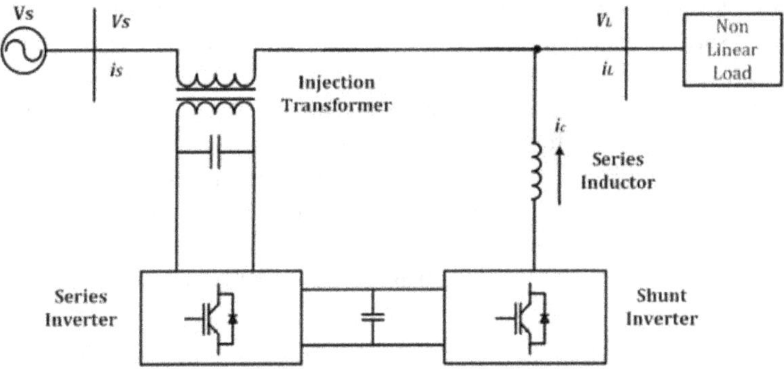

FIGURE 17.1 Schematic diagram of UPQC.

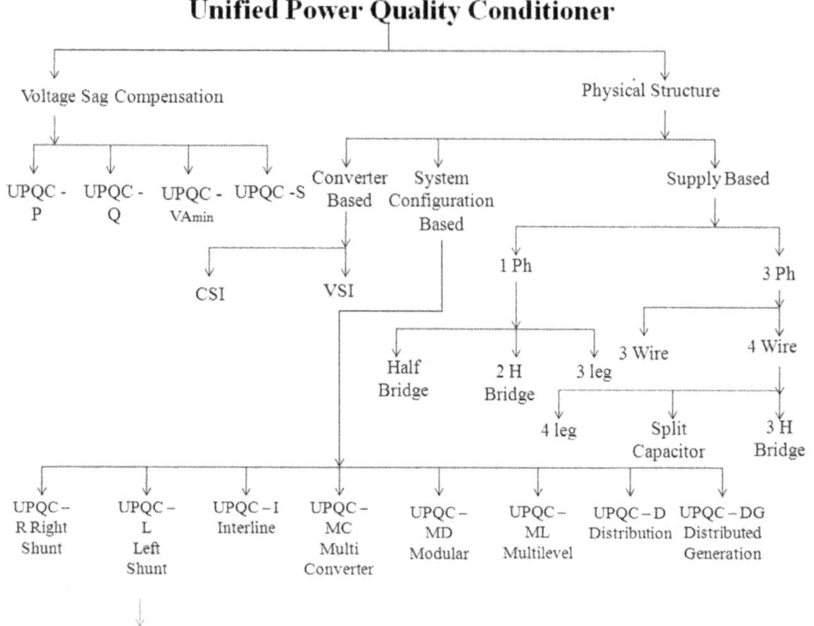

FIGURE 17.2 Classification of UPQC.

IEEE 519 power quality standards and maintains total harmonic distortion (THD) less than 5% in the load voltages and source currents, in addition to maintaining unity power factor operation. It is worthy to note that majority of the market versions of UPQC provide harmonic mitigation of almost less than 1% of % THD.

There is a variety of configurations possible for UPQC according to the applications, power rating, supply system, and topology [24]. As shown in Figure 17.2, UPQC is classified broadly into two categories, namely, physical structure and voltage sag compensation strategy. In the physical structure, it is further classified into three types on the bases of converter topology, supply system, and system configuration. There are two types converter topology: voltage source inverter and current source inverter. In supply system-based classification, there are two further categories: first is single phase with either two H-bridge or three-leg topology or half bridge; and second is three-phase with either three-phase three-wire or three-phase four-wire system. In the three-phase four-wire system, there are again three variations, namely, four-leg, split capacitor, and three H-bridges. In the system configuration-based classification, there are mainly eight variants, which are UPQC-R (right shunt), UPQC-L (left shunt), UPQC-I (interline), UPQC-MC (multiconverter), UPQC-MD (modular), UPQC-ML (multilevel), UPQC-D (distributed), and UPQC-DG (distributed generation) [25].

When any renewable source is operating in islanding mode supplying load through UPQC, it takes care of voltage sag, swell, flicker, and harmonics as well as load current harmonics, load voltage profile, and reactive power compensation of any nonlinear load [26]. The major lacuna of UPQC in such arrangement is its inability to compensate for voltage interruption from the source, as a result of which the load has to suffer blackout condition. Similar situation occurs when the load is fed electric power from a grid: if the grid fails, the load suffers from interruption. Even when there are multiple sources, such as combination of grid and one renewable energy source or multiple renewable energy sources supplying the power to the load with one of the sources acting as main and sharing full load and another is kept redundant or both renewable sources are supplying power to the critical load. In case of failure of main source in the first case or one renewable source in the latter case, the changeover to the other source takes some time to transfer, which may be highly undesirable for sensitive critical loads. Four of the UPQC configurations, namely, UPQC-MC, UPQC-DG with synchronous generator (SG)/induction generator (IG) as DG (UPQC-DG (SG/IG)), UPQC-DG with photovoltaic (PV) /fuel cell (FC)/ battery energy storage system (BS) as DG (UPQC-DG (PF/FC/BS)), and UPQC-I address such situations and provide solution for continuous power delivery in case of interruption of the main source, along with addressing all power quality issues.

17.3 UPQC-MULTICONVERTER (UPQC-MC)

The typical schematic arrangement of UPQC-MC configuration is shown in Figure 17.3(a), where two series active filters and one shunt active filter are present. In the main feeder feeder 1, the series active filter is formed with series inverter 1 (VSC1) and shunt active filter is formed with shunt inverter (VSC2), which are linked through a common DC bus and a capacitor C which forms UPQC for feeder 1 and supply to critical load L1. Here, load L1 is a nonlinear critical load that requires uninterrupted pure sinusoidal voltage supply, which must be fully protected against sag/swell and distortion for its proper operation, while its current is nonsinusoidal and contains harmonics. In feeder 2, series inverter 2 (VSC3) acts as series active filter, which is connected to the sensitive load L2 [27]. The load L2 is a sensitive load that requires sinusoidal voltage without sag/swell and harmonics. The series inverter 2 (VSC3) is connected to the same DC link capacitor C to which series inverter 1 (VSC1) and shunt inverter (VSC2) of feeder 1 are connected. Here, the objective of UPQC-MC is (1) to control the load voltage V_{Ll} and to prevent harmful effects of sag, swell, disturbance, and interruption in the system to the critical nonlinear load L1; (2) to control the load voltage $\mathrm{V_{L2}}$ and to prevent harmful effects of sag, swell and any disturbance in the system to the sensitive load L2; and (3) to compensate reactive power and load current harmonics produced by the nonlinear load L1. To achieve these objectives, series active filters (VSC1 and VSC3) operate as voltage controller and shunt active filter (VSC2) operates as current controller [28].

Here, both the feeders may be fed through grid, or the main feeder may be fed through grid and the other may be fed by renewable source, or both the feeders may be fed through renewable energy sources. The UPQC connected to feeder 1 looks after voltage-related issues such as sag, swell, flicker, and harmonics from main source V_{s1} and load current harmonics and reactive power compensation for critical load L1 [29]. Series active filter 2 connected to feeder 2 looks after voltage-related abnormalities such as sag, swell, flicker, and harmonics from source V_{s2}.

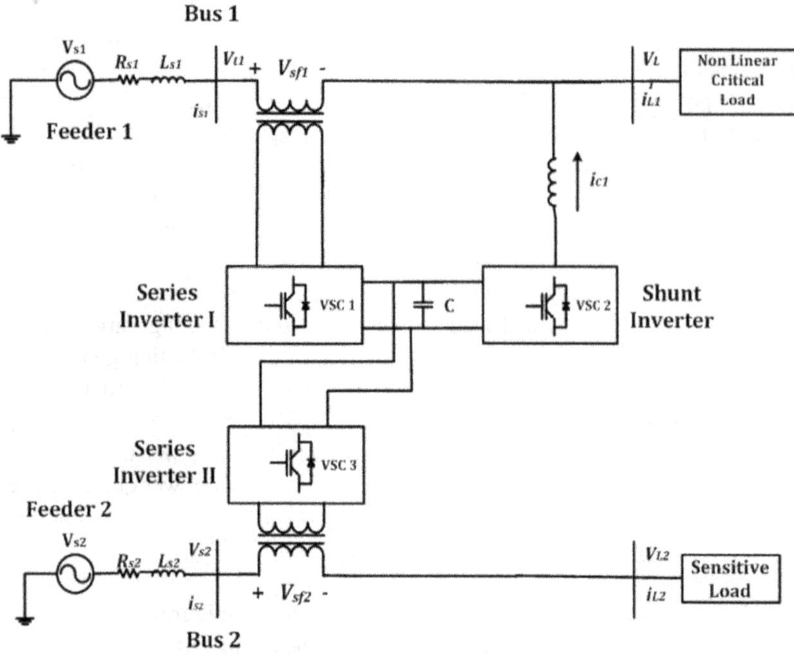

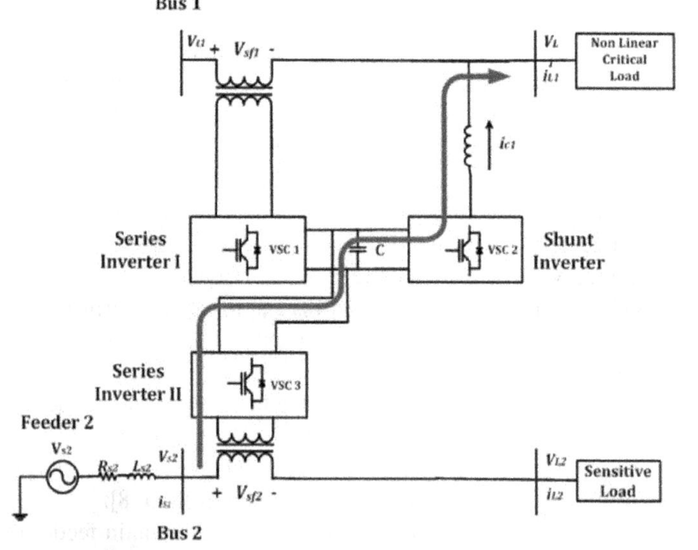

FIGURE 17.3 (a) Schematic arrangement of UPQC-MC (multiconverter). (b) Voltage supply in case of interruption at feeder 1.

In case of voltage supply failure of main feeder 1, as shown in Figure 17.3(b), the critical load L1 will be supplied power through series active filter 2 (VSC3) of feeder 2 and shunt active filter (VSC2) of feeder 1 through the common DC link capacitor C.

17.4 UPQC-DISTRIBUTED GENERATION

In UPQC-DG, the main feeder is supplying the critical load L through UPQC, as shown in Figure 17.4.

The series active filter (VSC1) mitigates source voltage V_s-related power quality issues such as sag/swell, flicker, and harmonics whereas shunt active filter (VSC2) tackle load current harmonics and reactive power compensation produced by critical load L. There can be different types of distributed generation sources, which can be directly connected to a common DC link of the UPQC through appropriate power electronic interface. In a broader classification, two types of categories of distributed generators can be assumed.

The first category may cover all renewable energy sources that produce DC electricity, namely, PV panels, FC, BS, etc. The other category may cover all generating sources that produce AC electricity with the help of either SG or IG. In case the voltage supply is interrupted in the main feeder, distributed generators can directly inject active power into the DC link and this injected power will be supplied to the critical load L through shunt active filter (VCS2), which will convert this DC power into required AC power. In the case of UPQC-DG, two modes of operations are possible: (1) interconnected mode of operation and (2) islanding mode of operation [30]. In *interconnected mode* of operation, there are further submodes of operation. One submode is known as *forward flow mode*, where the main source V_s supplies

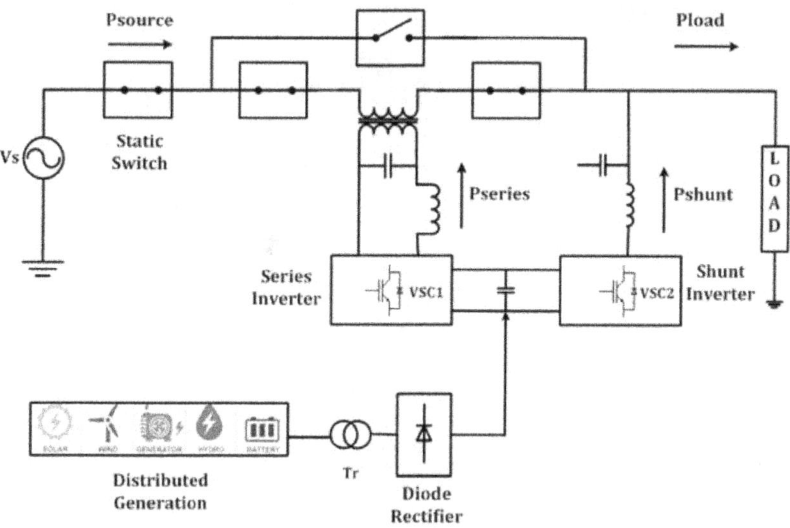

FIGURE 17.4 Schematic diagram of UPQC-DG (distributed generation).

the power to the load L and DG also feeds active power to the critical load L via shunt active filter (VSC2). In another submode known as reverse-flow mode, DG injects power to the load via shunt active filter together with the main source, and also injects power to the main source via DC link and series active filter (VSC1), as shown in Figure 17.5(a) [31]. In islanding mode of operation, the main source is interrupted and the critical load is supplied by DG source through DC link and shunt active filter (VSC2), as shown in Figure 17.5(b) [32].

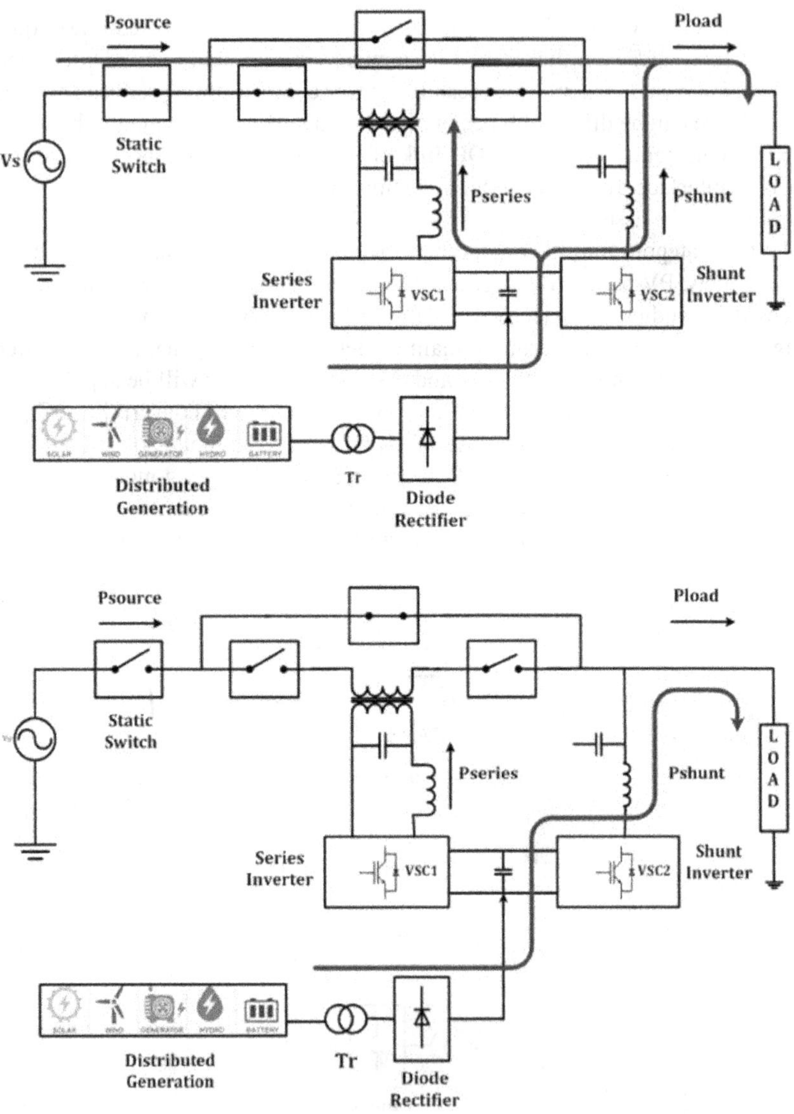

FIGURE 17.5 (a) Interconnected mode of operation of UPQC-DG with *forward flow* and *reverse flow*. (b) Islanding mode of operation of UPQC-DG.

In UPQC-DG, under normal condition, series inverter (VSC1) compensates voltage disturbance on the source side, which may due to various reasons. Series inverter control calculates the amount of compensation signal to be injected into the line by comparing positive sequence voltage with the disturbed source signal. Shunt inverter in this configuration covers two functions (1) current harmonic mitigation and reactive power compensation mode in normal operation and (2) output voltage control mode in case the main supply interrupts.

When DG source is generating AC power with the help of either SG or IG, it is connected to DC bus through transformer and a rectifier interface. This adds the extra cost to the overall implementation of the system. In case of DG generating DC power with the help of either solar PV, fuel cell or battery energy sources, etc., the system already consists of a DC-AC converter (inverter), whereas in the UPQC-DG implementation it will require DC-DC converter, which can be considered as an equal in cost and can be avoided if considered as extra cost.

17.5 UPQC-INTERLINE (UPQC-I)

The schematic arrangement of UPQC-I (interline) shown in Figure 17.6 depicts feeder 1 supplying power to load L1 and feeder 2 supplying power to critical load L2, which is very sensitive to voltage variations and hence must not be interrupted. In UPQC-I configuration, shunt active filter (VSC1) is connected with feeder 1 in parallel with the load L1 and series active filter (VSC2) is connected in series with load L2 through a common DC link capacitor C [33]. The main objective in this configuration is to maintain bus 1 voltage stable to provide sinusoidal voltages to load L1 and to protect critical load L2 from any supply interruption on feeder 2 along with the best quality voltage supply. To achieve this aim, the shunt active filter (VSC1) is

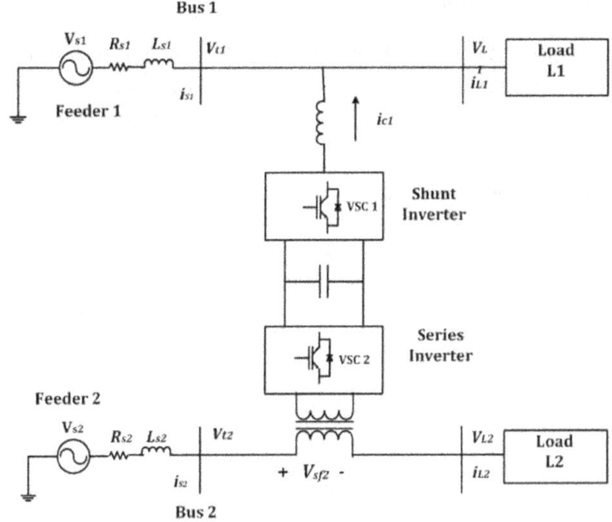

FIGURE 17.6 Schematic diagram of UPQC-I (interline).

operated in voltage controller mode, while series active filter (VSC2) regulates the voltage V_{L2} across the sensitive load. [34]

In normal situation, shunt active filter (VSC1) will take care of voltage-related issues such as sag/swell and distortion at bus 1 and supply sinusoidal voltages to load L1. Similarly, series active filter (VSC2) looks after sag/swell and distortion at bus 2 and supply sinusoidal voltages to the critical load L2. When feeder 2 trips due to any fault, the shunt active filter (VSC1) will supply power to the critical load through series active filter (VSC2) and DC link capacitor without any interruption. This configuration does not support current harmonics and reactive power compensation for load L1.

17.6 COMPARISON AND DISCUSSION

This section presents detailed comparison of UPQC-MC, UPQC-DG with PV/FC/BS as DG and SG/IG as DG, and UPQC-I with different power quality parameters such as shunt compensation, harmonic current compensation, reactive power compensation, sag/swell mitigation, voltage harmonic compensation, and power supply during main supply failure. It also evaluates these configurations for technical aspects such as switch ratings, cost economy, and control complexities, as well as operating aspects [35].

17.6.1 Shunt Compensation

UPQC-MC and both varieties of UPQC-DG provide shunt compensation as per the norms. At the same time, UPQC-I does not facilitate shunt compensation as shunt active filter (VSC1) connected to feeder 1 works in voltage controller mode.

17.6.2 Harmonic Current Compensation

UPQC-MC and both types of UPQC-DG mitigate current harmonics produced by nonlinear critical load through shunt inverter (VSC2), whereas in UPQC-I shunt inverter (VSC1) does not support current harmonic compensation.

17.6.3 Reactive Power Compensation

UPQC-MC and UPQC-DG offer unity power factor operation at source as their shunt active filter (VSC2) operates in current controller mode which compensates reactive power. UPQC-I does not compensate reactive power as shunt inverter (VSC1) cannot be configured for current controller mode of operation.

17.6.4 Sag/Swell Mitigation

All four configurations under considerations offer sag/swell mitigation as per power quality standards. Series active filter (VSC1) of UPQC-MC and both the variants of UPQC-DG will compensate source voltage sag/swell. In UPQC-I, it will be taken care by shunt active filter (VSC1).

17.6.5 Harmonic Voltage Compensation

All four configurations of UPQC offer voltage harmonic compensation which will be taken care by series active filter (VSC1) in UPQC-MC and in both the variants of UPQC-DG and by the shunt active filter (VSC1) in UPQC-I [36].

17.6.6 Power Supply during Interruption

All four configurations continuously supply power during interruptions. In UPQC-MC, when main feeder 1 interrupts, power is fetched through series active filter (VSC3) of adjacent feeder and shunt active filter (VSC2) of main feeder 1 through DC link capacitor to the critical load L1. In both types of UPQC-DG, in case of main source interruption, power is injected by DG to the DC bus through appropriate power electronic interface which is then supplied to critical load L1 via shunt active filter (VSC2). UPQC-I supplies power to critical load L2 through shunt active filter (VSC1) of feeder 1 and series active filter (VSC2) through DC link in case of interruption at feeder 2.

17.6.7 Number of Converters

In UPQC-MC, there are three converters: two for series active filters (VSC1 and VSC3) and one for shunt active filter (VSC2). In UPQC-DG where DC power is generated by DG, three converters are needed: the first as a series active filter (VSC1), the second as a shunt active filter (VSC2), and the third is a DC-DC converter, which works as an interface between DG source generating DC power such as PV/FC/BS and DC link bus. In the case of UPQC-DG, where DG is generating AC power through SG/IG, one is used in series active filter (VSC1), one is used in shunt active filter (VSC2), and one is converter rectifier. DG generating AC power will be injecting DC power into the DC link through the converter rectifier. UPQC-I requires two converters as it employs a simple UPQC between two feeders. One converter is used as shunt active filter (VSC1) in feeder 1 and the other is used as series active filter (VSC2) in feeder 2.

17.6.8 Number and Type of Switches

17.6.8.1 UPQC-MC

For UPQC-MC, a total of 18 controlled switches are required: 12 switches for series active filters 1 and 2 (VSC1 and VSC3) and 6 controlled switches for shunt active filter (VSC2).

17.6.8.2 UPQC-DG (SG/IG)

In UPQC-DG with AC power generators act as DG like SG/IG, 12 controlled and 6 uncontrolled switches are required. For series active filter (VSC1) six controlled switches, for shunt active filter (VSC2) six controlled switches and for converter rectifier six uncontrolled switches are employed.

17.6.8.3 UPQC-DG (PV/FC/BS)

UPQC-DG with DG as DC power-generating sources, such as PV/FC/BS, requires total 18 controlled switches, where 6 switches for series active filter (VSC1), 6 switches for shunt active filter (VSC2), and 6 switches for DC-DC (VSC3) interfacing converter are employed.

17.6.8.4 UPQC-I

In UPQC-I, 12 controlled switches are needed, where 6 switches are connected in shunt active filter (VSC1) and 6 switches are connected in series active filter (VSC2).

17.6.9 Switch Rating and Cost Consideration

17.6.9.1 UPQC-MC

In UPQC-MC, series active filter (VSC1) connected to feeder 1 looks after voltage-related issues of feeder 1 only; therefore, very low amount of compensating power may be required. Hence, power rating of six controlled switches connected to VSC1 is very low, which in turn costs less. In routine operation, shunt active power filter (VSC2) has to compensate load current harmonics and reactive power generated by nonlinear critical load, which demands substantial amount of compensating power. However, it is less than full load power requirement except in a few cases. Series active filter (VSC3) connected to feeder 2 also requires very less amount of compensating power to tackle voltage-related issues [37]. In case of interruption at feeder 1, considering the full load rated current of critical load to be carried by shunt active filter (VSC2) and series active filter (VSC3) switches supplied by feeder 2, the cost of these 12 controlled switches used in VSC2 and VSC3 will be comparatively very high.

17.6.9.1.1 Summary

A total of 6 controlled switches with low rating + 12 controlled switches with full load current rating are required.

17.6.9.2 UPQC-DG (SG/IG)

In the case of UPQC-DG with SG/IG as the energy source, six controlled switches of series active filter connected to feeder 1 will have low power rating as it has to compensate source voltage-related power quality issues, which poses very low power requirement. Shunt active filter (VSC2) will be carrying the full load current when feeder 1 fails and DG supplies power to the critical load; hence, six controlled switches connected to VSC2 will cost high. The converter rectifier will require six uncontrolled switched having full load rated current-carrying capacity of the critical load.

17.6.9.2.1 Summary

Six controlled switches with low rating + six controlled switches with full load current rating + six uncontrolled switches with full load current rating are required.

17.6.9.3 UPQC-DG (PV/FC/BS)

For UPQC-DG with PV/FC/BS as the energy source, as in the previous case, six controlled switches connected with series active filter (VSC1) will have low power rating and therefore the cost involved is less. Six controlled switches connected with shunt active power filter (VSC2) will carry full load rated current at the time of interruption and will cost very high. In addition, a DC power-generating source is connected to the DC bus through a DC-DC converter interface, which also uses six controlled switches having full load rated current-carrying capacity. Here, it is assumed that DG generating DC power through PV/FC/BS is already supplied with DC-DC converter as a standard component vcof DG setup and has not been accounted in the UPQC-DG system implementation cost. Usually, such DG systems are supplied with DC-AC converter to inject three-phase power into either the grid or three-phase loads, which also uses six controlled switches as is the case of DC-DC converter. Thus, the cost has been considered the same for both converters as the amount of power being handled is the same.

17.6.9.3.1 Summary

Six controlled switches with low rating + 12 controlled switches with full load current rating are required.

17.6.9.4 UPQC-I

UPQC-I employs two converters, namely, shunt active filter (VSC1) to take care of voltage-related power quality issues for load L1 connected at bus 1 and series active filter (VSC2) to take care of voltage-related power quality issues for critical load L2 connected at bus 2 which demands very low power rating. But in case of interruption at feeder 2, both the converters will have to carry the full load rated current of critical load L2, which considerably increases the cost of 12 controlled switches used in shunt active filter (VSC1) and series active filter (VSC2).

17.6.9.4.1 Summary

In total, 12 controlled switches with full load current rating are required. If the system is designed to facilitate reverse flow mode, where DG will be feeding power to the main source/grid through series active filter, the rating and cost will change accordingly.

17.6.10 Control Strategy

17.6.10.1 UPQC-MC

UPQC-MC employs three converters VSC1, VSC2, and VSC3 connected through a common DC-link capacitor sharing the energy for compensation for their own operation. Here, both the series active filters are acting as voltage controller mode and the shunt active filter operates in current controller mode. For each active filter, there is a separate controller controlling its own parameters for its assigned compensation along with operating in close coordination with the other two to maintain uninterrupted power supply in any situation. This makes the control strategy very complex; at the same time, it offers one of the best solutions for power quality enhancement in operating multiple sources supplying a single/multiple critical load.

17.6.10.2 UPQC-DG (SG/IG)

UPQC-DG (SG/IG) configuration functions as a simple UPQC in normal operating condition. The series active filter (VSC1) operates in voltage controller mode and the shunt active filter (VSC2) operates in current controller mode. In case of interruption at main feeder 1, DG generating AC power will start injecting power to DC link after rectification through a simple rectifier which does not require any separate controller. Yet, there is a scope for using a simple mechanical controller to modulate power output of SG/IG to maintain DC link voltage constant. The control strategy employed in this configuration is simple among all compared here. If the system facilitates the reverse power flow mode, the control strategy will become complex.

17.6.10.3 UPQC-DG (PV/FC/BS)

UPQC-DG (PV/FC/BS) employs three converters, one for series active filter (VSC1), one for shunt active filter (VSC2) and one for DC-DC conversion of the DC power generated by PV/FC/BS. The series active filter (VSC1) operates in voltage controller mode and the shunt active filter (VSC2) operates in current controller mode. Each converter will require its separate controller to fulfill its assigned function of compensation. Series and shunt active filter controllers will work in coordination with UPQC, whereas DC-DC converter controller may work independently as it has to modulate power injection into the DC bus and maintain DC bus voltages when required in case of interruption. It makes the control strategy of this configuration a bit complex. In case of system facilitating the reverse power flow mode, the control strategy will become more complex.

17.6.10.4 UPQC-I

UPQC-I employs two converters, one for shunt active filter (VSC1) and the other for series active filter (VSC2) which act as UPQC only. Here also it requires two separate controllers as in the case of UPQC, with the only difference being that the shunt active filter (VSC1) controller also works in voltage control mode in addition to current control mode. Yet, the control strategy is the simplest of all configurations discussed here.

17.6.11 Operating Scenario

17.6.11.1 UPQC-MC

This configuration offers the most reliable and optimum performance with a lot of flexibility to integrate any type of energy source to be operated as main or additional/backup mode. At the same time, the total numbers of controllers and number of controlled switches involved as well as their complex control strategies make the operation of this configuration the most complex.

17.6.11.2 UPQC-DG (SG/IG)

This scheme is comparatively simple to operate, as UPQC operation is simple in normal situation where the main source and DG are sharing the critical load, and in case of interruption of the main source, DG will start supplying the full load swiftly.

17.6.11.3 UPQC-DG (PV/FC/BS)

This is the most popular configuration in operation due to the popularity of renewable energy generation. This scheme also offers flexibility to integrate a variety of renewable energy sources. In this scheme, there are three different controllers working in tandem. It is little more complex in operation in comparison to its counterpart UPQC-DG (SG/IG).

17.6.11.4 UPQC-I

This configuration is the simplest of all discussed here as it employs a UPQC only for interline feeding. There are only two controllers and 12 controlled switches to be managed which makes it the simplest of all.

17.7 CONCLUSION

This chapter effectively compares various configurations, namely, UPQC-MC, UPQC-DG (SG/IG), UPQC-DG (PV/FC/BS), and UPQC-I, for uninterrupted power supply to the critical loads by integrating multiple energy sources, considering all the power quality parameters. From the points tabulated in Table 17.1 and the detailed comparison of different parameters and discussions in Section 17.6, it can be concluded that UPQC-MC offers very reliable operation along with flexibility to integrate any type of energy source. At the same time, cost, and control complexities of this configuration are the highest among all other configurations. UPQC-DG (SG/IG) and UPQC-DG (PV/FC/BS) offer reverse flow mode where power generated by DG can be injected to the main source along with supplying to critical load. In UPQC-DG (SG/IG), control complexity is less; at the same time, the overall cost of integrating SG/IG through transformer and rectifier may increase. UPQC-DG (PV/FC/BS) requires three converters, which increase control complexities. At the same time, the cost of PV/FC/BS has decreased drastically in the present times due to fast development and widespread use of the technology. Again, if DC-DC converter is assumed to be supplied as an integral component of PV/FC/BS system, the cost reduces drastically to declare this configuration the most suitable for the integration of multiple sources. UPQC-I is the simplest and cheapest of all the configurations; however, it has major drawbacks of inability to compensate for reactive power compensation, load current harmonic mitigation, and shunt compensation.

TABLE 17.1
Comparison of UPQC-MC, UPQC-DG (SG/IG), UPQC-DG (PV/FC/BS), and UPQC-I

	UPQC-Multiconverter	UPQC-Distributed Generation (SG[a]/IG[b])	UPQC-Distributed Generation (PV[c]/FC[d]/BS[e])	UPQC-Interline
Shunt compensation	Yes	Yes	Yes	No
Harmonic current compensation	Yes	Yes	Yes	No
Reactive power compensation	Yes	Yes	Yes	No
Sag/swell mitigation	Yes	Yes	Yes	Yes
Harmonic voltage compensation	Yes	Yes	Yes	Yes
Power supply during supply interruption	Yes	Yes	Yes	Yes
Number of converters	3	2 + 1	2 + 1f	2
Number and type of switches	18 (controlled)	12 (controlled) and 6(uncontrolled)	12 + 6[f] (controlled)	12 (controlled)
Switch rating	6 controlled switches (low rating) for VSC1, 12 controlled switches (full load rating) for VSC2 and VSC3	6 controlled switches (low rating) for VSC1,[g] 6 controlled switches (full load rating) for VSC2, 6 uncontrolled switches (full load rating)	6 controlled switches (low rating) for VSC1,[g] 6 controlled switches (full load rating) for VSC2 6 controlled switches (full load rating) for DC-DC converter	12 controlled switches (full load rating) for VSC1 and VSC2
Cost consideration	Costliest	Moderate	Lowest[f]	Low
Control strategy	Most complex	Simple	Moderate	Simplest
Operating scenario	Complex	Simple	Moderate	Simplest

[a] When DG employs synchronous generator (SG) with prime movers such as internal combustion engine, gas turbine, combined-cycle gas turbine, solar thermal, biomass, and gas turbine.

[b] When DG employs induction generator (IG) with prime movers such as wind energy system, microturbine, and small hydro generation.

[c] Photovoltaic cell.

[d] Fuel cell.

[e] Battery storage.

[f] DC-DC converter assumed an integral component supplied with PV/FC/BS system [35].

[g] When UPQC-DG facilitates *reverse power flow* mode, where DG injects power to the main source via series active filter VSC1, the switch rating will be high according to the amount of reverse power to be injected.

REFERENCES

1. Joseph, T., Ugalde-Loo, C. E., Liang, J., & Coventry, P. F. (2018). Asset management strategies for power electronic converters in transmission networks: Application to Hvdc and FACTS devices. IEEE Access, 6, 21084–21102
2. Cherian, E., Bindu, G. R., & Nair, P. S. C. (2016). Pollution impact of residential loads on distribution system and prospects of DC distribution. International Journal of Engineering Science and Technology, 19(4), 1655–1660.
3. Kazem, H. A. (2013). Harmonic mitigation techniques applied to power distribution networks. Advances in Power Electronics, 2013, 1–10
4. Grady, W. M., & Gilleskie, R. J. (1993). Harmonics and how they relate to power factor. In: Proceedings of the EPRI Power Quality Issues & Opportunities Conference (PQA'93), San Diego, CA.
5. Sharma, V. K. (2000). Power quality monitoring in industrial belt. In: Proceedings of the National Seminar (SASESC 2000), Agra, 103–107.
6. Al-Badi, A. H., Elmoudi, A., Metwally, I., Al-Wahaibi, A., Al-Ajmi, H., & Al Bulushi, M. (2011). Losses reduction in distribution transformers. In: International MultiConference of Engineers and Computer Scientists, Vol. II, Hong Kong.
7. Kung, P. (2018). Generator insulation-aging on-line monitoring technique based on fiber optic detecting technology. In: Sánchez, R. A. (Ed.), Simulation and Modeling of Electrical Insulation Weaknesses in Electrical Equipment. London: IntechOpen.
8. Owusu, P. A., & Asumadu-Sarkodie, S. (2016). A review of renewable energy sources, sustainability issues and climate change mitigation. Cogent Engineering, 3(1) https://doi.org/10.1080/23311916.2016.1167990.
9. Singh, M., Khadkikar, V., Chandra, A., & Varma, R. K. (2011). Grid interconnection of renewable energy sources at the distribution level with power-quality improvement features. IEEE Transaction on Power Delivery, 26(1), 307–315.
10. Bose, B. K. (2017). Power electronics, smart grid, and renewable energy systems. Proceedings of IEEE, 105(11), 2011–2018.
11. Elmoudi, A., Lehtonen, M., & Nordman, H. (2006). Effect of harmonics on transformers loss of life. In: Conference Record of the 1988 IEEE International Symposium on Electrical Insulation, 408–411.
12. Power Grid Corporation of India Limited. (2015). Swachh Power: A Glimpse of Power Quality in India. Gurugram: Power Grid Corporation of India.
13. Daut, I., Syafruddin, H. S., Rosnazri, A., & Zali, S. M. (2006). The effects of harmonic components on transformer losses of sinusoidal source supplying non-linear loads. American Journal of Applied Sciences, 3(12), 2131–2133.
14. El-Habrouk, M., Darwish, M., & Mehta, P. (2000). Active power filters: A review. IEE Proceedings – Electric Power Applications, 147(5), 403–413.
15. Stanciu, D., Teodorescu, M. A., Florescu, A., & Stoichescu. D. A. (2010). Single-phase active power filter with improved sliding mode control. In: Proceedings of the 17th IEEE International Conference on Automation, Quality and Testing, Robotics (AQTR), 15–19.
16. Patnaik, N., & Panda, A. K. (2014). Comparative analysis on a shunt active power filter with different control strategies for composite loads. In: Proceedings of the IEEE Region 10 Conference (TENCON 2014), 1–6.
17. Singh, B., Al-Haddad, K., & Chandra, A. (1999). Review of active filters for power quality improvement. IEEE Transactions on Industrial Electronics, 46(5), 960–971.
18. Newman, M. J., Holmes, G., Nielsen, J. G., & Blaabjerg, F. (2005). A dynamic voltage restorer (DVR) with selective harmonic compensation at medium voltage level. IEEE Transactions on Industry Applications, 41(6), 1744–1753.
19. Silva, S. A. O., & Campanhol, L. B. G. (2014). Application of shunt active power filter for harmonic reduction and reactive power compensation in three-phase four-wire systems. IET Power Electronics, 7(11), 2825–2836.

20. Fuchs, E. F., & Masoum, M. A. S. (2008). The roles of filters in power systems. In: Power Quality in Power Systems and Electrical Machines (359–395). Cambridge, MA: Academic Press.
21. Khadkikar, V., Chandra, A., Barry, A. O., & Nguyen, T. D. (2006). Conceptual study of unified power quality conditioner (UPQC). In: IEEE International Symposium on Industrial Electronics, Montreal, QC, 1088–1093.
22. Sant, A. V., Khadkikar, V., Weidong, X., & Zeineldin, H. (2013). Adaptive control of grid connected photovoltaic inverter for maximum VA utilization. In: Proceedings of the 39th Annual Conference of the IEEE Industrial Electronics Society (IECON 2013), Vienna, 388–393.
23. Fujita, H., & Akagi, H. (1998). The unified power quality conditioner: The integration of series active filters and shunt active filters. IEEE Transactions on Power Electronics, 13(2), 315–322.
24. Khadkikar, V. (2012). Enhancing electric power quality using UPQC: A comprehensive overview. IEEE Transactions on Power Electronics, 27(5), 2284–2297.
25. Ramanaiah, M. L., & Reddy, M. D. (2018). Moth flame optimization method for unified power quality conditioner allocation. International Journal of Electrical and Computer Engineering, 8(1), 530–537.
26. Anjana, P., Gupta, V., & Tiwari. H. (2014). Reducing harmonics in micro grid distribution system using APF with PI controller. In: IEEE PES T&D Conference and Exposition, Chicago, IL, 1–5.
27. Amirullah, Soeprijanto, A., Adiananda, & Penangsang, O. (2020). Power transfer analysis using UPQC-PV system under sag and interruption with variable irradiance. In: International Conference on Smart Technology and Applications (ICoSTA), Surabaya, Indonesia, 1–7.
28. Mohammadi, H. R., Varjani, A. Y., & Mokhtari. H. (2009). Multiconverter Unified Power Conditioning System: MC-UPQC. IEEE Transaction on Power Delivery, 24(3), 1679–1686.
29. Rahmani, S., & Al-Haddad, K. (2011). Filtering techniques for power quality improvement, IEEE Porto PowerTech, IEEE. doi:10.1201/b10643-45
30. Han, B., Bae, B., Kim, H., & Baek, S. (2006). Combined operation of unified power quality conditioner with distributed generation. IEEE Transaction on Power Delivery, 21(1), 330–338.
31. Khadem, S. K., Basu, M., & Conlon, M. F. (2013). A new placement and integration method of UPQC to improve the power quality in DG network. In: 48th International Universities' Power Engineering Conference (UPEC), Dublin, 1–6.
32. Karelia, N., & Pandya, V. (2015). Distributed generation and its role of UPQC-DG in meeting power quality criteria. Procedia Technology, 21, 520–525.
33. Jindal, A. K., Ghosh, A., & Joshi. A. (2005). The protection of sensitive loads from interharmonic currents using shunt/series active filters. Electric Power System Research, 73(2), 187–196.
34. Jindal, A. K., Ghosh, A., & Joshi, A. (2007). Interline unified power quality conditioner. IEEE Transaction on Power Delivery, 22(1), 364–372.
35. Karelia, N., Sant, A., & Pandya, V. (2019). Comparison of UPQC topologies for power quality enhancement in grid integrated renewable energy sources. In: IEEE 16th India Council International Conference (INDICON2019), Rajkot.
36. Akagi, H., Watanabe, E. H., & Aredes, M. (2007). Combined series and shunt power conditioners. In: Instantaneous Power Theory and Applications to Power Conditioning (265–374). Hoboken, NJ: John Wiley & Sons.
37. Jain, S. K., Agarwal, P., & Gupta, H. O. (2004). A control algorithm for compensation of customer generated harmonics and reactive power. IEEE Transactions on Power Delivery, 19(1), 357–360.

18 AI-Based Energy-Efficient Fault Mitigation Technique for Reliability Enhancement of Wireless Sensor Network

Syed Mufassir Yaseen[1], *Mithilesh Kumar Dubey*[1], *and Majid Charoo*[2]
[1]Lovely Professional University, Phagwara, Punjab, India
[2]Al-Jouf University, Al-Jawf, Saudi Arab

CONTENTS

18.1 INTRODUCTION

Wireless sensor network (WSN) consists of different autonomous nodes, which are linked together to form a network; in reality, it is simply a set of self-governing computer sensor devices used to monitor and control physical environments [1]. Each node inside a WSN has data sensing and processing capabilities. WSN technology has been attracting a lot of attention within the industry for many years. The self-setup, reliability, fast deployment, self-configuration, and easy update, as well as low operating costs, make WSN ideally suited for industrial use. It is also used in critical components, such as ball bearing motors, engines, and oil pumps [2]. A fault-tolerant WSN is configured to give a continuous supply of information despite disturbances to the sink terminal. Fault tolerance, the capacity to identify and diagnose unreliable nodes online, is an important feature. This research looks at the device detection of fault and mitigation issues in WSN. It is evident from the experiments that more than 80% of the errors occurring in actual systems are transient errors [3, 4]. When software is defective, an occasional fault originates from within the device. There will be no clear intermittent fault by its nature that makes the diagnosis of that type of fault a probabilistic event [5]. Ruling a successful trade-off can be developed in many ways by having a keen attention on the planned concluding performance. Thus, only the finest result cannot exist instead of a complete set of potential solutions of similar quality. This inspired us to use optimization algorithms that deal with several, probably conflicting, objective functions at such simultaneous optimization. Thus, a multiobjective algorithm was used for trade-off between energy, latency, and accuracy detection [6]. Also, for extraction of the best trade-off, Dhillon et al. [7] proposed a fuzzy-based mechanism called 2LB-MOPSO algorithm.

18.1.1 Architecture of WSN

It consists of multiple wireless sensor nodes capable of data processing, communication with other sensor nodes, and data storage. Because of its performance and cost-effectiveness, the WSN has gained wider applicability (Figure 18.1). Most applications of the WSN are related to sensing and monitoring of the environment. The sensing activity of the sensor nodes is achieved by the implanted microcontroller and radio transceivers. The microcontroller includes processor and memory that enable it to perform the computation and store the data, which is to be sensed by sensor nodes. The communication is achieved by the radio transceivers. These sensors are sensing the environment and processing and moving the sensed data to the next node.

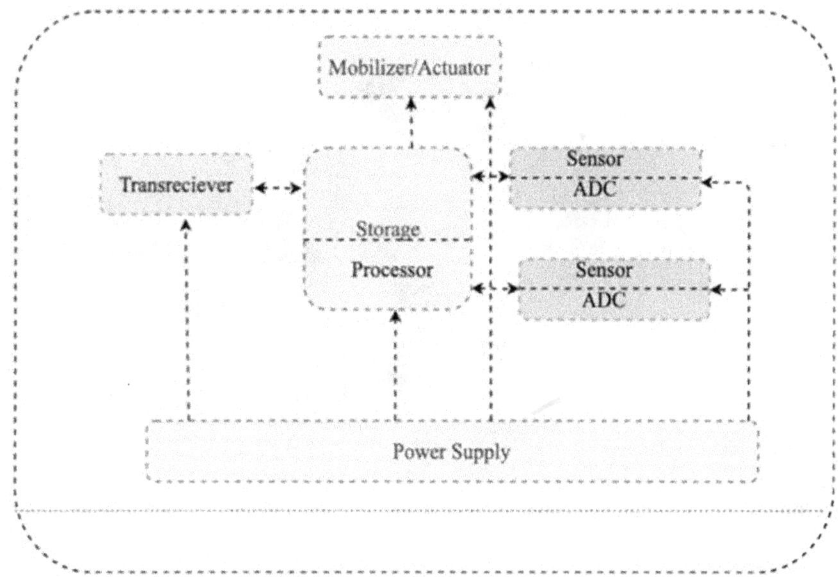

FIGURE 18.1 Typical sensor node.

There are two ways in which the communication between sensor nodes is done: direct and indirect. The connection includes two essential bodies, and they are the nodes of source and destination. The source node aims to forward the data to another node. The destination nodes are data receivers.

Figure 18.2 shows single-hop communication (direct communication), which is possible only when the receiving node lies in the contact range of source nodes. If the receiving node will not be within the source node's range of communication, then

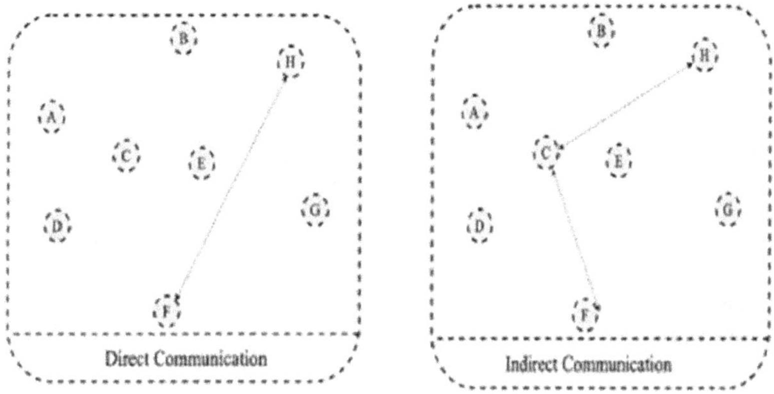

FIGURE 18.2 Communication types.

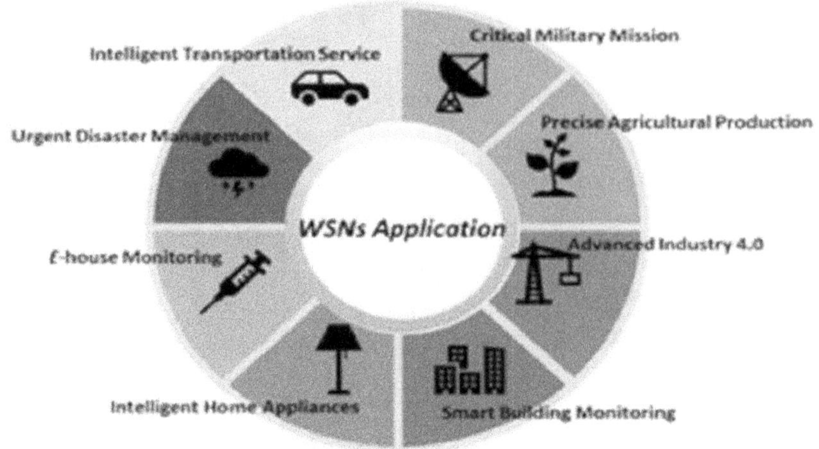

FIGURE 18.3 Overview of WSN applications.

the data is forwarded by using the intermediate to destination nodes. This is called indirect contact (or contact over multihop). WSN has a large number of applications in almost every field and some of them are described in Figure 18.3.

As described, the implementation of WSN technology will support a large number of applications, from habitat monitoring to battlefield surveillance [8–10]. Few benefits include fast deployment, high-fidelity sensing, low cost, WSN self-organization, and several other advantages. Despite many opportunities provided by the WSN, the technology also poses significant challenges. These problems are related to the features of WSN, namely:

- Reliability
- Scalability
- Inconsistent topology
- Hardware constraints
- Energy consumption
- Quality of services (QoS)

For efficient monitoring of an area, event data collected by nodes must be transferred efficiently to the next node within the sensor network. The most critical issue facing the sensor network is reliability. The sensor network can be claimed as secure, even when the nodes are compromised, if the network is not scattered. The sensor nodes are susceptible to failures and inadequate resources, unfavorable environmental conditions, and internal damage. The sensor network may experience a large number of node failures; however, the network's functionality must be met. Reliability is the main aspect of a wireless network, which depends on the complexity of the service.

When there are faults, the effects on human life, environmental effect, or fiscal loss may be severe. Faulty sensor outputs can result in misinterpretation or unwanted alarms. This could lead to serious incidents as a large number of WSNs would be interested in critical safety applications. For example, defective sensor nodes within a WSN embedded around railway bridge columns may not give any early sign of the danger. And for the effective utilization of bandwidth and energy, it is necessary to avoid the faults in a system. Thus, various faults in WSN are discussed in the next section.

18.1.2 Faults in WSN

A fault simply means an irregular substantial situation of a node that may occur due to temperature, humidity, power surge, erroneous deployment, radiations, etc. The term fault may in other terms also be defined as the deviations from expected algorithm or model [11]. Fault tends to lead to failure if the behavior of the sensor node deviates from its specific algorithm and thus does not perform its required job [12]. Categorization of faults depends on the part having fault, the perseverance of the fault, or the underlying cause [13, 14]. Faults can be categorized as soft fault and hard fault, depending on how a failed node behaves. A sensor node experiences soft faults with altered behaviors. The faults may be further categorized as permanent, temporary, or transient based on persistence. Permanent faults or hard faults are faults in hardware or software which always cause errors when completely exercised [5]. Temporary faults are further divided into internal (intermittent) faults and the external (transient). The former is soft faults triggered by events that come from the world of a sensor node, which does not mean that the node is defective. These are difficult to track because their unfavorable effects usually easily vanish [15]. The intermittent fault, which exhibits a fairly high rate of occurrence after its first occurrence, is a particularly problematic form of transient fault, and finally tries to become permanent [15, 16].

Figure 18.4 shows a usually observed order of faults in WSN applications at a physical layer [17]. In this way, early detection of fault is key to the efficient functioning of the sensor node. The basis for improvement is the ability to make significant

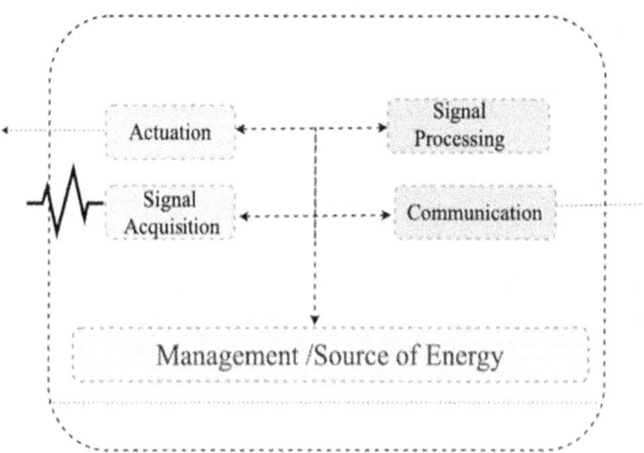

FIGURE 18.4 Block diagram of typical sensor node.

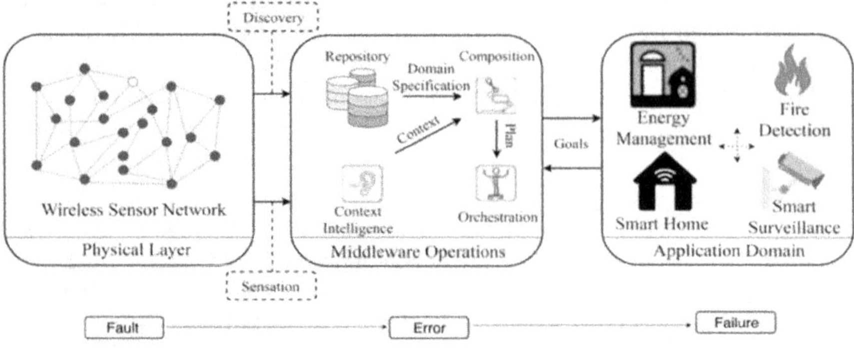

FIGURE 18.5 Architecture of sensor network applications [8].

and correct deductions from the information collected, which therefore includes high sensor information (Figure 18.5) [11]. Reliability of the program in this way is the main concern. We therefore must develop one of the best algorithms that can be used to identify faults in the network of wireless sensors.

18.1.3 Sources of Faults

Inherently, the collection and distribution of data in WSNs is incorrect and unpredictable [15, 16]. Key causes of failure include configuration error, hardware malfunction, harsh environmental conditions, failure to connect, and inadequate battery charging. As recorded in [18], the interaction of water molecules with the sensor parts can lead to short circuits and eventually to false readings. The primary cause of hardware failure may not be electrical malfunctions [19, 20]. Various sources can affect sensor readings, such as noise from external sources or some hardware sources.

18.1.4 Motivation

Since the late 1960s, researchers have recognized fault diagnosis as an important issue within wired interconnected networks. Although the basic guidelines are easily understood, their utilization to different areas is not well known, particularly in WSNs. Nevertheless, the recent neighboring coordination-based diagnosis breakthroughs in WSNs have opened up ample room for science. WSNs have the inherent property of deployment of low-cost sensor nodes at a large scale in unregulated or hostile environments. Sensor nodes are usually becoming inaccurate and unreliable. A WSN's regular operation suffers from inaccurate data as it decreases the base station's decision accuracy, increases the WSN's traffic, and wastes a great deal of available energy [21]. Diagnosis at device-level seems a feasible approach to such issues. The diagnosis becomes more difficult for intermittent, as in the intermittent defective node will pass a check and mark it as fault-free. Hence, it can take multiple test sessions to locate the faulty nodes. On the other hand, the influence of transient faults easily vanishes.

18.2 LITERATURE REVIEW

This portion deals with review of the literature for the system-level analysis algorithms that are designed for WSN. In this work, we are presenting the state of the art of system-level detection and mitigation algorithms. Previous algorithms for fault diagnosis in WSNs are broadly subdivided into two parts: distributed approach and centralized approach. A locally clustered node with a better storage capacity, higher processing strength, and continuous energy sources assumes duty for the diagnosis and identification of faults of the entire WSN in clustered approach. Central node sends diagnostic queries regularly to the WSN. Centralized approach is both effective and reliable in different ways, but for Linear Wireless Sensor Networks, implementation of these approaches might not be advocated.

18.2.1 Taxonomy Based on Techniques of Framework

Since WSNs are becoming very popular in scientific research, several diagnoses and detections of techniques of fault specifically developed for WSNs have emerged. Techniques for fault diagnosis of WSN can be generally classified into distributed approach and centralized approach. In the case of centralized strategies, a supervisory arbiter (sink node or base station) is often presumed to be available to interpret the diagnostic messages and disseminate diagnostic information. Implementing such an approach would put a lifelong bottleneck on the network. Diagnosis had been introduced and researched for these purposes. In distributed approaches, every node executes the detection algorithm and gives a local-view fault.

18.2.2 Distributed Approach

Specific applications require that the fault diagnosis is performed in a real-time mode with low latency, low overhead message, and high output. Thus, in addition to the aforementioned limitations of centralized methods, the design of diagnostic approaches will aim to address these issues. Depending on various aspects, distributed approach is further divided into various types, as described in the following.

18.2.2.1 Test-Based Approaches

In this strategy, tasks are allocated to sensor nodes and, on the basis of test results, defective nodes are classified. This form of approach tests the group of sensor nodes and transfers the test results onto other nodes. A system-level diagnosis was introduced in 1967 by Metze, Preparata, and Chien (PMC). Later on, Hakimi and Amin described the PMC model using another factor where the elements test each other. Chessa and Santi [22] suggested an algorithm that would diagnose the faults, namely, WSNDiag.

18.2.2.2 Majority-Voting Approaches

This method makes use of the fact that readings that are faulty are uncorrelated, while the usual ones are correlated spatially. Figure 18.6 shows readings from defective sensors that are geographically independent, while sensor readings are spatially associated in near proximity [23]. Miao et al. [24] suggested an agnostic

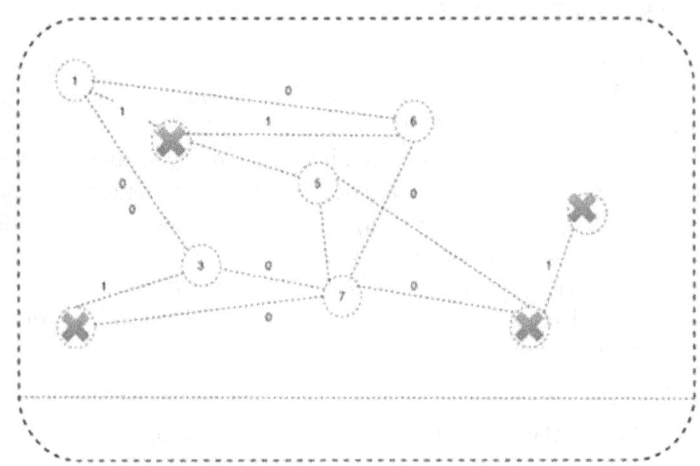

FIGURE 18.6 Comparison illustration of faults crossed.

diagnosis insufficiency identification system. This approach is inspired by the fact that sensors typically show such correlation patterns in the device parameters. This approach gathers from each sensor node 22 types of metrics, which are grouped into four categories. Luo et al. [25] suggested a system for adaptive detection of fault-tolerant cases. This technique makes use of a filter to accommodate transient faults. Confidence rates are utilized to control sensor node status. Whenever the fault is detected, the confidence levels are updated.

18.2.3 Weighted-Voting Approaches

This approach uses the weight properties such as physical distances from the case, measurements, and confidence. These measurements are used to make decisions about a sensor node's condition.

Guo et al. [26] proposed a diagnosis method to detect sensor nodes having faults, namely, FIND. The solutions to weighted-voting inherit the advantages of majority voting. The theoretical constraints of those methods, however, are additional. The weighted-majority-voting approaches show weak performance in sparse WSNs or WSNs with sparse regions, close to majority-voting approaches.

18.2.4 Hierarchal Detection Approaches

The fundamental concept behind hierarchical detection is that initially a spanning tree is identified with the root in the sink node or we can say a base station. The tree looks for all the nodes and then the faulty point is designated at each of the tree points. The tree then distributes the result that has been made at every node so as to detect each and every faulty node. Gheorghe et al. [27] put forward a protocol for adaptive trust management. Sensor nodes exchange reputation values during the exchange process and recalculate and evaluate them.

18.2.5 Self-Detection Approaches of Node

Here, the sensor node is capable of detecting status and it is done by having additional in the architecture.

I-Iarte et al. [28] suggested architecture to diagnose the faults in the elements of the nodes of the sensor. Here, both hardware and software interfaces are used, wherein the hardware part is made up of small accelerometers build on a circuit board.

Koushanfar et al. [29] suggested that sensor nodes be self-detected in WSNs. This method examines the sensor's binary outputs by contrasting them with the already set models having faults. Faults caused by the failure of batteries can be determined when the equipment is capable of calculating the voltage of the battery [2, 30].

18.2.6 Cluster-Based Approaches

These methodologies build a fundamental backbone of communication for sensor node groups and divide the WSN into various sets; in each of these sets, the process of fault detection is carried out. Typically, a cluster's lead node (e.g., the head of the cluster) conducts the process of detection of faults in groups using distributed or centralized method. Wang et al. [31] proposed a detection method based on an agreement to detect faults in cluster heads in underwater. It conducts a distributed recognition method periodically at every node of the cluster.

Sakib [32] suggested an asynchronous, ineffective identification of nodes. Here, the cluster heads and cluster leaders are given different detection protocols. A fault counter is used to monitor the data packets, which are then further transferred to another active node.

18.2.7 Watchdog Approaches

The most common cause of the firewall is to check the system for faults. The watchdog's basic working concept is to track whether the one-hop neighbor of a node is forwarding the packets. Within the fixed time, if the node fails to send the message, then it will be considered as the misbehaving node; if the misbehaving nodes reach an already defined threshold, then the source node is alerted and the packets are forwarded through other routes.

18.2.8 Soft-Computing-Based Approaches

This type of approach uses the characteristics of the nodes of the sensor and WSN for detection and diagnosis of faults. The neural network fiddles with its weights in the learning phase, which corresponds to the faulty models of healthy nodes. The development process compares sensor node current output to neural system output. The discrepancy between the two forms the basis for detecting the health status of a sensor. Barron et al. [33] applied this method with the operating system TinyOS on Moteiv's Tmote Sky platform.

From the literature review, it has been found that a large number of schemes for fault diagnosis have been proposed so far. Such current systems are, however, very costly in terms of time, communication, and energy perspectives.

18.3 CONTRIBUTION

1. To design a low-powered efficient algorithm for detection and mitigation of software faults in WSN.
2. To evaluate performance of the proposed methodology.

18.4 ROLE OF AI IN RENEWABLE ENERGY

The main objective of artificial intelligence (AI) is to develop a system that emulates the interaction capacities of a human being. The distributed AI pursues the same aim but stresses on human being societies [34].

Examples of renewable energy technologies incorporating AI are discussed in the following:

- **Energy accessibility**: AI is used for modeling the optimum costs and providing recommendations for smart home investments.
- **Energy efficiency**: AI is used for optimizing energy.
- **Forecasting of energy**: Data are used for training AI-based algorithms to help in making forecasting accurate and also informing power supply and its demand.

18.4.1 Energy Forecasting by AI

For renewable energy sources, unreliability is a persistent problem, such as solar power and wind. Sources of weather-based power also fluctuate in intensity. In Colorado Power Company, Xcel is introducing AI in an effort to resolve those challenges. By using the latest AI-based data mining system developed by the National Center for Atmospheric Research, Xcel was reportedly able to access weather forecasts with greater accuracy and detail.

18.4.2 AI for Energy Efficiency

Verdigris Technologies, which offers cloud-based software platform, aims to leverage AI so as to help customers optimize utilization of energy. Built for large commercial and corporate facilities, the cycle starts with IoT (Internet of Things) hardware installation.

18.4.3 AI for Energy Accessibility

PowerScout is currently using AI to design future optimization on electricity costs using industrial information in an attempt to enhance customer awareness and access to clean energy technologies.

18.5 SYSTEM MODEL

The n number of nodes is utilized by this framework with (x, y) position in a fixed area (2×2 m^2). Nodes like x_i and y_j are inside the scope of transmission TR. The suspicion arises if the Euclidean separation $d(x_i, y_j)$ is not as much as that of transmission extend, i.e., transmission range (TR). Here, we have $G = (V, E)$ which comprising vertices set like E and the set V as edges that conveyed each other with the connections in the middle of the hubs of the system.

18.6 DETECTION OF FAULT MODEL

Suggested algorithm for diagnosis of the fault comprises four different stages, namely, communication phase, phase of identification and classification, clustering phase of faults, and phase of isolation. Figures 18.7 and 18.8 show stages of identification and

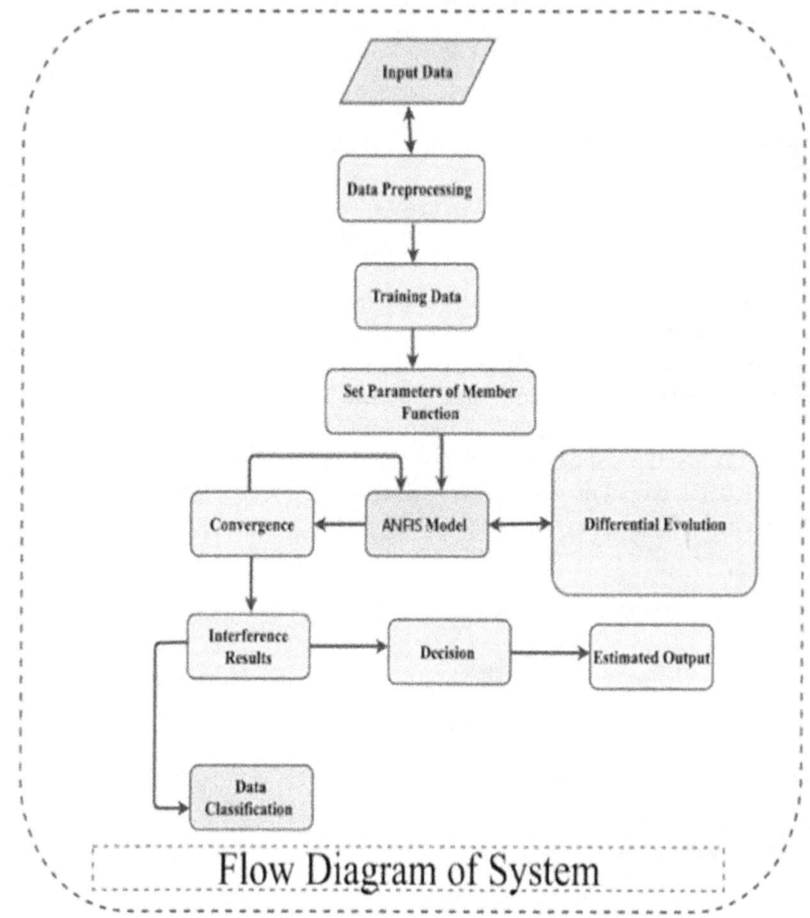

FIGURE 18.7 Flow diagram.

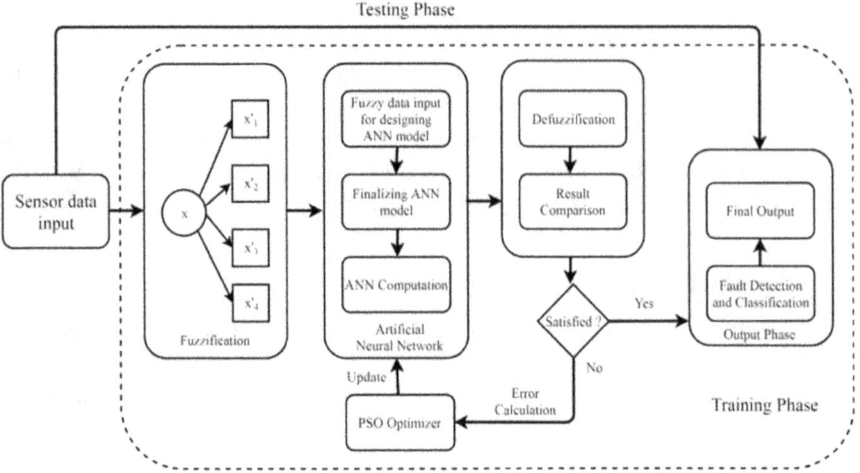

FIGURE 18.8 Fault detection [8].

classification of faults, defined in four steps: (i) input data, (ii) artificial neural network (ANN) layout, (iii) ANN training, and (iv) ANN testing:

1. Input data: Within a set range of 0_1 and 0_2 sensor nodes, produce the data by using the usual distribution.
2. ANN design: In NMO, M is the hidden node number, N is the node number, and O is the output.
3. ANN training: The NN is instructed on how to update the understanding-based scheme. To update weights, we use the backpropagation neural network evolutionary approach to the proposed work.
4. Test ANN: Once the training dataset has been trained by ANN, in the phase of testing, output or selection will be compared with a match that is closest to any of fault or may be of fault class.

$$\text{Mean squared error} = \frac{1}{2}\text{N}\left(\text{Target}(i) - \text{output}\,(i)\right)^2.$$

18.7 METHODOLOGY WORKFLOW DIAGRAM

18.7.1 Algorithm for Mitigation of Faults

There are various sequential steps that are to be performed to incorporate improvised fault mitigation techniques.

$$\textbf{Software faults} = \{\text{s1, s2, s3, …, s10}), \textbf{Hardware faults} = \{\text{h1, h2, h3, h4, h5}\}$$

Step 1: Establish the network in the smaller area consisting of sensor nodes. These sensor nodes are randomly distributed nodes.

Step 2: All the sensor nodes collect the data for some fixed parameter and send it back to the base station. The base station collects the data and either processes the data or forwards the data to some remote location for further analysis.

Step 3: On the base station, there is an improvised component that detects the fault in both the hardware and software categories.

Step 4: There can be either one fault or multiple faults at the sensor node or base station itself.

Step 5: In multiple fault conditions, some optimization criteria needed to be done, which will identify the optimized fault within the set of recognized faults at the base station.

Step 6: Activate the mitigation process for the optimized fault.

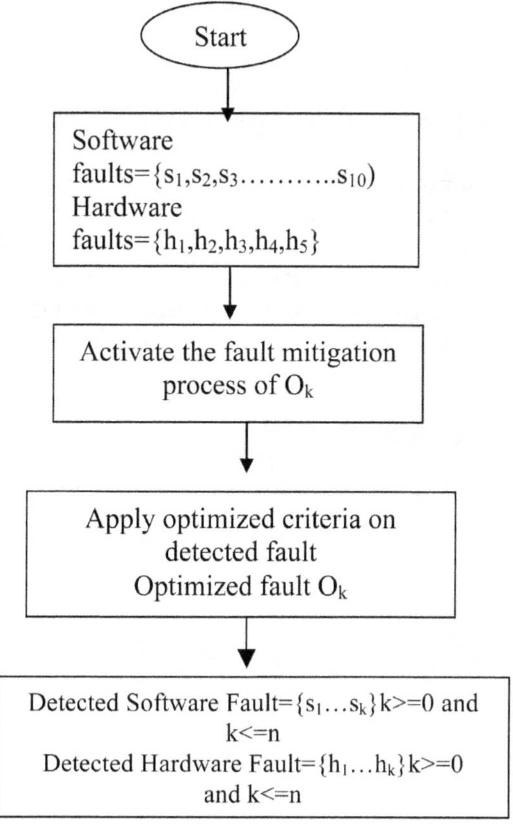

Flow Chart 1

18.7.2 An Energy-Efficient Algorithm

Input: Generate chromosomes and crossover.
Initialization: Weights and biases with random chromosomes = 0 and MSE.
Output: Faults mitigation and detection.
Step 1: Normalization: Generation of input

$$N = \begin{cases} 2, & d_0 \le d \\ 3, & d_0 > d \end{cases}.$$

Input: Values of the sensor node in the input_layer;
while (Not satisfying condition) **do**
for each set of sensor node values, n_i $(1 \le i \le n)$ **do**
For calculation of nodes by utilization of function activation, $z_j = f(z_{inj})$;
end for
for in output layer **do**
Calculate net input to the output nodes (y_{ink}); Calculate output of the output nodes $y_k = f(y_{ink})$;
Do denormalization;
end for
To calculate Error, E = Target – *Y*; Mean Squared Error = Mean Squared Error + E^2;
end for
Mean Squared Error = MSE/instance;
Step 2: **for** every node in hidden_layer and output_layer **do**
Update Weights; For each Epoch calculate fitness Mean Squared Error;
Sort the population according to the fitness of their MSE in ascending order;
Replacement of lowest fit chromosomes with highest fit chromosomes; Crossover preformation done;
end for
end while
Step 3: Senor node test values;
for input_node i.e, n_i **do**
for neuron_Hidden **do**
Hidden nodes input calculation; Output Calculation for hidden-node using activation function.
end for
Step 4: Calculate input to the output-node;
Step 5: output generation by using activation-function; Find the node index with highest valued y ;
end for

After executing the algorithm, Figure 18.9 shows us the best fitness and mean fitness.

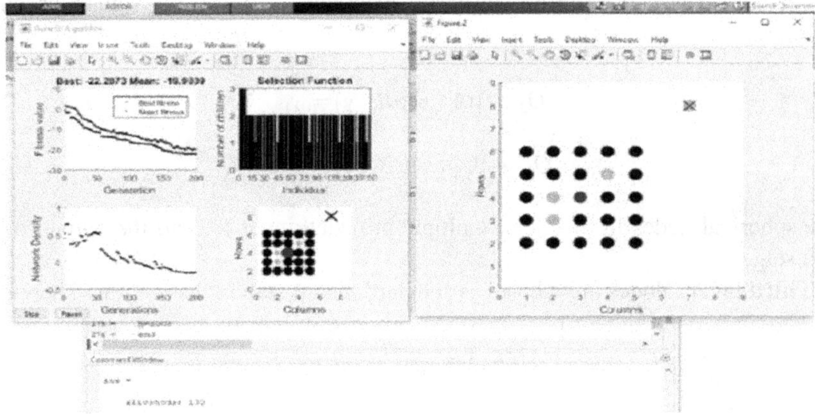

FIGURE 18.9 Output of algorithm.

18.8 ANFIS

A construct known as fuzzy neural system is produced by combining the ANN with the fuzzy inference system. Fixing prediction problems is prevalent in many areas, such as an anticipated average temperature, inventory systems, and flow of traffic.

Adaptive neuro-fuzzy inference system (ANFIS) layers: The ANFIS structure shown in Figure 18.10 consists of five layers. The input node is I and the respective node is A_i or B_{i-2}.

First layer: Let $O_{1,n}$ be the *n*th layer output_node yield 1. For these adaptive nodes, node features are described as follows:

$$O_{1,1} = \mu A_1(x),\ O_{1,2} = \mu A_2(x), O_{1,3} = \mu B_1(x),\ O_{1,4} = \mu B_2(x) \tag{18.1}$$

Thus, here $O_{1,1}$, $O_{1,2}$, $O_{1,3}$, $O_{1,4}$ are the membership grades of $\{A_1, A_2, B_1, B_2\}$ fuzzy set and $\mu(x) = (1 + |x - c_i| / a_i \times 2b_i)$, where a_i, b_i, c_i are called premise parameters and, usually, $\mu Ak(x)ax$, is elected as bell-shaped fuzzy membership function.

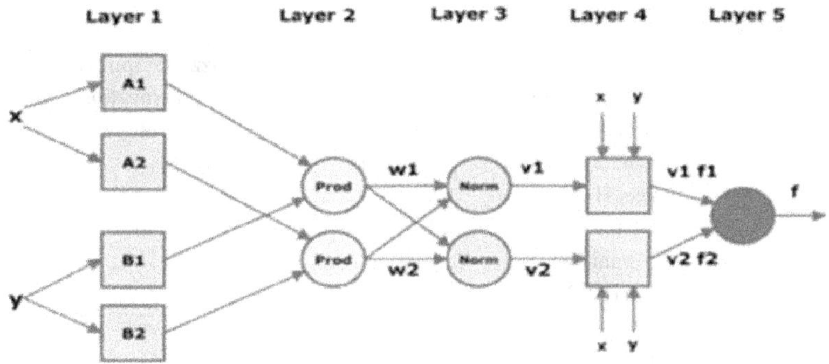

FIGURE 18.10 ANFIS structure.

Second layer: Nodes in this layer are fixed and are incoming signals product; therefore,

$$O_{2,1} - \mu A_1(x)\mu B_1(y) = w_1$$

$$O_{2,2} - \mu A_2(x)\mu B_2(y) = w_2. \tag{18.2}$$

The spherical nodes in this layer multiply provided input & send the output to the next stage.

Third layer: Nodes are labeled as standard means standard strengths. Therefore,

$$\left(O_{3,1}\right) = \frac{w_1}{w_1 + w_2} = w_1. \tag{18.3}$$

Fourth layer: Let $\{p_k, q_k, r_k\}$ be a collection of parameters called consequent variables. The adaptive feature are as:

$$O_{4,k} = W_k \times f_k = wk \times \left(p_k x + q_k y + r_k\right),\ k = 1,\ 2. \tag{18.4}$$

Fifth layer: Single nodes aggregate signals globally.

$$O_{5,k} = \sum_k w_k f_k,\ k = 1,\ 2. \tag{18.5}$$

18.8.1 ANFIS Modeling of Wireless Sensor Network

MATLAB Fuzzy Logic Toolbox and Sugeno-type are used to design the ANFIS model. Seventy percent of the information set has been randomly allocated as the

TABLE 18.1
Possible If-Else Rule

	Fuzzy Input Variables		
Rule Number	Residual Energy	Distance to BS	Fuzzy Output Variable Communication Range
1.	Low (L)	Close (C)	Very small
2.	Medium (M)	Close (C)	Small (S)
3.	High (H)	Close (C)	Rather small
4.	Low (L)	Medium (M)	Medium small
5.	Medium (M)	Medium (M)	Medium (M)
6.	High (H)	Medium (M)	Medium large
7.	Low (L)	High (H)	Rather large
8.	Medium (M)	High (H)	Large (L)
9.	High (H)	High (H)	Very large

TABLE 18.2
FIS Structure Parameters

Parameter Name	Parameter Value	Parameter Name	Parameter Value
FIS structure	Sugeno	Squash factor	1.25
No. of input	3	Accept ratio	0.5
No. of output	1	Optimization method	Hybrid learning algorithm
Rules count	625	Rejection rate	0.15
Influence of range	0.5	Epoch number	1000

network operation instruction set. The remaining 30% was used to test the ANFIS estimator's efficiency. The placed WSN network parameter relies on four parameters that will be taken as input, namely, packet delivery ratio, fault ratio, number of re-transmissions, and residual node energy. The node status is the parameter of one output. The training for the proposed algorithm is given by a proper set of training information; therefore, input and output can be assessed based on the dataset. The data is to be trained to describe the Sugeno-type FIS hybrid algorithm-based parameters. The criteria used to evaluate network performance were false alarm rate, power cost, fault detection accuracy, probability of failure, and network life.

Rates of error for each feature were calculated and are given in Tables 18.1, 18.2 and 18.3. Gaussian membership function (gauss2mf) was recognized as ANFIS model membership function because it gives the lowest rates of error. The blurred laws created using the Gaussian membership function architecture were automatically generated by the ANFIS model. For the first three inputs, five membership functions are defined as low (L), medium (M), high (H), very small (VL), and high (H). For the last known critical factor input, two membership functions have been defined: low and high. The number of laws was determined automatically by ANFIS using the following equation:

$$\text{Number of rules} = \left(\text{Number of membership function value}\right)^{\text{Number of input}}$$

Here, number of rules is given as: 250 ($5^3 \times 2^1$).

TABLE 18.3
Different Function Types' Error Rate

Type of Function	Rate of Error	Type of Function	Rate of Error
Trimf	0.0059457	Gaussmf	0.0058196
Trapmf	0.0059457	Pimf	0.0059457
Gbellmf	0.0058196	Psigmf	0.005861
Gauss2mf	0.005853	Dsigmf	0.0058561

18.9 PERFORMANCE EVALUATION OF THE PROPOSED MODEL

To evaluate the efficiency of the ANFIS model, the model was contrasted with the ANN model. A set of information was used to build the ANN model. The ANN model consisted of five layers and ten concealed layers. The training algorithm was established as Trainlm in this model and was based on backpropagation. Moreover, forecast outcomes are collected in Table 18.4 with the results ranging from 1 to 3. Cut-off values were determined for these outcomes to be classified. The outcomes ranged from 1 to 3. Depending on the choice, they were split into three equal components.

ANN and ANFIS methods give different separate predictive values, mean squared error (MSE), and classification accuracy values. The classification accuracy and MSE are given by the following equations:

$$\text{Classification accuracy} = \text{Corrected number predictions/Total predictions.}$$

$$\text{Mean squared error} = \frac{1}{n}\sum_{k=0}^{n}\left(n_{i=1}\right)\left(f_i - f_i{}^*\right)^2,$$

where f_i is the actual output value of data and f_i is the predictive values predicted by the models ANFIS and ANN. The MSE value of the ANFIS model is approximately 0.056 and that of the ANN model is approximately 0.146. The ANFIS model has categorized the fault as 92.7, whereas the complete precision of the ANN model was 86.9. The ANFIS prediction capacity, based on (fuzzy C-means) FCM and (differential evolution) DE, is very effective than ANN, according to these outcomes.

Design of proposed work:

```
%position of sink node
sink_x=Xm/2; sink_y=Ym/2; Xpos=zeros(1,N); Ypos=zeros(1,N); basexposition=50;
baseyposition=50;
temp = [39.45,29.32,34.23,46.4,34.43,39.45,29.32,34.23, 39.45,29.32,4.43,39.45,29.3
2,34.23,46.]
y=zeros (1, 1500); Eo=2;
%deployment in first grid
```

TABLE 18.4
Accuracy and Mean Squared Error Value Classification

Parameters	ANN	ANFIS
ANFIS	0.146	0.056
Incorrect forecast	8	5
Total classification accuracy	0.845	0.916

```
ma=zeros(100,3); flag=zeros(N1);
for i=1:N1
flag(i)=0;
end
source=1; figure(1); count = 1; entries = 0; error = 0;
        for i=1:60
            for j=1:17
            x_new(1)=Xpos(101); y_new(1)=Ypos(101); x_new(2)=Xpos(i);
y_new(2)=Ypos(i);
            line(x_new,y_new, 'Color','r','Line-Style','--','linewidth',2);
            pause (0.1) count = count+1;
            if (count == 50)
               count = 1;
            end
            hold off; end
        end         if (entries>=40)
    disp ('All temp reading has been transmitted'); else
    disp ('Some integration fault has arisen due to the faulty condition');
end
```

a. **Software fault for the wrong/incorrect processing:** Figure 18.11 shows software fault when wrong processing is done such that some wrong condition is put up on the data. This type of fault may also be named white box fault.
b. **Integration fault:** There can be integration error in the system such that the data collected by some sensor node is not in the format in which it is supposed to be. Base station will consider this as the integration error.

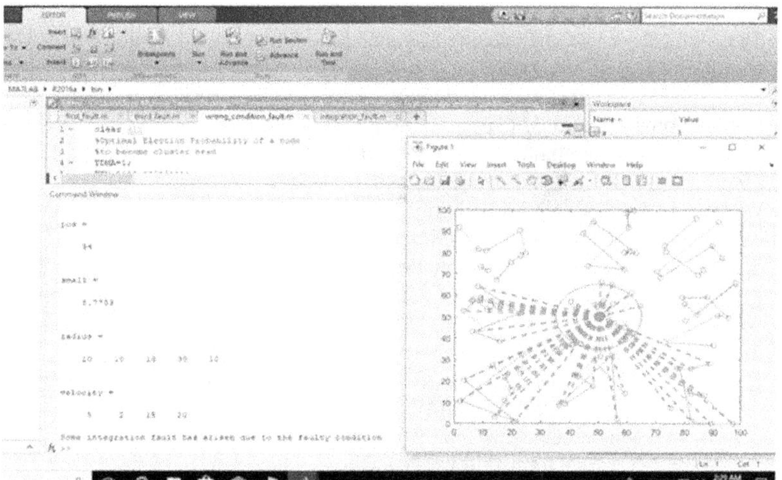

FIGURE 18.11 Integration fault.

18.10 CONCLUSION AND FUTURE RESEARCH

In this work, we discussed the state-of-the-art techniques for software fault detections in WSN and gave updated fault categorization techniques. The completeness of the fault diagnosis system is said to be achieved if sensor nodes could be diagnosed accurately by the fault detection computations along with the mitigations of the software faults. We have designed a low computational algorithm which makes our system more energy efficient.

The full procedure for the analysis of faults is presented in this chapter with different phases. In future work, we can extend our work to various different applications of real life and also for IoT devices with different wireless sensor multifunctional networks so that we will achieve better operability in networking applications. For future research, from the study we summarized the areas that need attention:

- Differentiation of error and event.
- Fault tolerance, fault replacement, and fault recovery.
- Topology independence for mobile nodes with fault detection and selection of parameters.

REFERENCES

1. Abdul-Rahman, A., & Hailes, S. (2000, January). Supporting trust in virtual communities. In: *Proceedings of the 33rd Annual Hawaii International Conference on System Sciences* (9 pp.).
2. Benini, L., Castelli, G., Macii, A., Macii, E., Poncino, M., & Scarsi, R. (2000, January). A discrete-time battery model for high-level power estimation. In: *Proceedings of the Conference on Design, Automation and Test in Europe Conference and Exhibition 2000 (Cat. No. PR00537)* (pp. 35–41).
3. Horst, R., Jewett, D., & Lenoski, D. (1993). The risk of data corruption in microprocessor-based systems. In: *The Twenty-Third International Symposium on Fault-Tolerant Computing (FTCS-23)* (pp. 576–585).
4. Siewiorek, D., & Swarz, R. (2017). *Reliable Computer Systems: Design and Evaluation.* Digital Press, Bedford, MA.
5. Barborak, M., Dahbura, A., & Malek, M. (1993). The consensus problem in fault-tolerant computing. *ACM Computing Surveys*, *25*(2), 171–220.
6. Zhao, S. Z., & Suganthan, P. N. (2011). Two-*lbests* based multi-objective particle swarm optimizer. *Engineering Optimization*, *43*(1), 1–17.
7. Dhillon, J., Parti, S. C., & Kothari, D. P. (1993). Stochastic economic emission load dispatch. *Electric Power Systems Research*, *26*(3), 179–186.
8. Syed, M., & Dubey, M. (2019, December). A novel adaptive neuro-fuzzy inference system-differential evolution (Anfis-DE) assisted software fault-tolerance methodology in wireless sensor network (WSN). In: *2019 International Conference on Computational Intelligence and Knowledge Economy (ICCIKE)* (pp. 736–741).
9. Sohraby, K., Minoli, D., & Znati, T. (2007). *Wireless Sensor Networks: Technology, Protocols, and Applications.* John Wiley & Sons.
10. Dargie, W., & Poellabauer, C. (2010). *Fundamentals of Wireless Sensor Networks: Theory and Practice.* John Wiley & Sons.
11. Ni, K., Ramanathan, N., Chehade, M. N. H., Balzano, L., Nair, S., Zahedi, S., … & Srivastava, M. (2009). Sensor network data fault types. *ACM Transactions on Sensor Networks*, *5*(3), 1–29.

12. Jalote, P. (1994). *Fault Tolerance in Distributed Systems.* Prentice-Hall, Inc., Upper Saddle River, NJ,
13. Elhadef, M., Boukerche, A., & Elkadiki, H. (2008). A distributed fault identification protocol for wireless and mobile ad hoc networks. *Journal of Parallel and Distributed Computing, 68*(3), 321–335.
14. Siewiorek, D. P., & Swarz, R. S. (1982). *The Theory and Practice of Reliable System Design.* Digital Press, Bedford, MA.
15. Gobriel, S., Khattab, S., Mossé, D., Brustoloni, J., & Melhem, R. (2006, September). Ridesharing: Fault tolerant aggregation in sensor networks using corrective actions. In: 2006 3rd Annual IEEE Communications Society on Sensor and Ad Hoc Communications and Networks (pp. 595–604).
16. Zhao, J., & Govindan, R. (2003, November). Understanding packet delivery performance in dense wireless sensor networks. In: *Proceedings of the 1st International Conference on Embedded Networked Sensor Systems* (pp. 1–13).
17. Babaie, S., Khadem-zadeh, A., & Badie, K. (2012). Distributed fault detection method and diagnosis of fault type in clustered wireless sensor networks. *Life Science Journal, 9*(4), 3410–3422.
18. Szewczyk, R., Osterweil, E., Polastre, J., Hamilton, M., Mainwaring, A., & Estrin, D. (2004). Habitat monitoring with sensor networks. *Communications of the ACM, 47*(6), 34–40.
19. Ramanathan, N., Balzano, L., Burt, M., Estrin, D., Harmon, T., Harvey, C., ... & Srivastava, M. (2006). Rapid deployment with confidence: Calibration and fault detection in environmental sensor networks. UCLA: Center for Embedded Network Sensing. Available at: https://escholarship.org/uc/item/8v26b5qh
20. Li, H., Price, M. C., Stott, J., & Marshall, I. W. (2007, September). The development of a wireless sensor network sensing node utilising adaptive self-diagnostics. In: Hutchison, D., Katz, R. H. (eds) *Self-Organizing Systems. IWSOS 2007. Lecture Notes in Computer Science, vol 4725* (pp. 30–43). Springer, Berlin, Heidelberg.
21. Lee, M. H., & Choi, Y. H. (2008). Fault detection of wireless sensor networks. *Computer Communications, 31*(14), 3469–3475.
22. Chessa, S., & Santi, P. (2002). Crash faults identification in wireless sensor networks. *Computer Communications, 25*(14), 1273–1282.
23. Vuran, M. C., Akan, Ö. B., & Akyildiz, I. F. (2004). Spatio-temporal correlation: Theory and applications for wireless sensor networks. *Computer Networks, 45*(3), 245–259.
24. Miao, X., Liu, K., He, Y., Papadias, D., Ma, Q., & Liu, Y. (2013). Agnostic diagnosis: Discovering silent in wireless sensor networks. *IEEE Transactions on Wireless Communications, 12*(12), 6067–6075.
25. Luo, X., Dong, M., & Huang, Y. (2005). On distributed fault-tolerant detection in wireless sensor networks. *IEEE Transactions on Computers, 55*(1), 58–70.
26. Guo, S., Zhong, Z., & He, T. (2009, November). FIND: faulty node detection for wireless sensor networks. In: *Proceedings of the 7th ACM Conference on Networked Sensor Systems* (pp. 253–266).
27. Gheorghe, L., Rughiniş, R., Deaconescu, R., & Ţăpuş, N. (2010). Adaptive trust management protocol based on fault detection for wireless sensor networks. In: *Proceedings of the Second International Conferences on Advanced Service Computing* (pp. 216–221).
28. Koushanfar, F., Potkonjak, M., & Sangiovanni-Vincentell, A. (2002, June). Fault tolerance techniques for wireless ad hoc sensor networks. In: SENSORS, *2002 IEEE* (Vol. 2, pp. 1491–1496).
29. Luo, H., Wu, K., Guo, Z., Gu, L., & Ni, L. M.(2011). Ship detection with wireless sensor networks. IEEE Transactions on Parallel and Distributed Systems, 23(7), 1336–1343.

30. Rakhamtov, D., & Vrudhula, S. (2001, August). Time-to-failure estimation for batteries in portable electronic systems. In: *Proceedings of the 2001 International Symposium on Low Power Electronics and Design* (pp. 88–91).
31. Wang, P., Zheng, J., & Li, C. (2007, November). An agreement-based fault detection mechanism for under water sensor networks. In: *IEEE GLOBECOM 2007-IEEE Global Telecommunications Conference* (pp. 1195–1200).
32. Sakib, K. (2012). "Asynchronous failed sensor node detection method for sensor networks. *International Journal of Network Management*, 22(1), 27–49.
33. Barron, J. W., Moustapha, A. I., & Selmic, R. R. (2008, April). Real-time implementation of fault detection in wireless sensor networks using neural networks. In: *Fifth International Conference on Information Technology: New Generations (ITNG 2008)* (pp. 378–383).
34. O'Hare, G., O'Grady, M., & Marsh, D. (2006). Autonomic wireless sensor networks: Intelligent ubiquitous sensing. In: *Proceeding of ANIPLA 2006, International Congress on Methodologies for Emerging Technologies in Automation.*

19 AI Techniques Applied to Wind Energy

Swagat Kumar Samantaray[1] *and*
Shasanka Sekhar Rout[2]
[1]National Institute of Science and Technology, Berhampur, Odisha, India
[2]Gandhi Institute of Engineering and Technology University, Gunupur, Odisha, India

CONTENTS

19.1 INTRODUCTION

The global energy demand and abrupt escalation of gas emission due to the use of fossil fuels have given us opportunity to support huge investment into renewable energy. Over the decades, the evolution of renewable energy technologies has transcended all expectations. The availability of appropriate technologies such as hydro, solar, biomass, and wind energy sources has enable us to properly tap the renewable energy sources. The sources of renewable energy have significant impact on economic development, mainly wind energy, which has been in limelight due to its eco-friendly and clean nature. The installed capacity and production from all renewable sources have increased extensively and have also spread to more countries. Recently, wind energy has turned out to be one of the most favorable renewable sources of energy and is utilized worldwide due to its sustainability and, most importantly, its lowest adverse effects on environment. Wind power capacity

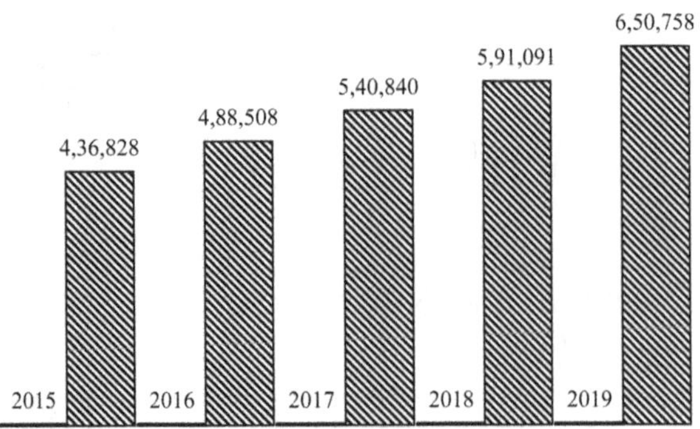

FIGURE 19.1 Wind power installation capacity.

worldwide reaches 650,758 MW according to statistics provided by the World Wind Energy Association (WWEA). Figure 19.1 gives an idea about the popularity of wind power in provisions of the installation capacity from 2015 to 2019.

The mechanical design and functioning of wind turbines are the key building blocks for harvesting wind energy. A significant amount of efforts has been applied into producing the resourceful design of wind turbine system. The wind turbine structure comprises rotor blade, pitch system, yaw system, driven train, gear box system and turbine, which is driven by generator. Because a variety of components are associated with wind turbine, around 64% of fault occurs due to blades of rotor, drive train, nacelle and generator. However, the maintenance cost in offshore or mountainous locations to ensure that the turbines operate at their finest level over their lifespan (20–25 years) is significantly high (25% of the installation). As a result, monitoring of wind turbine structure is most essential for identifying developing fault and reducing maintenance cost.

Machine learning (ML) technique and its performance are extensively useful for various structural monitoring processes. Several optimistic outcomes have pointed out that ML techniques are helpful for monitoring structure and providing better and efficient result compared to traditional method of monitoring. Artificial intelligence (AI) techniques play a significant role in monitoring of turbine structure. There are a large number of sensors associated with structural health monitoring (SHM) process and thus it is very difficult to take decision by the conventional human intelligence method. Hence, ML techniques such as regression, support vector machine (SVM), K-nearest neighbors and decision tree are able to provide solutions to complex problems and provide better decision.

This chapter gives a general outline in the thrust area of AI paradigm for SHM process for damage detection used for wind turbine and allied areas. The need for ML boosts the investment opportunities to provide the state of wind-turbine blade along with notification about the promising failure. In addition to this, some fundamental and future scopes of the ML approach for SHM have been discussed. This chapter briefs about applying ML techniques for fault assessment such as detection of damage and its localization and quantification for different parts associated with wind turbine tower. Different ML models and their prominences in competency of damage detection motivate to utilize for wind turbine structural monitoring application. Thus, this chapter focuses on the different AI techniques and acquires significant consideration and are established themselves as a novel group of intelligent methods for use in structural engineering for wind turbine.

19.2 MOTIVATION AND BACKGROUND

Attention toward the utilization of renewable energy has increased in recent years, and wind energy has received a substantial consideration due to global warming, issues related to environmental and reduction in fossil fuels [1]. The wind turbine installation capacity has increased continuously [2, 3]. However, faults also take place due to age of the structure, wear and tear of moving parts or unpredicted events, which may possibly lead to breakdown. Subsequently, the issues related to wind turbines can be overcome or condensed by implementing appropriate SHM techniques. Over the last two decades, most of the investigations were carried out toward different nondestructive evaluation technologies [4]. However, the advance in technology has been toward the development of wireless and autonomous system with large number of sensors for monitoring in real time. A huge number of sensors data, types of damage and its information will be useful to get better life safety and to trim down the maintenance cost, which leads to the use of ML techniques for efficiently scrutinizing components associated with wind turbine system, as shown in Figure 19.2.

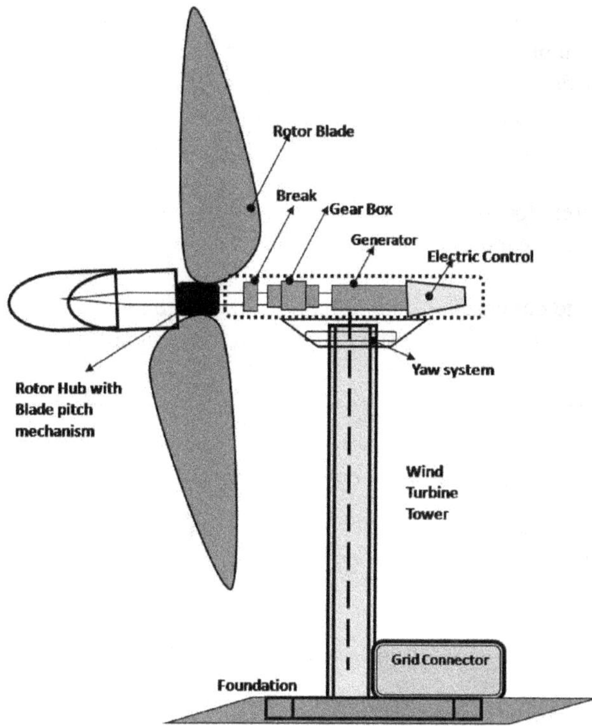

FIGURE 19.2 Components of a wind turbine [5].

In the case of wind farm, power losses may be due to inappropriate placement of wind turbine, which raises alarming situations of instability of structure. The routine scheduled maintenance of turbines, which are usually situated at remote areas such as seashore and mountain, becomes a difficult task. Hence, to ensure a safe operation condition, it uses SHM methods to judge whether any damage is present in a structure. SHM refers to the use of continuous or scheduled measurements and analysis of the key parameters of structure, which provide warnings about any kind of defect, thus avoiding fatalities [5].

19.3 STRUCTURAL DAMAGE AND ITS TYPES IN WIND TURBINE

A wind turbine functions continuously and most of the components are moveable or difficult to access when damaged. Some of the difficulties are listed below.

- Due to the height of the turbine, it is difficult to do inspection on a regular basis.
- Due to delay in inspection, there might be serious accidents reported.
- Wind turbine is usually located in remote areas such as mountainous area or seashore; thus, it is more challenging to repair and maintain.
- Lifting and handling of large equipment during installation and operation of turbine systems are also a huge concern.
- The risk occurs to be high due to its increases in capacity and expensive installation price associated with wind turbine.
- To get a better safety throughout its operation, minimize its breakdown time and provide reliable power generation, the wind turbine system must be monitored routinely to ensure its good condition of operation [6].

There are different location and parts where damage can happen, starting from concrete base to bolt shears. Some other possible damages with their location are listed in Table 19.1.

Among all the damages, the common type of failure is rotor or blade failure [8]. Thus, most of the time attention should be given to the monitoring of blade because

TABLE 19.1
Some Possible Defects and Its Location in Wind Turbine [7]

Damage Location	Possible Defects in Wind Turbine
Turbine rotor blade	Cracks, damage on the surface of blade
Hydraulic system, pneumatic system	Leakages, corrosion
Drive train	Corrosion and leakages
Nacelle and force and moment transmitting components	Cracks and corrosion in components
Wind tower and its foundation	Crack on the foundation of wind tower
Braking system and safety devices	Break and wear
Transformer station, control system and switchgear	Corrosion, dirt, loosening

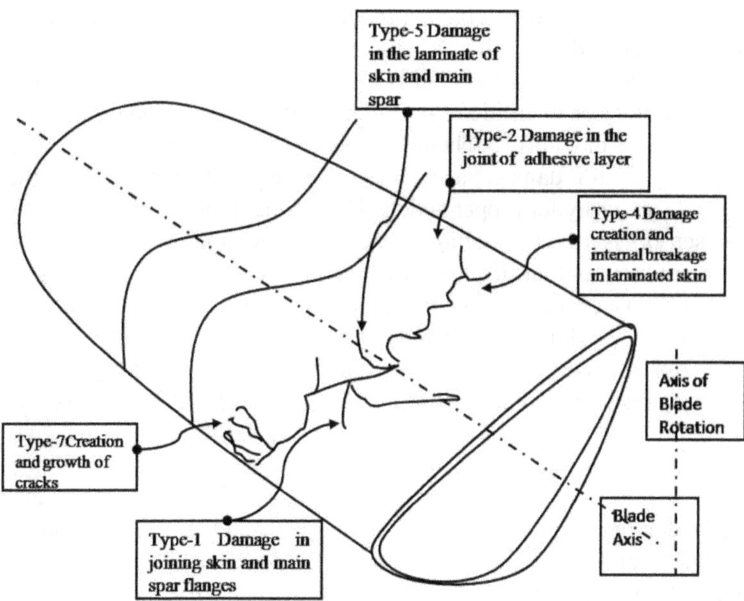

FIGURE 19.3 Common types of damage found in blade of wind turbine system [10].

it is one of the key elements for power generation. Regular maintenance is needed to avoid any unexpected failure, as the cost of the blade is around 15–20% of the total cost of turbine [9].

Different types of failure take place near the turbine blade, for example, damage formation and growth in adhesive layer, interference face, splitting, bucking, etc. Figure 19.3 shows the different damage types and their locations in a wind turbine blade. The root causes for this type of failure may be due to reduced quality control, improper installation and, most importantly, moving components, which are largely responsible for failure. Some natural calamities and lightning can cause severe damage and destruction to wind towers and its blade. Figure 19.3 shows the damage: type 1, which is found in main spar flange/adhesive layer deboning; type 2, between the skins of adhesive joint on the leading edge; and type 4, at delimitation by buckling load and internal breakage in laminated skin [10].

Recent requirements and development of power systems, rotating machinery, and offshore oil platforms are heavily dependent on structural and mechanical systems. Several of these presented systems are presently nearing the end of their unusual design life.

19.4 STRUCTURAL HEALTH MONITORING BACKGROUND

SHM is one of the greatest techniques for scientific community. Because of ageing as well as effects of environmental condition on structure, it is necessary to have continuous monitoring to avoid unwanted failure. Thus, the schedule maintenance procedure can be replaced by efficient monitoring strategy. The wind turbine size is physically large and it is difficult to do maintenance and repairing works. As a result, an efficient monitoring system is needed for early recognition of the structural degradation prior to failure to prevent any type of catastrophic collapse of the wind turbine. The monitoring scheme should have the capability to locate, detect and evaluate structural damage, as well as to subsequently communicate the status to responsible authority for proper action. The information collected through different smart sensors is used for finding the damage state of the structure to reduce the inspection duration and ensure the uninterrupted service for wind energy service providers. Also, it can deal with the fatigue fractures and calculate the remaining life of components in wind turbine on a regular basis.

The benefits of having a monitoring system for detection of damage are avoidance of premature breakdown, reduction of maintenance cost and supervision at remote sides. To make wind turbine more affordable, an SHM system needs to be integrated with wind turbine system which could be used to assess different components to prevent any failure. Various components of wind turbine which need to be monitored are rotor blade, turbine gear box, yaw system, and wind tower [5]. For online monitoring of blades and for identifying remaining life, mechanical errors and structural condition, different sensors need to be deployed at essential location.

19.4.1 Damage Detection Techniques

Power losses in wind energy occur due to improper installation of wind turbine, which can raise a frightening condition of increased turbulence, and the situation becomes worse by the structural condition of blade and tower. A continuous monitoring is necessary in order to identify the damage and to notify different type of damages with respect to SHM. Damage detection techniques such as acoustic emission, thermal imaging, ultrasonic and strain monitoring, etc., need some complex instruments and algorithms to recognize, localize and estimate the damage severity [11–14]. Also, the methods need to be capable of prediction of failure and to estimate the residual service life of the structure.

19.5 INTRODUCTION TO MACHINE LEARNING

ML is a study to develop different algorithms which build a functional relation between partial data without specialist intervention. Frequently, the learning techniques in AI deal with three different types of problems: classification, regression and density estimation. Commonly, there are two type of learning techniques: supervised and unsupervised. Supervised learning techniques deal with the regression and classification, whereas unsupervised learning techniques deal with density estimation [15]. In supervised learning process, observational evidence constructs computational models, and in unsupervised learning processes, it is inside a measurement dataset to determine the underlying pattern or configuration. The structural behavior learns by learning algorithms from the past data and carrying out pattern recognition for damage identification [16]. Commonly used ML approaches are shown in Figure 19.4.

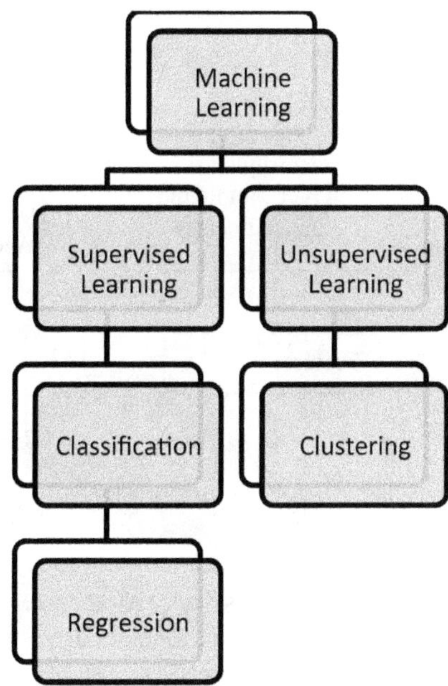

FIGURE 19.4 Commonly used machine learning approaches.

In SHM and its aligned areas, it is expected to come across some application with reference to ML, how it can be used to identify the problems in structure and provide the damage index parameters for further plan of action [17]. In addition, future behavior of a structure is also provided in extreme measures or natural calamities [18]. The ML algorithms can be divided into either supervised or unsupervised based on the data collected. First, the acquired data need to be grouped and then interpreted using the input data; then, second, it requires the information in relation to the output data. The supervised learning predicts an output variable using labeled input data, whereas unsupervised learning draws interpretations from data without labeled inputs..

The ML method's sequence of steps is given as follows:

- Data are acquired by different input. Then, data are preprocessed by performing integration and are cleaned by finding the outliers.
- Different input feature selection is done and characteristics are extracted.
- Selection of appropriate model: an appropriate model needs to be chosen.
- Validation: finally evaluate on validation set of data for performance measure.

19.5.1 Common Machine Learning Tasks

Two frequently used ML tasks are classification and regression. Classification consists of data processing and then equalization of classes to ensure that the model has

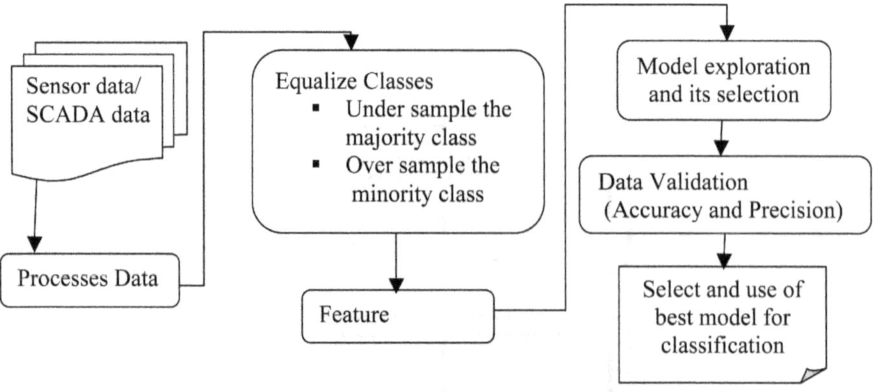

FIGURE 19.5 Workflow of classification learning process.

equal distribution of data. After that, to select and keep relevant features, suitable extraction process is used. Finally, model parameters are estimated and tested. The workflow of classification is shown in Figure 19.5.

In regression-based method, the task is to find out how signals and its features are related to output during different components. The relationship among different components is used by fitting a regression model when the system is in healthy state. Any new data that come up are compared with base state which is healthy; if any kind of divergence is found for a number of time interval, an alert message or signal is raised to indicate damage. The behavior of any components can be captured during the regression method [19]. The workflow of regression is shown in Figure 19.6.

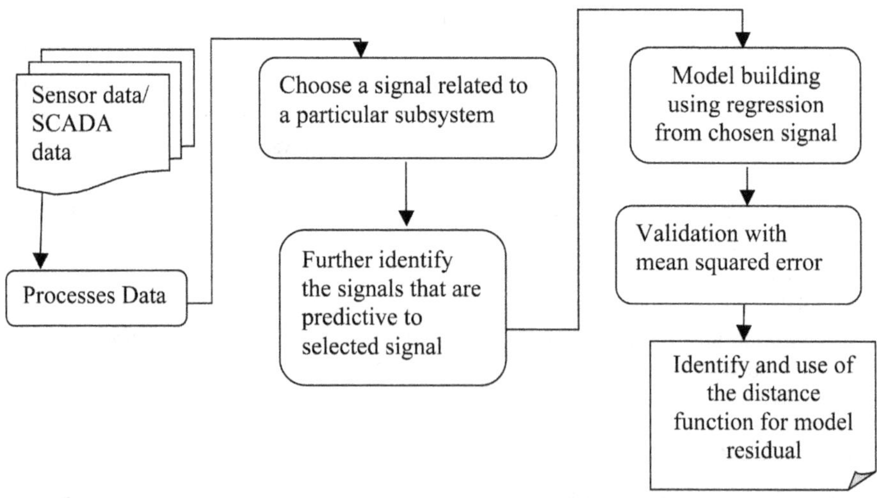

FIGURE 19.6 Workflow of regression learning process.

19.5.2 Machine Learning in Allied Areas of Wind Energy

The power of wind is always directly proportional to cube of the wind speed, and wind speed also has the essential information with regard to its variability. Thus, by forecasting the wind speed, we may be able to save not only the human hours but also the operation cost along with the maintenance cost. As wind energy is concentrated, forecasting of wind, wind wakes and ramp events is significant for the operation of wind turbine. ML has been used with several functions such as regression, classification, pattern recognition and clustering. Most known techniques such as artificial neural network (ANN) also have been used for predicting weather parameters such as the speed of the wind and its direction, which are needed for wind farm's operation. Table 19.2 lists some ML methods applied for evaluating different tasks of wind turbine.

Commonly used AI techniques comprise different feature-based algorithms that basically use sensors data mounted over wind turbine components. Important variables such as acoustic information, temperature, speed of rotor, and blade condition are frequently monitored through different sensors. For wind farms, data

TABLE 19.2
Machine Learning Methods Applied for Evaluating Performance of Wind Turbine

Task	Method	Evolution	SCADA/Simulation Data
Performance assessment of turbine	Regression	Gaussian mixture models (GMMs)	SCADA data [20]
Lubricant pressure monitoring	Regression/normal behavior mode	Mean absolute percentage error (MAPE) calculated between 2.9 and 14.25	SCADA data [21]
Temperature monitoring of generator	Regression/normal behavior mode	Nonlinear state estimation technique (NSET) achieves better modeling accuracy	SCADA data [22]
Turbine blade faults detection	Classification	RIPPER (frequent incremental pruning to create error reduction) (rule-based algorithm) with accuracy of 85.5%	SCADA data [23]
Wind turbine gearbox fault detection	Classification	MSCNN(multiscale convolution neural network for crowd counting) obtained 98.53% F1	Simulated/experimental data [24]
Fault detection, diagnosis and prediction of generator faults	Classification	Support vector machine (SVM) obtained high recall but low precision on a number of scenarios	SCADA data [25]

from supervisory control and data acquisition (SCADA) system, maintenance data and failure history may be used for damage sensitive analysis [26]. Hence, the evaluation of wind turbine structures for the finding, and quantification, of damages is essential for secure and dependable operation to avoid any catastrophic failure [27].

For recognizing the damage existence and its location, unsupervised learning techniques can be used. However, the severity of damage can be found using supervised learning techniques. Most importantly, the sensor network is only used to measure different parameters; therefore, it needs feature extraction and statistical classification to convert the sensor data into damage information.

Damage assessment involves monitoring the various components of a wind turbine to make out every change in its operation that can be investigative of an emergent fault. In the process of monitoring, analysis of specific measurements such as vibration, strain and acoustic emission is needed. Latest technology development in smart sensors, Big Data, ML algorithms and Internet of Things improves the data acquisition capability and opportunities to utilize it in reliable and robust decision making in monitoring structural damage [28].

19.5.3 Data Set Selection and Preprocessing

Any condition monitoring system for wind farm may depend on several types of statistics and datasets. Most of the data sets are based on its operation or any particular event data which may be provided by SCADA systems or different time-series data collected through sensors from parts of wind turbine. The raw data collected from SCADA system (temperature, vibration, wind speed, etc.) are a wide range of data collected over different operating conditions. Furthermore, it is very difficult to identify the actual fault from the data collected, so an appropriate data analysis tool is needed to understand the condition of wind turbine.

With the aim of extracting the accurate information from SCADA data, a preprocessing algorithm needs to be implemented which helps us to understand the behavior of data and their correlation. The algorithm is as follows:

- Step I: Identify the SCADA data of concern parts/section.
- Step II: Filter out the data collected and compare with pristine condition of wind turbine data.
- Step III: Find minimum and maximum of different parameter (for example, for wind speed it is V_{min} and V_{max}).
- Step IV: The value of N can be estimated by $N = \frac{V_{max} - V_{min}}{0.5}$.
- Step V: Find the indices for different parameters within a certain range.
- Step VI: Conduct the statistical analysis and calculated expected value.
- Step VII: Number of iterations for different range of data.

The benefit of the algorithm is that it reduces the error caused by outliers in the data collected from SCADA system. Another advantage is that the correlation between data can be extracted successfully.

19.6 ARTIFICIAL NEURAL NETWORK

ANN is one of the widely used techniques in energy sector for forecasting speed of wind, power control of turbine and, in many cases, identification of damage and its evaluation. ANN is multilayered fully connected neural nets that consist of an input layer, single or multiple numbers of hidden layers and an output layer. It has a number of elements which are processed and connected to form layers of neurons. In neural network (NN) approach, it has the capability to model a complex relationship between input and output for finding suitable patterns. It is effective enough to capture complex nonlinear relationship between parameters, without deriving any numerical equations.

The NN is trained with the data collected from a wind turbine's SCADA system. Three sets of data can be used to avoid any kind of overfitting. In wind turbine system, firstly different data sets are to be taken for testing and training and a different set of data is to be taken for validation. For example, in order to model a wind turbine rotor system, sets of both inputs and outputs are taken. Different input parameters such as wind speed, direction of wind and pitch angle (in all the three directions) and output parameters such as rotor speed can be applied to the NN model and trained and validate [29].

The neuron acts on the input data (X_i) and sends the calculated weight forward to summing nodes, as shown in Figure 19.7. To determine the internal activity of a neuron, a nonlinear transfer function is derived from each neuron with an activation function that decides whether it fires an output or not by a threshold. All the neurons are interconnected through the weights (W_{ik}), which make a decision of next level. At the junction, all the input signals from neurons are added with associated weights (W_{1k} to W_{nk}).Thus, the degree of connectivity can be determined by different value of weights. The activation function (θ) which bounded bounds the output with in a finite range of [0 to 1] or [−1 to +1] and produces an output (Y_k).

Through a supervised training process, the experimental knowledge is stored in the weights to find the connectivity among the neurons. To start the process, the weights in the neurons are set with a random value and transmitted through the network. This gives an error signal, which is a difference between the original input

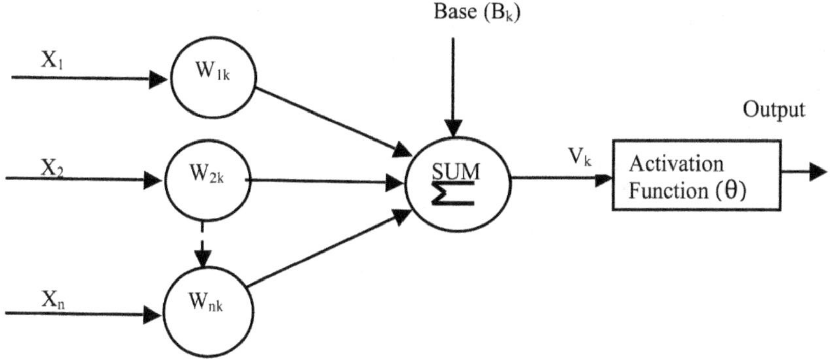

FIGURE 19.7 The neuron internal model.

and the desired output. Afterward, the error signal is feedback to the network and the weights are adjusted to minimize the scale of error. The most suitable algorithm used for adjusting the weight to minimize the error is back-propagation algorithm. This is an iterative process until the weights are constant. After the network is stabilized, it used the testing data set for estimating the as close as possible to the desired output values. This decides how well the network has learned the unknown input. An NN technique can be used to estimate the power output of the turbine with key parameters selected from SCADA data. For better output and accurate estimation, a data set of sufficient parameters needs to be modeled [30].

19.7 LOGISTIC REGRESSION TECHNIQUE FOR DAMAGE ASSESSMENT

A binary classification logistic regression algorithm is used to monitor the damage and its existence on wind turbine, especially on turbine blade. As the name suggests, the system only shows the threshold which indicates the damage in the form of binary and can determine its severity. In this algorithm, a set of training samples (n) which consists of the reduced feature vector and the information about healthy and damage conditions are provided. To develop a suitable hypothesis function, it is imitated with weighted parameter θ.

As shown in Figure 19.8, the algorithm can further categorize the input into its respective classes. The output may be recorded visually and decision may be taken to repair or replace the components under inspection [31].

The selected input features have both time and frequency domain components associated with parameters. First, the time domain data are put in a vector form of 'n' features to produce a feature matrix of $m \times n$ with 'm' number of training example.

$$X = \begin{bmatrix} x_1^1 & \cdots & x_1^n \\ \vdots & \vdots & \vdots \\ x_m^1 & \cdots & x_m^n \end{bmatrix}. \tag{19.1}$$

After selection of the feature matrix, the hypothesis is redefined to produce a discrete output which uses the most common learning algorithm logistic regression and classification. For damage detection, only two states are considered, either 'healthy state' or 'damaged state', which are noted as $Y = 1$ for healthy and $Y = 0$ for damaged.

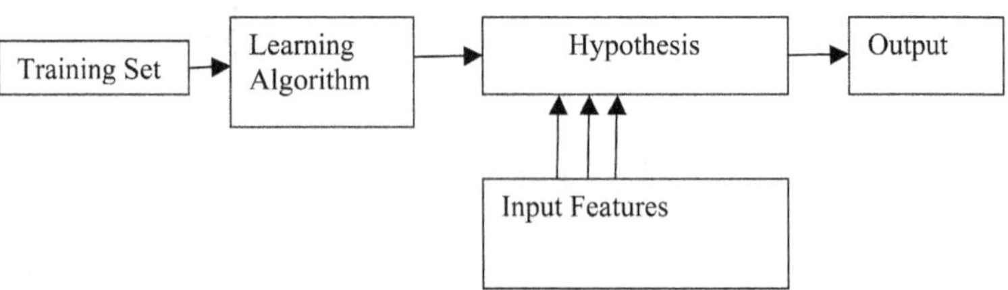

FIGURE 19.8 The supervised learning algorithm process.

The hypothesis is between 0 and 1 represented by sigmoid function. The hypothesis representation using sigmoid is specified in equation 19.2.

$$h_\theta(x) = g(\theta^T x) \rightarrow g(z) = \frac{1}{1+e^{-z}} \rightarrow h_\theta(x) = \frac{1}{1+e^{-\theta^T}x} \tag{19.2}$$

The algorithm predicts the output value of $Y = 1$ when $\theta^T x \geq 0$ and predicts $Y = 0$ when $\theta^T x \leq 0$. The hypothesis corresponds to the probability estimation of output (Y) which is equal to input (x). Thus, the hypothesis can be represented as cost function, $\text{cost}(h_\theta(x), y)$, represented in equation 19.3.

$$\text{Cost}(h_\theta(x), y) = \begin{cases} -\log(h_\theta(x)), \text{if } Y = 1 \\ -\log(1 - h_\theta(x)), \text{if } Y = 0 \end{cases}. \tag{19.3}$$

For 'm' number of training sets, the equation can be written as shown in equation 19.4.

$$j(\theta) = -\frac{1}{m}\left[\sum_{i=1}^{m} y^i \log\left(h_\theta(x^i)\right) + (1 - y^i)\log\left(1 - h_\theta(x^i)\right)\right]. \tag{19.4}$$

Several optimization algorithms are available which need to be employed to minimize the cost function, to get the best possible values to fit for the proposed hypothesis. To check the algorithm capability, an iterative process needs to be implemented with suitable parameter. Based on the training data set, the parameters which generate the lowest cost are preferred and used in hypothesis for prediction of future output. By comparison of hypothesis and cost function, it is able to find the damage and undamaged instances. Using the logistic regression algorithm for monitoring the wind turbine system has a large potential and promising future toward damage diagnosis.

19.8 PRINCIPAL COMPONENT ANALYSIS (PCA)

Principal component analysis (PCA) is one type of unsupervised, nonparametric statistical technique mostly used for dimensionality reduction in ML. This dimensionality reduction provides an opportunity to visualize the clusters of damage sensitivity features and interesting results related to classification of different damage. Consequently, the technique uses the data taken from different states of structure and creates pattern which acts as an input to the machine. For further analysis, a small set of uncorrelated variables is used for understanding the structural behavior [32]. For data comparison and noise elimination, PCA is one of the most useful methods used in SHM. The training process is shown in Figure 19.9.

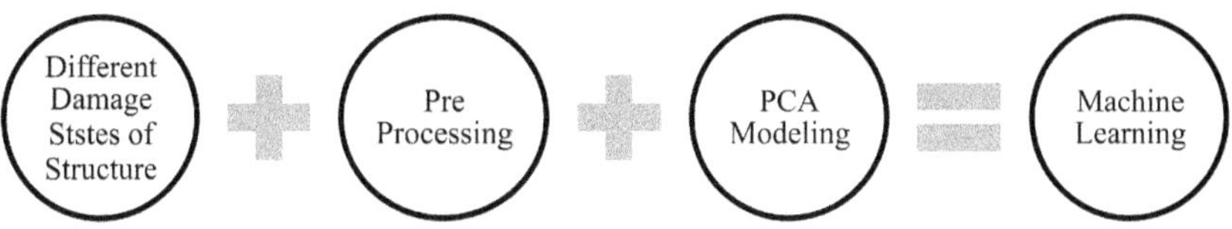

FIGURE 19.9 Data processing through training.

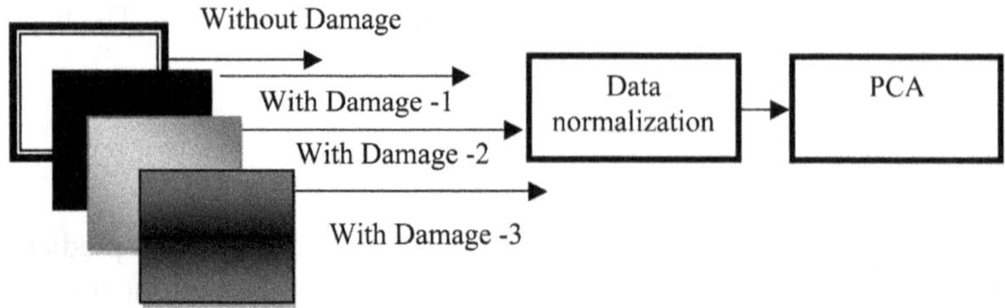

FIGURE 19.10 Organization of the input data and normalization for processing.

Commonly, the process includes number of a preprocessing step and provide with PCA modeling algorithm. The actuation phases are organized in a matrix and the data normalization techniques need to be applied which described in Figure 19.10 [33].

The training methodology consists of the following steps:

- The acceleration time histories' data are used to calculate the input data matrix.
- The data are leveled as 'without damage' and 'with damage-1', 'with damage-2' and 'with damage-3', which are shown in Figure 19.10. Further, it can be used randomly also.
- Then SVM needs to be trained with a part of the dataset.
- The rest of the data are used to test the model efficiency, by indicating different structural conditions.

After all the training processes are completed, the input data sets need to be used for principal component analysis. Moreover, it incorporates in the well-trained machine to determine the condition of the structure [34].

19.9 SUMMARY

For structural analysis of wind tower and to find the damage sensitive features by varying the operational and environmental conditions is one of the major challenges for researchers to put into practice. The recent progress in computational world has opened up many opportunities for researches and wind energy investors. Analysis of the state of wind turbine and in-depth examination to find failure in real time can provide a cost-effective solution. Improved monitoring practices by using different ML techniques can notify about the damaged condition of wind turbine components, as a result of which the maintenance engineers can plan accordingly. In the era of Big Data, the ML techniques and algorithms have the potential to use AI approaches such as ANN, logistic regression and PCA. Some of these methods have been explained in brief with a view of monitoring the wind turbine structure. The data can be a simple damage indictor, such as looseness or crack, or major damage, such as turbine blade or rotor fault. By employing the well-built mathematical formation, some of the algorithms can be used for representing damage statistics and characterizing structural behaviors and its useful parameters.

REFERENCES

1. Fthenakis, V., and H. C. Kim. *Land use and electricity generation: A life-cycle analysis.* Renewable and Sustainable Energy Reviews, 2009. **13**(6–7): pp. 1465–1474.
2. Leung, D. Y., and Y. Yang. *Wind energy development and its environmental impact: A review.* Renewable and Sustainable Energy Reviews, 2012. **16**(1): pp. 1031–1039.
3. Premalatha, M., T. Abbasi, and S. A. Abbasi. *Wind energy: Increasing deployment, rising environmental concerns.* Renewable and Sustainable Energy Reviews, 2014. **31**: pp. 270–288.
4. De Souza Rabelo, D., et al. *Impedance-based structural health monitoring and statistical method for threshold-level determination applied to 2024-T3 aluminum panels under varying temperature.* Structural Health Monitoring, 2017. **16**(4): pp. 365–381.
5. Ashley, F., et al. Bethany Wind Turbine Study Committee Report, 2007.
6. Hameed, Z., et al. *Condition monitoring and fault detection of wind turbines and related algorithms: A review.* Renewable and Sustainable Energy Reviews, 2009. **13**(1): pp. 1–39.
7. Germanischer Lloyd. *Wind energy, GL wind.* Possible Wind Turbine Damage, 2007. **30**: p. 11. http://www.glgroup.com/industrial/glwind/3780.htm
8. Ciang, C.C., J.-R. Lee, and H.-J. Bang. *Structural health monitoring for a wind turbine system: A review of damage detection methods.* Measurement Science and Technology, 2008. **19**(12): p. 122001.
9. Jureczko, M., M. Pawlak, and A. Mężyk. *Optimisation of wind turbine blades.* Journal of Materials Processing Technology, 2005. **167**(2–3): pp. 463–471.
10. Sørensen, B.F., et al. Improved design of large wind turbine blade of fibre composites based on studies of scale effects (Phase 1). Summary report. Riso-R1390 (EN), Risø National Laboratory, Denmark, 2004.
11. Joosse, P., et al. *Acoustic emission monitoring of small wind turbine blades.* Journal of Solar Energy Engineering, 2002. **124**(4): pp. 446–454.
12. Avdelidis, N., et al. Structural integrity assessment of materials by thermography. In: Conference on Damage in Composite Materials (CDCM), Stuttgart, Germany. 2006.
13. Raišutis, R., E. Jasiūnienė, and E. Žukauskas. *Ultrasonic NDT of wind turbine blades using guided waves.* Ultragarsas "Ultrasound", 2008. **63**(1): pp. 7–11.
14. Schroeder, K., et al. *A fibre Bragg grating sensor system monitors operational load in a wind turbine rotor blade.* Measurement Science and Technology, 2006. **17**(5): p. 1167.
15. Farrar, C. R., and K. Worden. Structural Health Monitoring: A Machine Learning Perspective. 2012. Hoboken, NJ: John Wiley & Sons.
16. Freudenthaler, B., et al. Case-based decision support for bridge monitoring. In: Third International Multi-Conference on Computing in the Global Information Technology (ICCGI 2008), Athens, 2008.
17. Worden, K., and G. Manson. *The application of machine learning to structural health monitoring.* Philosophical Transactions of the Royal Society A: Mathematical, Physical and Engineering Sciences, 2007. **365**(1851): pp. 515–537.
18. Singh, S., W. K. Seah, and B. Ng. Cluster-centric medium access control for WSNs in structural health monitoring. In: 13th International Symposium on Modeling and Optimization in Mobile, Ad Hoc, and Wireless Networks (WiOpt), Mumbai, 2015.
19. Stetco, A., et al. *Machine learning methods for wind turbine condition monitoring: A review.* Renewable Energy, 2019. **133**: pp. 620–635.
20. Lapira, E., et al. *Wind turbine performance assessment using multi-regime modeling approach.* Renewable Energy, 2012. **45**: pp. 86–95.
21. Wang, L., et al. *Wind turbine gearbox failure identification with deep neural networks.* IEEE Transactions on Industrial Informatics, 2016. **13**(3): pp. 1360–1368.

22. Guo, P., D. Infield, and X. Yang. *Wind turbine generator condition-monitoring using temperature trend analysis.* IEEE Transactions on Sustainable Energy, 2011. **3**(1): pp. 124–133.
23. Godwin, J. L., and P. Matthews. *Classification and detection of wind turbine pitch faults through SCADA data analysis.* International Journal of Prognostics and Health Management, 2013. **4**: p. 90.
24. Jiang, G., et al. *Multiscale convolutional neural networks for fault diagnosis of wind turbine gearbox.* IEEE Transactions on Industrial Electronics, 2018. **66**(4): pp. 3196–3207.
25. Leahy, K., et al. *Diagnosing and predicting wind turbine faults from SCADA data using support vector machines.* International Journal of Prognostics and Health Management, 2018. **9**(1): pp. 1–11.
26. Yang, W., R. Court, and J. Jiang. *Wind turbine condition monitoring by the approach of SCADA data analysis.* Renewable Energy, 2013. **53**: pp. 365–376.
27. Dao, P. B., et al. *Condition monitoring and fault detection in wind turbines based on cointegration analysis of SCADA data.* Renewable Energy, 2018. **116**: pp. 107–122.
28. Moness, M., and A. M. Moustafa. *A survey of cyber-physical advances and challenges of wind energy conversion systems: Prospects for internet of energy.* IEEE Internet of Things Journal, 2015. **3**(2): pp. 134–145.
29. Tian, Z., et al. *Condition based maintenance optimization for wind power generation systems under continuous monitoring.* Renewable Energy, 2011. **36**(5): pp. 1502–1509.
30. Nithya, M., S. Nagarajan, and P. Navaseelan. Fault detection of wind turbine system using neural networks. In: IEEE Technological Innovations in ICT for Agriculture and Rural Development (TIAR), Chennai, 2017.
31. Regan, T., et al. Wind turbine blade damage detection using various machine learning algorithms. In: International Design Engineering Technical Conferences and Computers and Information in Engineering Conference, Charlotte, NC, 2016.
32. Vitola, J., et al. Data-driven methodologies for structural damage detection based on machine learning applications. In: Ramakrishnan, S. (Ed.), *Pattern Recognition: Analysis and Applications.* 2016. London: IntechOpen.
33. Bulut, A., et al. Real-time nondestructive structural health monitoring using support vector machines and wavelets. In: Advanced Sensor Technologies for Nondestructive Evaluation and Structural Health Monitoring. 2005, 5770: pp. 180–189. International Society for Optics and Photonics.
34. Santos, P., et al. *An SVM-based solution for fault detection in wind turbines.* Sensors, 2015. **15**(3): pp. 5627–5648.

20 Comparative Performance Analysis of Multi-Objective Metaheuristic Approaches for Parameter Identification of Three-Diode-Modeled Photovoltaic Cells

Saumyadip Hazra and Souvik Ganguli
Department of Electrical and Instrumentation Engineering, Thapar Institute of Engineering and Technology, Patiala, Punjab, India

CONTENTS

20.1 INTRODUCTION

For coping with the problem of environmental pollution, increase in the shortage of energy and change in climate, many attempts have been made which focused on the research and development of the renewable energy resources [1]. Renewable energy resources have received large interest throughout the world because of the presence of many issues such as cost of fossil fuel, its depletion and political concerns [2]. Energy calamity and fuel collapse are crucial challenges which signify the vitality of renewable resources. Among all renewable energy resources present currently such as tidal wave, wind, nuclear, biomass, geothermal, and many others, solar energy is considered as the one with the highest potential because of the unlimited availability and its nonpolluting nature [3]. Leaving aside the initial cost of generation of power through photovoltaic (PV) and the limitation that only during certain times of day power can be generated and its seasonal dependency, it is known to be a popular method for generation of power. Its eco-friendly nature has given a big advantage for power generation, and it is usually used to generate power for all area sizes, ranging from small to large. It also plays a crucial role in the electric power sector and its dispersion in electric power production sector is increasing day by day [2]. New technologies are being used in PV and are encouraged as they are helping in cutting the costs by implementing less complex production methods and the materials are cheaper. Monocrystalline and polycrystalline are the two major types of PV cells which are manufactured by sawing thin wafers from cautiously cooled and hardened silicon slabs acquired from a method of crystal growth. Thin films are another type of PV cell, which are manufactured by depositing one or more thin films of PV material on lightweight and stretchy substrate. They are easier to manufacture and are more resistant to different factors than the delicate monocrystalline and polycrystalline cells [4, 5].

For the purpose of controlling the PV systems, it becomes important to have knowledge about the characteristic behavior of the PV cell based on their correct mathematical model and also by plotting the voltage-current relationship. The accuracy of PV cells or arrays largely depends on their model considerations and parameters. However, these parameters generally are not provided by the manufacturer and change gradually due to use of the cell, faults and various extreme operating conditions [6]. Still, they always have a significant impact on the efficiency of the cell throughout their life. Hence, a challenge is presented when power is to be produced from the PV cell without knowing its performance parameters. Therefore, a model is required for determining the characteristics of the PV cell accurately so that its nonlinear curves can be plotted successfully and further analysis on its control and evaluation can be carried out. From the knowledge of accurate mathematical model of PV cell, the cell parameters can be determined which are important because the output power is directly dependent on the solar irradiance. These parameters either can be estimated from the manufacturer's datasheet or can be determined experimentally [7]. This has inspired the advancement of different types of parameter extraction methods over the last years. Hence, it is necessary to propose an efficient, practicable and reliable method for accurately calculating the parameters of solar PV cells [3, 6–9].

The three-diode model (TDM) for a PV cell is a recently formulated model in which all the losses occurring in a diode are represented and hence it is more accurate than the other models. However, it is completely true that the single-diode model of the cell gives very dynamic and fast response and also has a simple structure, whereas the double-diode model of a cell lies between the single-diode model and the three-diode model. The single-diode model is a compromise between accuracy and simplicity with only five parameters to be estimated. The double-diode model of a cell has a total of seven parameters to be determined which makes it achieve higher accuracy as compared to single-diode model of a PV cell. The three-diode model of a PV cell is considered for this study of the parameters of the PV cell. The extraction of accurate parameters for three-diode model is critical in control, design and performance estimation. Also, new challenges are introduced as increased accuracy of three-diode model also increases the number of unknown parameters to be extracted and, as a result of this, the complexity of the problem for parameter estimation also increases. Further, the three-diode model is considered as the most accurate model, especially in the case of low-radiation operating conditions, and has nine unknown parameters [10–13]. Therefore, the conclusion that can be drawn is that a reliable method for extraction of parameters is required.

The analytical methods are usually employed for solving the implicit equations. The analytical methods include equations which are formulated through basic relations among the variables. These relations are then used for the determination of characteristic points of the current-voltage (IV) and PV curves or they use some means of approximations to convert equations into explicit form to make them solvable easily. However, it is true that calculations in these methods are quite quick and simple, but several methods are nonlinear and have multiple variables in them and some may be multimodal equations where there are many maxima and minima present. In these cases, the solution becomes very tough and tedious to obtain [4]. Due to the limitations, the analytical methods are not used for the extraction of parameters. Another approach for the solution is the numerical-based methods which employ the numerical solutions of the problem and calculate the result of objective function at each step [14]. The numerical-based methods are further extended as deterministic heuristic methods and stochastic heuristic methods. The deterministic methods include Newton-Raphson, Nelder-Mead simplex and Levenberg-Marquardt. Generally, these methods show a very quick response toward the convergence and the results obtained are also correct but they suffer from a limitation which is their solution is entirely dependent on the initial value chosen and many times suffer to give good and correct results for multimodal problems [15].

Artificial intelligence (AI) is a field of study that deals with the special capabilities shown by machines, which allows it to perceive the processes happening in its surroundings and allows it to adopt the necessary steps for ensuring the maximum possible output for reaching the targeted goal. AI, in some place or the other, requires optimization for effective decision-making regarding the right step to be taken next for maximizing the output. The actual mathematical optimization processes also employ AI techniques and fall under the category of heuristics, known as metaheuristic algorithms, which are much popular these days as they provide much better results than the other methods. These algorithms are known for their capabilities of finding the optimum solutions under the given conditions for the given set of problem and are population-based methods where the whole population is randomly assigned with positions initially. During the course of iterations, the solution is improved using the decision-making capability till the desired accuracy result is obtained. Metaheuristic algorithms have been used extensively for the estimation of parameters of a PV cells as good results are provided by them. Therefore, stochastic heuristic methods are being employed to calculate the parameters of the PV model and have been improved constantly in the past decade for obtaining better results. The main advantage of these algorithms is that they impose no restriction on the total number of variables in the problem as well as the total number of constraints as they are completely population based [2, 16].

In this chapter, modeling of the PV cell is done using the three-diode model and results are calculated using four metaheuristic algorithms. The results have been tabulated and corresponding IV and PV curves are also plotted. The future scope for this chapter lies in the use of other types of PV cell models and use of other metaheuristic algorithms.

The rest of the chapter has been presented in the following manner. Section 20.2 presents the problem statement in which the mathematical model and the corresponding equations are derived. Section 20.3 is the methodology section where a description is given on the different metaheuristic algorithms employed for the study. Section 20.4 represents the results section, where the results obtained are tabulated along with their corresponding IV and PV curves and are interpreted. Finally, Section 20.5 draws the conclusion from the whole chapter.

20.2 PROBLEM STATEMENT

In three-diode model of PV cell, three diodes are connected in parallel representing three different types of recombination losses in the cell. The ideal mathematical modeling of the diode is done with the help of Shockley's equation for determining the current passing through each diode. Hence, the diode current is given as:

$$I = I_0\left[\exp\left(\frac{q(V+IR_s)}{aKT}\right)-1\right], \tag{20.1}$$

where I_0 signifies the reverse saturation current, a represents the ideality factor of diode, R_s is the resistance connected in series, K is the Boltzmann constant ($1.3806503 \times 10^{-23}$ J/K), T is the cell temperature in kelvin (K) and q is the electronic charge ($1.60217646 \times 10^{-19}$ C). The series resistance represents the obstruction caused in the path of the current due to the various elements connected, and in the modeling another resistance, known as parallel resistance, R_p, is present which represents the recombination due to electron hole pair. Generally, all the calculations regarding the cell are carried out at STC. In this chapter as well, all the analysis regarding the parameter extraction is carried out in STC.

Now, the results obtained after applying Kirchhoff's current law (KCL) in Figure 20.1 are:

$$I = I_{pv} - I_{D1} - I_{D2} - I_{D3} - \frac{V+IR_s}{R_p} \tag{20.2}$$

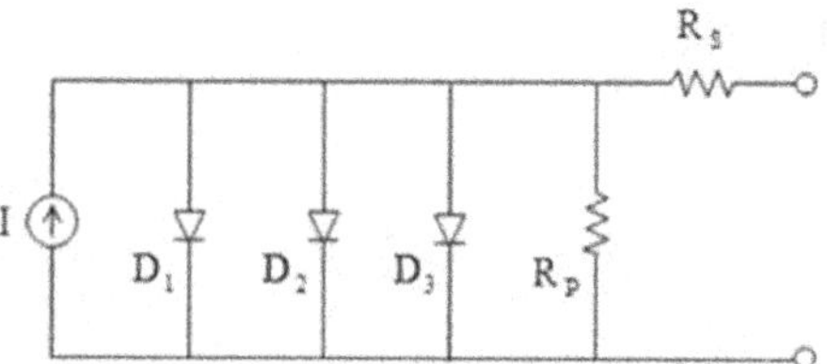

FIGURE 20.1 Three-diode model of PV cell [17].

In the above equation, the values of the reverse saturation currents are substituted from equation 20.1. The results obtained are:

$$I = I_{pv} - I_{01}\left[\exp\left(\frac{q(V + IR_S)}{a_1 KT}\right) - 1\right] - I_{02}\left[\exp\left(\frac{q(V + IR_S)}{a_2 KT}\right) - 1\right]$$
$$I_{03}\left[\exp\left(\frac{q(V + IR_S)}{a_3 KT}\right) - 1\right] - \frac{V + IR_S}{R_p} \quad (20.3)$$

The nine unknown parameters in the above equation are I_{pv}, a_1, a_2, a_3, R_s, R_p, I_{01}, I_{02}, I_{03}. The equation is implicit; therefore, some conditions are to be applied for the generation of constraints [18]:

1. Open circuit condition, $V = V_{OC},\ I = 0$,

$$I_{pv} = I_{01}\left[\exp\left(\frac{qV_{oc}}{a_1 KT}\right) - 1\right] + I_{02}\left[\exp\left(\frac{qV_{oc}}{a_1 KT}\right) - 1\right] + I_{03}\left[\exp\left(\frac{qV_{oc}}{a_1 KT}\right) - 1\right] + \frac{V_{oc}}{R_p} \quad (20.4)$$

2. Short circuit condition, $V = 0,\ I = I_{SC}$,

$$I_{sc} = I_{ph} - I_{01}\left[\exp\left(\frac{q(IR_S)}{a_1 KT}\right) - 1\right] - I_{02}\left[\exp\left(\frac{q(IR_S)}{a_2 KT}\right) - 1\right]$$
$$- I_{03}\left[\exp\left(\frac{q(IR_S)}{a_3 KT}\right) - 1\right] - \frac{IR_s}{R_p} \quad (20.5)$$

3. Maximum power point (MPP) condition, $V = V_{mp},\ I = I_{mp}$,

$$I_{mp} = I_{pv} - I_{01}\left[\exp\left(\frac{q(V_{mp} + I_{mp}R_S)}{a_1 KT}\right) - 1\right] - I_{02}\left[\exp\left(\frac{q(V_{mp} + I_{mp}R_S)}{a_2 KT}\right) - 1\right]$$
$$- I_{03}\left[\exp\left(\frac{q(V_{mp} + I_{mp}R_S)}{a_3 KT}\right) - 1\right] - \frac{V_{mp} + I_{mp}R_S}{R_p} \quad (20.6)$$

4. The slope of the PV curve at MPP is given as $\frac{dI}{dV} = -\frac{I_{mpp}}{V_{mpp}}$,

$$I_{mp} = \left(V_{mp} - I_{mp}R_s\right)\left[\left\{\frac{I_{01}q}{a_1KT}\exp\left(\frac{q\left(V_{mp}+I_{mp}R_s\right)}{a_1KT}\right)\right\} + \left\{\frac{I_{02}q}{a_2KT}\exp\left(\frac{q\left(V_{mp}+I_{mp}R_s\right)}{a_2KT}\right)\right\}\right.$$
$$\left. + \left\{\frac{I_{03}q}{a_3KT}\exp\left(\frac{q\left(V_{mp}+I_{mp}R_s\right)}{a_3KT}\right)\right\} + \frac{1}{R_p}\right] \tag{20.7}$$

5. The slope of the IV curve is given as $\frac{dI}{dV} = -\frac{1}{R_{par}}$,

$$\frac{R_s}{R_p} = \left(R_p - R_s\right)\left[\left\{\frac{I_{01}q}{a_1KT}\exp\left(\frac{qI_{sc}R_s}{a_1KT}\right)\right\} + \left\{\frac{I_{02}q}{a_2KT}\exp\left(\frac{q(I_{sc}R_{se})}{a_2KT}\right)\right\}\right.$$
$$\left. + \left\{\frac{I_{03}q}{a_3KT}\exp\left(\frac{qI_{sc}R_s}{a_3KT}\right)\right\}\right] \tag{20.8}$$

The errors that are generated because of the above equations can be calculated as:

$$Err_{OC} = I_{01}\left[\exp\left(\frac{qV_{OC}}{a_1KT}\right) - 1\right] + I_{02}\left[\exp\left(\frac{qV_{OC}}{a_2KT}\right) - 1\right] + I_{03}\left[\exp\left(\frac{qV_{OC}}{a_3KT}\right) - 1\right]$$
$$+ \frac{V_{OC}}{R_p} - I_{pv} \tag{20.9}$$

$$Err_{SC} = I_{SC} + I_{01}\left[\exp\left(\frac{q(IR_S)}{a_1KT}\right) - 1\right] + I_{02}\left[\exp\left(\frac{q(IR_S)}{a_2KT}\right) - 1\right] + I_{03}\left[\exp\left(\frac{q(IR_S)}{a_3KT}\right) - 1\right]$$
$$+ \frac{I_{SC}R_S}{R_p} - I_{pv}. \tag{20.10}$$

$$Err_{mp} = I_{pv} - I_{01}\left[\exp\left(\frac{q\left(V_{mp}+I_{mp}R_S\right)}{a_1KT}\right) - 1\right] - I_{02}\left[\exp\left(\frac{q\left(V_{mp}+I_{mp}R_S\right)}{a_2KT}\right) - 1\right]$$
$$- I_{03}\left[\exp\left(\frac{q\left(V_{mp}+I_{mp}R_S\right)}{a_3KT}\right) - 1\right] - \frac{V_{mp}+I_{mp}R_S}{R_p} - I_{mp} \tag{20.11}$$

$$E_{PV} = \left(V_{mp} - I_{mp}R_s\right)\left[\left\{\frac{I_{01}q}{a_1KT}\exp\left(\frac{q\left(V_{mp}+I_{mp}R_s\right)}{a_1KT}\right)\right\}+\left\{\frac{I_{02}q}{a_2KT}\exp\left(\frac{q\left(V_{mp}+I_{mp}R_s\right)}{a_2KT}\right)\right\}\right.$$
$$\left.+\left\{\frac{I_{03}q}{a_3KT}\exp\left(\frac{q\left(V_{mp}+I_{mp}R_s\right)}{a_3KT}\right)\right\}+\frac{1}{R_p}\right]-I_{mp} \tag{20.12}$$

$$E_{IV} = \left(R_p - R_s\right)\left[\left\{\frac{I_{01}q}{a_1KT}\exp\left(\frac{qI_{sc}R_s}{a_1KT}\right)\right\}+\left\{\frac{I_{02}q}{a_2KT}\exp\left(\frac{q(I_{sc}R_s)}{a_2KT}\right)\right\}\right.$$
$$\left.+\left\{\frac{I_{03}q}{a_3KT}\exp\left(\frac{q(I_{sc}R_s)}{a_3KT}\right)\right\}\right]-\frac{R_s}{R_p} \tag{20.13}$$

The final compilation of error is done on the basis of square error as it is able to produce accurate results [19]:

$$E = E_{oc}^{\ 2} + E_{sc}^{\ 2} + E_{mp}^{\ 2} + E_{PV}^{\ 2} + E_{IV}^{\ 2} \tag{20.14}$$

20.3 METHODOLOGY

In engineering, there is a major role played by the optimization for the solution of problems. Optimization is the process of selecting the best alternative of the given set of alternatives for a particular problem(s). In the simplest words, they maximize or minimize a particular problem in order to calculate its global maxima or minima, respectively, by checking that all the variables in it are safely inside the determined limits. The optimization problems are divided into two categories: single objective optimization and multi-objective optimization problems. In single objective optimization, there is a single objective function which needs to be optimized. The results obtained are the global maxima and the results are generally easy to interpret and compare as there is only one objective function involved. But most of the problems in engineering are multi-objective, i.e., they include two or more objectives to be optimized. In this case, more than one maximum and minimum are found and out of them the best result has to be chosen. There are two ways for solving a multi-objective problem: priori method and posteriori method. In priori methods, the multi-objective is converted into single objective by applying some weights. The weights are decided on the basis of the importance of the objective on the problem and then the converted single objective is optimized. Whereas, in posteriori method, the multi-objective nature of the problem is maintained and the variables are searched over a wide range based on the conditions applied to it [20–22].

20.3.1 Multi-Objective Gray Wolf Optimization (MOGWO)

The gray wolf is an animal which belongs to the canine family and are generally found in Eurasia and North America. The gray wolves are known for their hunting method and are considered as very good hunters. They are generally found in a small group of 5–12 members, and among all the members, they maintain a social relationship, out of which one is the leader, known as alpha, and is always the best in management and planning. The next post is beta, which acts as informers to the alpha and also checks the discipline among the remaining wolves. After beta, there is delta who follows the orders of beta and alpha, and finally there is omega, which is the lowest post the group. After the wolves locate their prey, they start encircling them. The alpha leads the hunt and is participated by the beta and delta. After encircling, they continuously update the position of themselves based on the position of their leader and prey. The change in position of the wolves is guided by the alpha. This is the exploration phase where the wolves make their prey tired. Once the prey stops moving, the exploitation phase starts and they attack the prey to complete their hunt [23]. For performing multi-objective calculations with the help of GWO, firstly an archive is created to store the nondominated Pareto optimal solutions. It acts as a tank for the nondominated solutions. An archive controller is made which is responsible for controlling the solutions entering the archives. A solution is only allowed to enter the archive when it dominates the existing ones. When the archive is full, then first it creates a hypercube, which stores the most crowded region of the archive and a solution is deleted from it randomly. There is also a leader selection technique in which the new leaders are selected from the archive unlike the GWO, where the solutions obtained by alpha, beta and delta were assumed to be the best solution obtained till then. The selection is done from the least crowded area of the archive where the probability is maximum. Roulette wheel selection method is applied to find the least crowded area and randomly they are assigned as alpha, beta and delta, respectively [24].

20.3.2 Multi-Objective Grasshopper Optimization Algorithm (MOGOA)

Grasshoppers are small insects which eat plants and other vegetation. They take part in their swarm, which can be very large, for getting food and migration. Both the adult and the larvae take part in the swarm. The larvae are not able to move very fast and take small steps, whereas the adult grasshopper can jump from one plant to another rapidly. During the exploration phase, the grasshoppers move abruptly. This algorithm includes three main features of the grasshopper, out of which the first is social interaction. The social interaction is largely dependent on the distance between two neighboring grasshoppers and also on a term known as social factor. The grasshoppers move freely when it is in the comfort area; otherwise, it may get attracted or repelled from the other grasshopper. The next is the effect of gravitational forces. These forces are experienced by them when they jump. The third important feature is the effect of wind. Their movement, especially for the nymph grasshopper which does not have any wings, depends on the direction of wind. The position of the grasshoppers is updated at the end of every iteration, based on the effect of these factors, and during the exploitation phase, the algorithm is made to converge toward the food source [25]. For the multi-objective operation of the algorithm, firstly to save the nondominated results obtained so far from the Pareto front, an archive is created. The mathematical equations are the same as that of the original algorithm. It is assumed that the solution is present in the archive. The key task in MOGOA is to track down the target, as target is the optimal solution of the optimization problem. In order to achieve this, the distance of a solution is calculated from all the neighboring solutions. This result is anticipated to be equal to the number of neighbors and is used for finding the crowded areas of the Pareto solutions. The maximum probability of finding a best solution lies in the least crowded areas and roulette wheel selection is used for the same. The solutions are deleted from the archive when it is full and the most crowded one is deleted. This process is repeated till the desired result is obtained [26].

20.3.3 Multi-Objective Salp Swarm Optimization (MOSSA)

Salps are marine organisms which have a transparent body. They fill water in them and thrust out the water to move forward. They form up a long chain and travel in that chain, which is called their swarming behavior. The leading salp is known as the leader salp and all the other are the followers. The optimum solution is location of the food particle. The algorithm initiates with randomizing the positions of all the salps and calculating the fitness value for each salp. The one with the best fitness is considered as the leader and the chase for the food particle starts. After each iteration, the fitness of each salp is calculated for determining if there is any other salp which has become fitter. If any salp during their hunt goes out of the food chain, then it is brought back into the chain. The best solution or the position of the food particle is updated during the course of iteration as there is a possibility that the salps may find a better food particle. This process is done till the salps find the optimum solution. For the multi-objective approach of this algorithm, first an archive is created in order to store the nondominated Pareto solutions as the original algorithm saves only the best solution that is obtained till then and not all of them. Each salp is compared with the ones present in the archive. If it is fitter than any of them inside the archive, then that salp is deleted and the new one is added. This is how the archive only stores the nondominated solutions. When the archive becomes full, then, in place of randomly removing the salps from the archive, the distance of a particular salp with all the other neighboring salp is calculated and is then used in the roulette wheel selection method to find the most crowded area of the archive. Then the most crowded area is deleted. One problem faced by the original algorithm is its inability to select the best solution. For solving this issue, the best solution is selected from the archive where there is least crowded area. Then the probability is calculated of that area and the best solution is obtained [27].

20.3.4 Multi-Objective Dragonfly Algorithm (MODA)

Dragonflies are small insects which can fly and live by eating other small insects. The nymph phase of dragonflies also hunts other small insects and eats them. There are two types of swarming behavior shown by dragonflies: static and dynamic. In static swarming behavior, the flies travel in small groups and search for insects for their foods. In dynamic swarming behavior, they form groups and migrate from one place to another. The behaviors shown by dragonflies in their swarm are of five types:

separation, which includes aversion of collision from other dragonflies from their neighborhood; alignment, which includes the matching of speed of one dragonfly with others; cohesion, which includes the movement along the center of mass of the swarm; attraction toward the food source; and distraction from the enemy. The algorithm starts with the assumption of positions of the dragonflies at random position within the bounds as defined. At the end of each iteration, the positions of the dragonflies are updated with the help of the five types of forces. This process is done until the satisfied results are obtained. For solving multi-objective optimization problems with the help of this algorithm, first the algorithm is equipped with an archive which is added to save nondominated Pareto solutions. For the selection of food source, the solutions present in the archive are used. For the proper segmentation of solutions of archive, hypersphere is created. Then the least populated area or the least populated hypersphere is calculated with the help of roulette wheel selection which is based on calculating the probability of the area with the least number of solutions [28].

20.3.5 Multi-Objective Antlion Optimization (MOALO)

Antlions are small insects which hunt for smaller insects, mostly ants. They are known for their unique hunting technique in which they dig a cone-shaped pit and they lie in the bottom of it with their wide jaws opened. When any ant falls in the pit, they engulf it and further process takes place. The ants are supposed to move randomly over the search space. Hence, the population of the ants is determined randomly initially. Since their walk is random in nature, they are normalized at each iteration to keep them within the bounded limits. It is considered that antlions build their tarps based on their fitness. Antlions who have greater hunger build larger traps. The fitness of antlions is calculated at each iteration. The final step of the algorithm is when the antlion catches its prey. When an ant falls in the pit of the antlion, the antlion shoots sand toward it in order to make it slip and fall. Once it falls, the antlion eats the ant. The fittest antlion as calculated at each iteration gives the best solution obtained till then and is known as the elite. This process is repeated till the desired result is obtained [29]. The multi-objective version of the antlion algorithm is done first by employing an archive for the storage of the nondominated Pareto solutions. The best solution is picked from the archive and then it is optimized with the help of ALO. In order to manage the distribution of the archive and select the best solution, niching process is used. In this, all the neighbors of a particular solution are calculated and then are assumed to be equal to the distribution. This is then used to find the one with lowest crowdedness and for that roulette wheel selection is used. As the size of the archive is limited, solutions need to be deleted from it when it becomes full. The deletion of solution is done from the one with the largest crowdedness [30].

20.4 SIMULATION RESULTS AND DISCUSSIONS

As described in the preceding section, there are five equations whose error is minimized and the power from the cell is maximized with the help of square error equation. This is only possible with help of multi-objective optimization where the conflicting constraints are dealt with based on the posteriori multi-objective optimization.

TABLE 20.1
Datasheet Information of the Various PV Modules Incorporated

	Company			
	SolarWorld Pro SW255 Polycrystalline	AS-6P30 255W Polycrystalline	SolarWorld Plus SW280 Monocrystalline	Nemy JB270M-60 Monocrystalline
Maximum power P_m (W)	255	255	280	270
Voltage at nominal power V_{mpp} (V)	30.9	30.5	31.2	31.1
Current at nominal power I_{mpp} (A)	8.32	8.37	9.07	8.68
Open circuit voltage V_{oc} (V)	38	38.1	39.5	38.6
Short circuit current I_{sc} (A)	8.88	8.83	9.71	9.2
Number of cells per module	60	60	60	60

The results are obtained based on the manufacturer's datasheet for the four commercially available solar panels. The details about the panels considered for the study are shown in Table 20.1. Results are obtained for all the algorithms considering 50 search agents and 500 iterations. Based on the parameters, the upper and lower limits of the variables were selected accordingly and are shown in Table 20.2. A comparison is made among the algorithms presented in Section 20.3 and is tabulated along with their comparative PV and IV curves. A total of 20 runs are taken and the best result out of them is reported in the table for each of the iteration and model of PV panel.

As there are four panels, there are four tables which have been drawn comparing all of the algorithms. The comparison yields that the algorithms have given good results for all the panels and the variation of results among them is quite less. Table 20.3 shows the parameters extracted from SolarWorld Pro SW255.

The values obtained from SolarWorld Pro for different algorithms are well within the limits. The value of PV current is greater than the minimum current required and hence the results are correct. The value of the ideality factors is not much high, representing the quite ideal nature of the cell. The values of the series resistance

TABLE 20.2
Bounds of the Variables

Parameter	Lower Bound	Upper Bound
Photovoltaic current, I_{pv} (A)	5	12
Ideality factor, a	0.5	2
Series resistance, R_s (Ω)	0.001	0.1
Parallel resistance, R_p (Ω)	25	500
Reverse saturation current, I_0 (A)	0	1e-06

TABLE 20.3
Parameters Estimated for SolarWorld Pro SW255 TDM Model

Method	I_{pv}	a_1	a_2	a_3	R_S	R_P	I_{01}	I_{02}	I_{03}
MOGWO	8.9580	1.5050	1.5673	1.4771	0.0539	62.9413	4.6671e-07	4.4082e-07	3.923e-07
MOGOA	9.1252	1.5106	1.6969	1.6641	0.0025	31.9541	5.9142e-07	3.1283e-07	5.8845e-07
MOSSA	9.0942	1.5326	1.7532	1.4848	0.0016	38.4756	8.2327e-07	8.5405e-07	7.2900e-07
MODA	8.9046	1.6441	1.5743	1.3765	0.0474	63.1919	9.9481e-07	2.7689e-07	75613e-07
MOALO	9.1973	1.5363	1.7912	1.5727	0.1364	76.0239	8.8679e-07	4.3350e-07	7.846e-07

are very less, which are true and they represent the resistance generated due to the obstructions in the path. Coming to the reverse saturation currents, they are of the order of 10^{-7}, representing quite ideal values. Their corresponding PV and IV curves are represented in Figure 20.2, where it can be seen that all the algorithms have an overlapping effect over each other, showing the convergence of the algorithms.

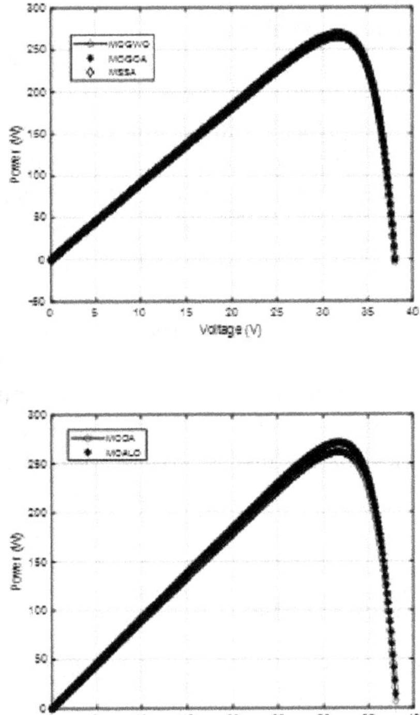

FIGURE 20.2 PV and IV curves obtained for SolarWorld Pro SW255.

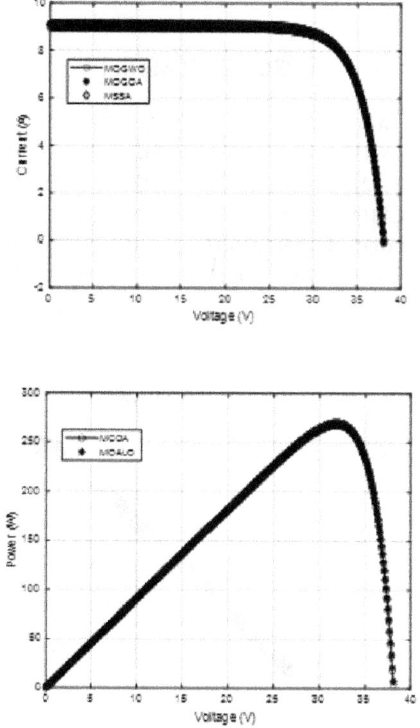

FIGURE 20.2 (Continued)

The results are also verified from these curves as the maximum power obtained from each of the algorithms is just near the value proposed by the manufacturer.

Table 20.4 shows the estimated parameters for AS-6P30 255W, where all the values are within the limits after 20 runs for each iteration. The variation in the values of the parameters is also very less, which determines the correctness of the results. The ideality factor obtained in the case of MOSSA is quite large as it crosses the value of 1.7; for the rest of the algorithms, all the ideality factors mimic nearly the same as the exact model. The series and parallel resistances have quite good values, which should be normal for the cells. The reverse saturation currents for each of the algorithm are again of the order of 10^{-7}, which is very less and represents the ideal behavior.

The results obtained in Table 20.4 are verified as shown in Figure 20.3, where all the curves have smooth texture and are overlapping in nature, proving that all the algorithms produce nearly same values and have good convergence characteristics. Also, the maximum powers for all the algorithms are nearly equal to the actual value given in the datasheet. The results obtained by all the algorithms are good enough as all of them have nearly same values.

TABLE 20.4
Estimated Parameters for AS-6P30 255W Model

Method	I_{pv}	a_1	a_2	a_3	R_S	R_P	I_{01}	I_{02}	I_{03}
MOGWO	8.9722	1.5236	1.4919	1.6202	0.0189	50.3572	3.7305e-07	2.9772e-07	2.8623e-07
MOGOA	8.8654	1.5965	1.5221	1.6116	0.3027	30.1753	7.880e-07	4.2765e-07	6.2538e-07
MOSSA	9.0955	1.7526	1.4814	1.4115	0.1435	81.06453	3.1077e-07	4.9686e-07	5.666e-07
MODA	9.1346	1.4685	1.6115	1.4009	0.0167	73.1191	7.29414e-07	2.9257e-07	7.1368e-07
MOALO	9.0562	1.5230	1.5608	1.6965	0.4512	45.7455	3.608e-07	6.414e-07	5.695e-07

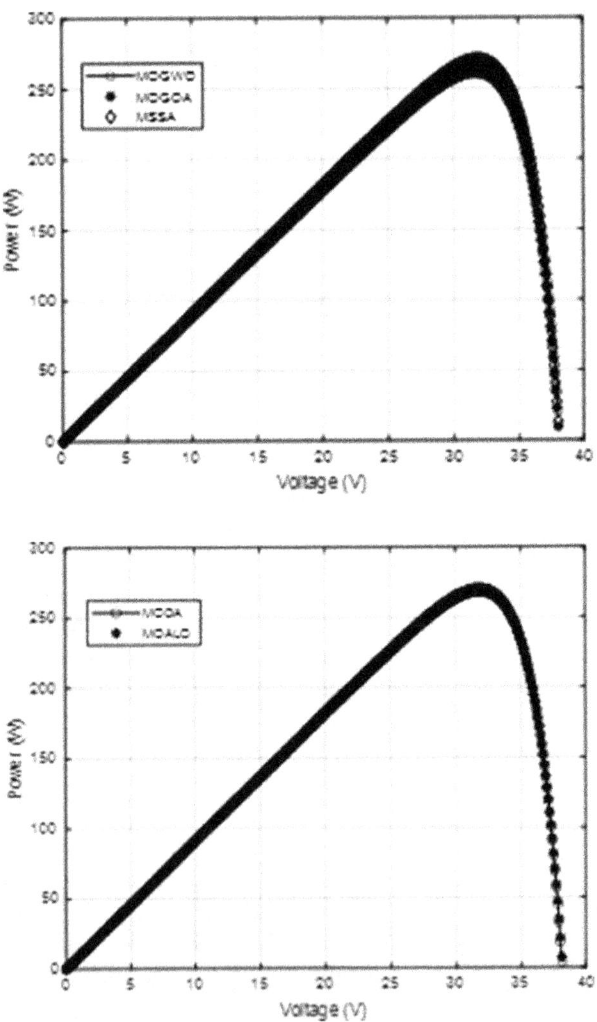

FIGURE 20.3 PV and IV graphs for AS-6P30 255W.

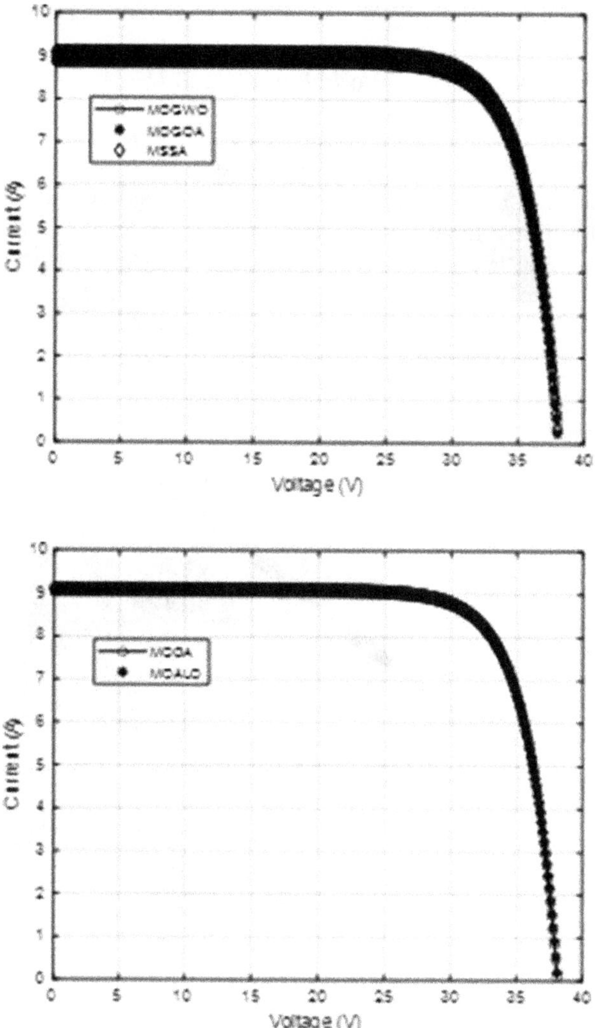

FIGURE 20.3 (Continued)

Table 20.5 shows the values of the parameters for SolarWorld Plus SW280, in which the value of PV current is more than the short-circuit current of the cell and the ideality factors have good values, except for two cases of MOGOA and MSSA, where the values cross 1.7. The reverse saturation currents are of the order of 10^{-7}, representing quite the ideal behavior of the cell.

The PV and IV curves obtained for SolarWorld Plus SW280 are shown in Figure 20.4. The figure shows that the best result is shown by MOGOA algorithm as the maximum power and short-circuit current are much nearer to the actual values than in the other algorithms. Other than that, all the algorithms have good overlapping nature and have smooth texture.

TABLE 20.5
Parameters Extracted for SolarWorld Plus SW280 Model

Method	I_{pv}	a_1	a_2	a_3	R_S	R_P	I_{01}	I_{02}	I_{03}
MOGWO	10.0917	1.5833	1.5663	1.3266	0.0397	40.8597	5.4205e-07	3.4403e-07	3.3747e-07
MOGOA	9.7468	1.7296	1.5517	1.3251	0.2446	95.5771	5.8862e-07	5.4779e-07	5.0115e-07
MOSSA	10.0414	1.7054	1.5695	1.6855	0.0012	75.3169	8.6330e-07	5.7342e-07	7.2595e-07
MODA	10.0708	1.6021	1.6173	1.5471	0.3384	128.3208	7.4630e-07	4.9169e-07	7.0448e-07
MOALO	9.9737	1.5532	1.7574	1.5667	0.3170	59.9874	6.2715e-07	2.597e-07	6.4039e-07

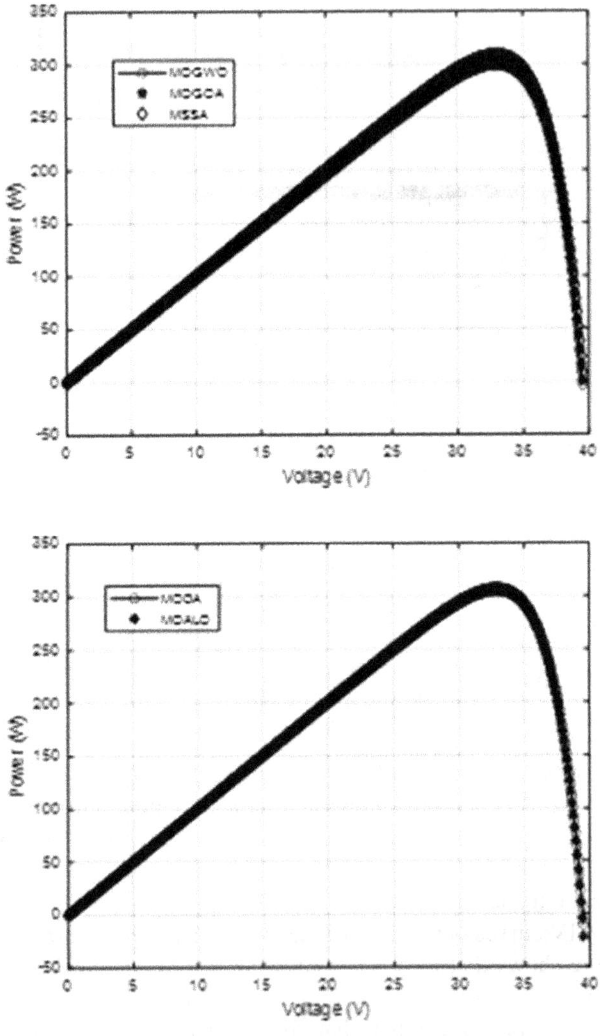

FIGURE 20.4 PV and IV characteristics of SolarWorld Plus SW280 model.

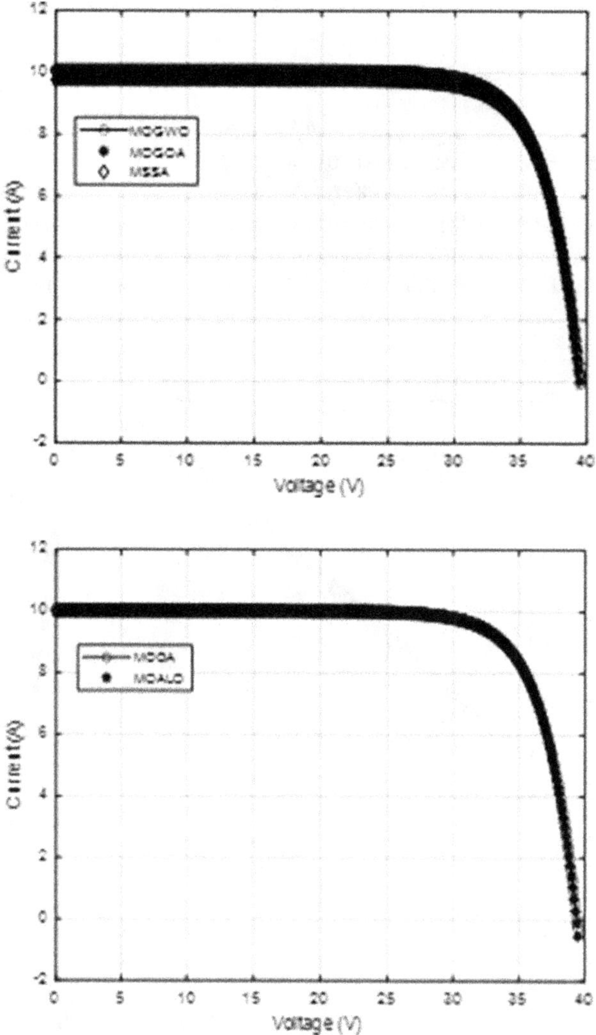

FIGURE 20.4 (Continued)

Table 20.6 shows the parameters extracted for Nemy JB 270M60, where the values of the PV current and series and parallel resistance are quite good as their values show nearly ideal values. The value of ideality factor for MOGOA is greater than 1.8 and is a large value. Other than that, all the values obtained for different algorithms show very less variation among them.

Figure 20.5 shows the curves obtained for Nemy JB-270M60, where the worst result is obtained by MSSA as the short circuit current for it is higher than the other algorithms. But the variation among the curves is still quite less. The maximum power for all the algorithms is quite good and resembles the original model.

TABLE 20.6
Extracted Parameters for Nemy JB-270M60 Model

Method	I_{pv}	a_1	a_2	a_3	R_S	R_P	I_{01}	I_{02}	I_{03}
MOGWO	9.4653	1.7980	1.5452	1.4684	0.0050	46.9958	2.7880e-07	5.7102e-07	4.0562e-07
MOGOA	9.3819	1.5244	1.8428	1.4644	0.1532	40.6132	4.2264e-07	1.9724e-07	6.9227e-07
MOSSA	9.6725	1.7409	1.5798	1.4779	0.2832	29.4308	7.1880e-07	7.4398e-07	8.6751e-07
MODA	9.4347	1.5792	1.6862	1.4679	0.0143	60.5504	7.8352e-07	1.858e-07	2.237e-07
MOALO	9.2392	1.7701	1.4577	1.7571	0.0047	38.5322	2.2180e-07	2.1036e-07	4.3747e-07

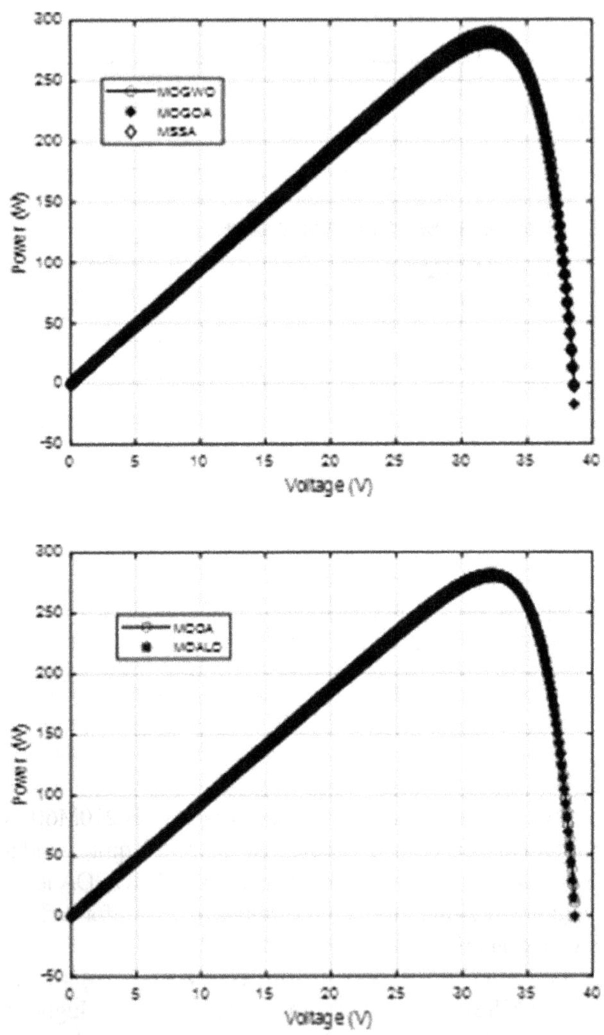

FIGURE 20.5 PV and IV characteristics for Nemy JB-270M60.

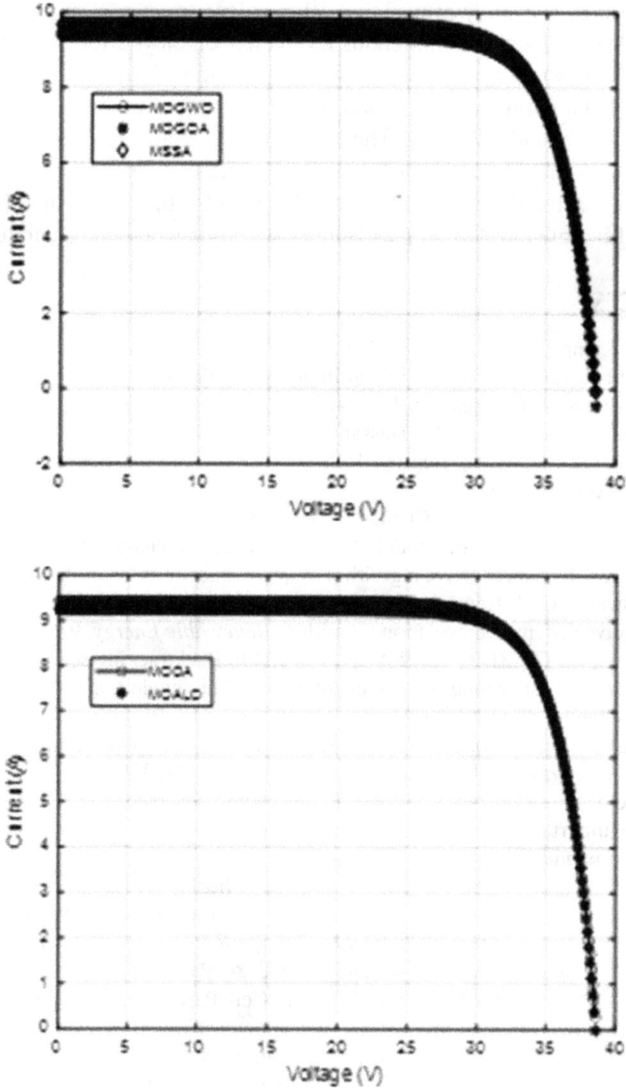

FIGURE 20.5 (Continued)

20.5 CONCLUSION

The chapter presented a comparative study of the various types of metaheuristic algorithms for the parameter extraction of a PV cell using the three-diode model. Three-diode model is the most recent model and is able to give the accurate image of a PV cell by representing all types of losses. Besides, this chapter presented the use of multi-objective method of optimization where two conflicting objectives were considered, namely minimization of error and maximization of power. Multi-objective optimization has not been used much for the extraction of parameters as per the literature. The study proved that all the algorithms used produces satisfying results for all the types of commercial PV panels employed for the study, as their results' variation was very less, which is evident from the plotted PV and IV curves. The values of the parameters obtained for all the algorithms were quite good and were within the bounds decided. The best results for all the cases were given by MOGWO, MODA and MOALO, while the others had a little variation. Hence, it can be correctly said that these algorithms can be used for the determination of parameters of any kind of commercial panels and the results would be accurate.

REFERENCES

1. Yu, K., Liang, J. J., Qu, B. Y., Chen, X., & Wang, H. (2017). Parameters identification of photovoltaic models using an improved JAYA optimization algorithm. *Energy Conversion and Management*, *150*, 742–753.
2. Oliva, D., El Aziz, M. A., & Hassanien, A. E. (2017). Parameter estimation of photovoltaic cells using an improved chaotic whale optimization algorithm. *Applied Energy*, *200*, 141–154.
3. Wu, L., Chen, Z., Long, C., Cheng, S., Lin, P., Chen, Y., & Chen, H. (2018). Parameter extraction of photovoltaic models from measured IV characteristics curves using a hybrid trust-region reflective algorithm. *Applied Energy*, *232*, 36–53.
4. Chellaswamy, C., & Ramesh, R. (2016). Parameter extraction of solar cell models based on adaptive differential evolution algorithm. *Renewable Energy*, *97*, 823–837.
5. Zhang, Y., Jin, Z., Zhao, X., & Yang, Q. (2020). Backtracking search algorithm with Lévy flight for estimating parameters of photovoltaic models. *Energy Conversion and Management*, *208*, 112615.
6. Jordehi, A. R. (2016). Parameter estimation of solar photovoltaic (PV) cells: A review. *Renewable and Sustainable Energy Reviews*, *61*, 354–371.
7. Gnetchejo, P. J., Essiane, S. N., Ele, P., Wamkeue, R., Wapet, D. M., & Ngoffe, S. P. (2019). Important notes on parameter estimation of solar photovoltaic cell. *Energy Conversion and Management*, *197*, 111870.
8. Abbassi, R., Abbassi, A., Jemli, M., & Chebbi, S. (2018). Identification of unknown parameters of solar cell models: A comprehensive overview of available approaches. *Renewable and Sustainable Energy Reviews*, *90*, 453–474.
9. Mehta, H. K., Warke, H., Kukadiya, K., & Panchal, A. K. (2019). Accurate Expressions for Single-Diode-Model Solar Cell Parameterization. *IEEE Journal of Photovoltaics*, *9*(3), 803–810.
10. Qais, M. H., Hasanien, H. M., Alghuwainem, S., & Nouh, A. S. (2019). Coyote optimization algorithm for parameters extraction of three-diode photovoltaic models of photovoltaic modules. *Energy*, *187*, 116001.
11. Yousri, D., Thanikanti, S. B., Allam, D., Ramachandaramurthy, V. K., & Eteiba, M. B. (2020). Fractional chaotic ensemble particle swarm optimizer for identifying the single, double, and three diode photovoltaic models' parameters. *Energy*, *195*, 116979.
12. Elazab, O. S., Hasanien, H. M., Alsaidan, I., Abdelaziz, A. Y., & Muyeen, S. M. (2020). Parameter estimation of three diode photovoltaic model using grasshopper optimization algorithm. *Energies*, *13*(2), 497.
13. Qais, M. H., Hasanien, H. M., & Alghuwainem, S. (2020). Parameters extraction of three-diode photovoltaic model using computation and Harris Hawks optimization. *Energy*, *195*, 117040.
14. Muangkote, N., Sunat, K., Chiewchanwattana, S., & Kaiwinit, S. (2019). An advanced onlooker-ranking-based adaptive differential evolution to extract the parameters of solar cell models. *Renewable Energy*, *134*, 1129–1147.

15. Jordehi, A. R. (2018). Enhanced leader particle swarm optimisation (ELPSO): An efficient algorithm for parameter estimation of photovoltaic (PV) cells and modules. *Solar Energy, 159*, 78–87.
16. Yu, K., Qu, B., Yue, C., Ge, S., Chen, X., & Liang, J. (2019). A performance-guided JAYA algorithm for parameters identification of photovoltaic cell and module. *Applied Energy, 237*, 241–257.
17. Qais, M. H., Hasanien, H. M., & Alghuwainem, S. (2019). Identification of electrical parameters for three-diode photovoltaic model using analytical and sunflower optimization algorithm. *Applied Energy, 250*, 109–117.
18. Rawat, N., Thakur, P., & Jadli, U. (2019). Solar PV parameter estimation using multi-objective optimisation. *Bulletin of Electrical Engineering and Informatics, 8*(4), 1198–1205.
19. Biswas, P. P., Suganthan, P. N., Wu, G., & Amaratunga, G. A. (2019). Parameter estimation of solar cells using datasheet information with the application of an adaptive differential evolution algorithm. *Renewable Energy, 132*, 425–438.
20. Marler, R. T., & Arora, J. S. (2004). Survey of multi-objective optimization methods for engineering. *Structural and Multidisciplinary Optimization, 26*(6), 369–395.
21. Branke, J., & Deb, K. (2005). Integrating user preferences into evolutionary multi-objective optimization. In: *Knowledge Incorporation in Evolutionary Computation* (pp. 461–477). Springer, Berlin, Heidelberg.
22. Deb, K. (2012, September). Advances in evolutionary multi-objective optimization. In: *International Symposium on Search Based Software Engineering* (pp. 1–26). Springer, Berlin, Heidelberg.
23. Mirjalili, S., Mirjalili, S. M., & Lewis, A. (2014). Grey wolf optimizer. *Advances in Engineering Software, 69*, 46–61.
24. Mirjalili, S., Saremi, S., Mirjalili, S. M., & Coelho, L. D. S. (2016). Multi-objective grey wolf optimizer: A novel algorithm for multi-criterion optimization. *Expert Systems with Applications, 47*, 106–119.
25. Saremi, S., Mirjalili, S., & Lewis, A. (2017). Grasshopper optimisation algorithm: Theory and application. *Advances in Engineering Software, 105*, 30–47.
26. Mirjalili, S. Z., Mirjalili, S., Saremi, S., Faris, H., & Aljarah, I. (2018). Grasshopper optimization algorithm for multi-objective optimization problems. *Applied Intelligence, 48*(4), 805–820.
27. Mirjalili, S., Gandomi, A. H., Mirjalili, S. Z., Saremi, S., Faris, H., & Mirjalili, S. M. (2017). Salp swarm algorithm: A bio-inspired optimizer for engineering design problems. *Advances in Engineering Software, 114*, 163–191.
28. Mirjalili, S. (2016). Dragonfly algorithm: A new meta-heuristic optimization technique for solving single-objective, discrete, and multi-objective problems. *Neural Computing and Applications, 27*(4), 1053–1073.
29. Mirjalili, S. (2015). The ant lion optimizer. *Advances in Engineering Software, 83*, 80–98.
30. Mirjalili, S., Jangir, P., & Saremi, S. (2017). Multi-objective ant lion optimizer: A multi-objective optimization algorithm for solving engineering problems. *Applied Intelligence, 46*(1), 79–95.

21 Artificial Intelligence Techniques in Smart Grid

Irtiqa Amin and Mithilesh Kumar Dubey
School of Computer Science and Engineering, Lovely Professional University, Phagwara, Jhalandar, Punjab, India

CONTENTS

21.1 INTRODUCTION

A smart grid is an electricity grid that incorporates a wide range of functional and power interventions as well as intelligent meters, smart devices and green reusable energy infrastructure and energy capable services. A smart grid is a power supply network capable of incorporating the behavior with activities of its linked utilities and customers, in order to ensure a cost-effective, reliable, low loss, quality and supply protection network. Creative products and services are implemented by an intelligent grid with smart surveillance, power control, transmission and self-repairing technologies offering improved information and suggestions to clients on how to use their inventory [1]. However, the total electricity supply network helps reduce the environmental consequences and ensures proper connection and handling of generators of all sizes and technologies [2]. It also allows clients to contribute toward optimizing the system operation and maintaining or increasing current highest levels of device reliability, efficiency and supply protection [3].

21.1.1 Features of Smart Grid

i. It provides efficient maintenance and improvement of current infrastructure. The intelligent grid uses technologies such as city estimation to enhance failure prediction. It allows the infrastructure to self-heal without the involvement of specialists. It would make electricity more stable and less vulnerable to natural disasters or attacks, thus ensuring the reliability of smart grids.
ii. Power transmission infrastructure of the next generation will be better able to manage possible multi-way power streams, enabling distributed generation of electric car batteries [4], wind turbines (WTs), hydroelectric pumped power and the utilization of fuel cells and other resources, for instance, the photovoltaic (PV) cells on roofs of buildings for better flexibility in network topology [5].
iii. The development of smart grid technologies is expected to lead to overall improvements in energy infrastructure efficiency [6]. This involves management on demand, such as switching air conditioners off in short-term electricity price fluctuations and decreasing the voltage on distribution lines through the voltage/VAR optimization (VVO), thus removing the threat [7]. The net impact of the transmission and distribution lines is lower reliability as well as increased usage of generators, which result in lower energy prices, thus providing better efficiency.
iv. The increased versatility of the smart grid allows for better infiltration of extremely large variable renewable resources of energy, including solar and wind, even without energy storage [8, 9]. Speedy variabilities in centralized production such as overcast or rainfall are major challenges for electrical engineers who are required to achieve stable levels of power by differing the outputs of the more controllable generators; thus, smart grids provide an improved sustainability to these issues.

21.1.2 Artificial Intelligence in Smart Grids

Artificial intelligence (AI) usage in the power sector is rapidly entering emerging markets where it can make a major contribution to clean, affordable and reliable energy development. Through the transfer of knowledge from the power sector to IT companies, challenges can be addressed over time. AI systems can be especially useful when specifically built for routine and organized task automation, leaving people to deal with the energy challenges of tomorrow [9, 10]. The core of power development is access to energy. The lack of access to power is therefore a crucial barrier to change the impacts on health, education, food security, equality between the sexes, livelihoods and poverty reduction [7, 8, 11]. One of the sustainable development goals (SDGs) is universal access to affordable, secure and sustainable renewable energy [9, 12]. Nevertheless, it will remain a target if new approaches and emerging technology cannot be overcome. Furthermore, diversified and decentralized energy production with the advent of new technology and changing patterns of demand, production, transmission, distribution and consumption of power in all nations create complicated challenges. The potential of AI is to reduce energy waste and power expenses and accelerate the use of clean renewable power sources worldwide in electricity grids [10, 11]. AI may also enhance energy system's planning, functioning and control. As a result, AI technologies are closely related to the development capability of providing clean, cost-effective energy.

21.1.2.1 AI in Emerging Markets in India

According to the International Energy Agency report in November 2019 [13], there is a worldwide shortage of access to electricity in about 860 million people. Around three billion people cook in their homes and heat them with open fires and stoves, which are fed with biomass, liquid kerosene or coal. Over four million people die prematurely due to air pollution in household diseases. Therefore, energy supplies go beyond merely providing energy as they become necessary for human health and security. Renewable energy on power development, one of the UN SDGs, would play a significant role in increasing access to electricity. By improved planning, operation and control of the power system, AI tool's speed and robustness lead to relative insensitivity to noisy or missing data. AI can thus promote the integration of renewable energy into electric design to develop low-carbon hybrid energy systems [14]. The shift to renewable energies can take place much quicker by using AI [15].

India was particularly acknowledged for its efforts to expand renewable energy production. India currently has an installed capacity of 75 GW [13] from various sources of renewable energy (wind, solar, etc.), and its target is of 175 GW from renewable energy sources by 2022. Despite regulatory initiatives to promote clean energy investments, the spread and extension of renewable energy has remained a challenge. AI is seen as a possible solution to accelerate the adoption of renewable energies. AI-based smart grid can learn and adjust according to the load and the variable volume of renewable energy that flows through the grid network.

21.1.2.2 AI Application in Smart Grid

i. Fault prediction.
ii. Maintenance facilitated by image processing.
iii. Energy efficiency decision making.
iv. Disaster recovery.
v. Prevention of losses due to informal connections.

AI is basically classifiable into four separate fields: expert systems (ESs), fuzzy logic (FL), artificial neural network (ANN), and genetic algorithms or evolutionary computing (EC) [16]. AI has become nearly identical with neural network applications in recent years. If a system's regulation, estimate or diagnosis is usually based on AI [17], it is usually called a smart intelligent machine. The conventional operation of a dynamic model is based on a PI (proportional-integral) or PID (proportional-integral-derivative) controller, whose specifications are calculated from the numerical model of the system. As compared to the conventional systems, the smart intelligent control does not need a mathematical model. This control is mostly called self-study, self-order or self-adaptive control. It is said that the integration with the smart grid of new power electronics, computers, transmission and artificial information and web technologies has already formed a demanding baseline for implementing smart network outline [7, 16]. With the new AI techniques, this challenge will be more intimidating. Smart grid is strictly an interdisciplinary complex dimension with multi-faceted technology [10, 12]. This chapter will first discuss the basic concepts of AI (i.e., ES, FL and NN) in detail and then help to understand the applications of AI in renewable energy systems (RESs) and smart grid and NN of AI will be discussed in detail as it can be used to build these smart intelligent grids. Lastly, some of the novel applications are discussed.

21.2 EXPERT SYSTEMS

The ES is primarily an "intelligent and smart" application software based on Boolean logic (i.e., 0's and 1's), designed to embed a mortal being's knowledge and experience into a definite framework so that it can handle the individual expert's replacement problems. ESs are developed to fix complicated issues using information bodies, which are primarily represented as IF-THEN measures rather than traditional native programming. The human professionals possess the knowledge of the domain over an extended period of time, through education and training. The purpose of an ES is to replace the human expertise with machine-embedded expertise to solve the specific issue. The basic components of ES smart grids consist of:

- User interface.
- Rule engine.
- Knowledge base with expertise and available databases.
- User system, i.e., expert and nonexpert explanation.

The basic cornerstone of an ES is the knowledge base, i.e., knowledge engineer, who acquires intelligence from the ES and translates into a computer program. The domain expert is a power system engineer who does not require the software expertise needed to organize and interpret their knowledge into reliable ES software platforms. Expert knowledge is the key to knowledge base, consisting of a matrix of IF-THEN rules endorsed by the database. Expertise is often described as "shallow" or "deep." Shallow knowledge is derived straight from the expert domain, and deep knowledge is derived from the model system and simulation response, which is purely based on the expertise of the scientists. Basically, the inference or rule engine [18] is the core controller of the system. It evaluates the knowledge base rules in pattern, and attempts to make conclusions or inferences. The testing of the rules may be through either "forward chaining" or "backward chaining." First, the conditional part is evaluated in the forward chaining principle, and if it is accurate, then the rule is ejected. The inference engine decides that the action part is accurate in backward chaining, and then tests the validity of the conditional element before triggering the rule backwards. There are different software that can be utilized in the field of ES, such as LISP, PROLONG and C. The inference engine translates with users in a highly sociable way and helps them to solve any problem and tries

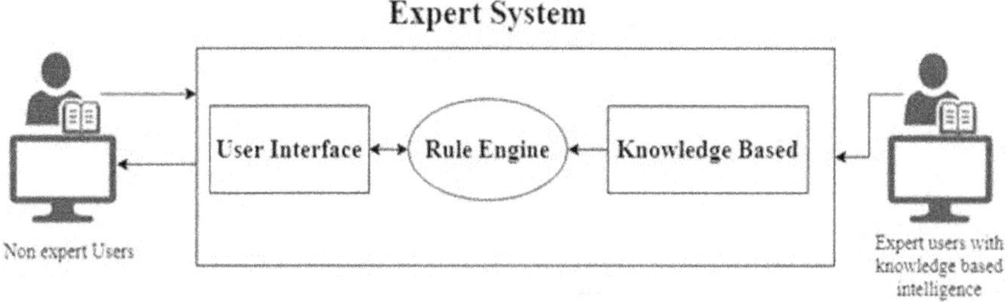

FIGURE 21.1 Block diagram of expert system.

to give answer of how and why and help responses about it [18]. These users are experts who provide best knowledge to nonexpert users. The mechanism of ES is depicted in Figure 21.1.

Basic applications of ES [19] are as follows:

i. Classification of an entity that is based on the defined properties.
ii. Infer system failure from observable data through control system.
iii. Comparing data from a program that is continuously observed to recommend actions by monitoring devices.
iv. Process management is performed by controlling a processed mathematical model.
v. Plan and set up a device to match requirements.
vi. Updating an action plan by scheduling and planning.
vii. Creation of alternative solutions to a problem.

21.2.1 Expert System in Intelligent Smart Grid

An intelligent smart grid is a perception of tomorrow's sophisticated and fully developed electrical grid using futuristic technologies in electric systems, such as power electronics, information systems, servers, networking, AI and web, to enhance the system accessibility, dependability, electrical excellence, power competence and defense, with optimal use of resources with cost-effectiveness. The block diagram of ES in smart grids is shown in Figure 21.2. The grid comprises a huge segment of green renewable resources along with fossil and nuclear power plants generating excess electricity, and these bulk energy storage facilities (mainly for RES support, HVDC (high voltage dc) systems, VAR, etc.) [20] are geographically spread all across the grid. The entire network is partitioned into a variety of domains which are under the control of centralized master controller. The master controller is designed with an ES-oriented knowledge base in which the rules are formulated based on broad offline investigation, design and simulation of the entire system. It is an enormous task requiring a lot of iterations. Usually, the ES rules have a great set of variables with varying parameter values, and the knowledge base can be continually modified; as a result, more system knowledge is obtained. The master

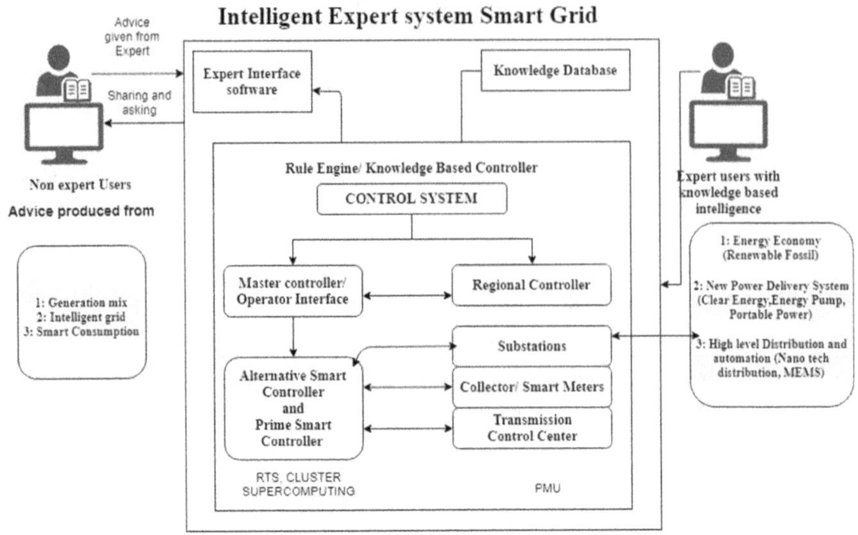

FIGURE 21.2 Block diagram of expert system in smart grid.

controller has a real-time simulator (RTS) that controls all the regional controllers, substations, collectors, distributed systems, transmission controlling centers and smart metered devices associated with it. The RTS is based on a group of supercomputers that incorporate the entire grid's conceptual model including the electronic quick response power systems. It acts as a virtual machine, operating independently of the physical network. The RTS in master controllers obtains all of the system's signals, while the regional controllers only obtain signals from the regions [21]. A great number of grid signals, either sensor-based or sensor less, can be obtained via system-wide phasor measuring units through master controller, which is also known as operator interface.

21.3 FUZZY SYSTEM

The term "fuzzy" applies to details that are ambiguous or uncertain. Every behavior, procedure, or activity that is continually evolving cannot always be classified as either true or false, which means we have to describe these activities in a fuzzy way. FL is a form of multi-valued logic in which variable truth values can be any relevant figure consisting of either 0's and 1's or between them and can be considered as Boolean logic. It is used to handle the idea of half-truth, where the value of truth can range from complete true to complete false. FL emulates the methodology used for making human decisions. It helps in understanding the simplification of the real-world problems based on degrees of truth rather than normal true/false or 1/0 interpretation based of Boolean. Lofti Aliaskar Zadeh is considered the father of fuzzy systems. His research lets us understand the concepts of fuzzy systems/Boolean logic. The framework of Boolean logic in fuzzy systems

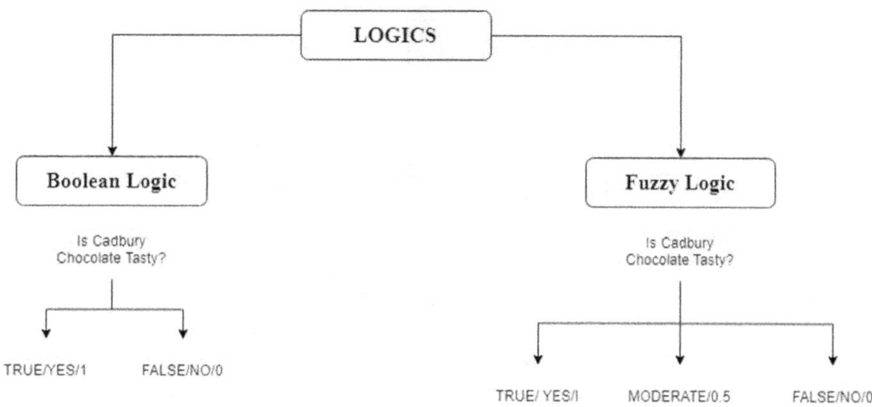

FIGURE 21.3 Boolean logics in fuzzy system.

is shown in Figure 21.3. Knowingly the values are given by a number within the range from 0 to 1: 1 stands for absolute truth and 0 stands for complete false. The number that shows the value is called the truth value in fuzzy systems and the logic is often described as fuzziness. Fuzzy control system is the core unit of an FL system with decision making as its primary function [22, 23]. It uses the rules of "IF-THEN" along with the "OR" or "AND" connectors to create essential rules for decision making to produce expected outputs [24].

The characteristics of fuzzy system are as follows:

i. Fuzzy inference system (FIS) output is often a fuzzy collection or set, regardless of its data, which can be fuzzy or crisp [25].
ii. When used as a controller, it is important to have fuzzy output.
iii. When fuzzy variables are converted into crisps, it is necessary that de-fuzzification process is used with this system.

21.3.1 Block Function or Working of Fuzzy System

The working model of fuzzy system is described in Figure 21.4. The functional block fuzzy system has five basic components:

1. Rule base includes fuzzy IF-THEN rules.
2. Database defines the proper membership functions of fuzzy sets used in fuzzy rules.
3. Fuzzification control unit helps to convert the crisp quantities into fuzzy quantities.
4. Decision-making unit operates on required guidelines to provide approximate decisions by applying different methods to it.
5. De-fuzzification control unit helps to convert fuzzy quantities back into crisp quantities.

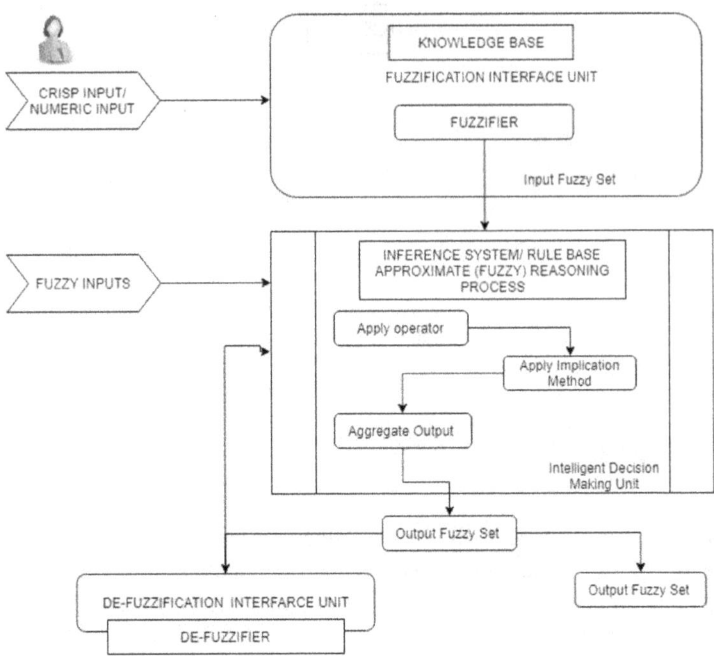

FIGURE 21.4 Working diagram of fuzzy system.

The process undergoes three stages: fuzzification, execute and de-fuzzification.

- **Fuzzification**: This process allocates a system's numerical input to fuzzy or uncertain sets including some degree of membership. The degree can occur anywhere between the interval of [0, 1]. If the value of the degree is 1, then it is included in a given fuzzy set, and if the value of degree is 0, then it is not included in it. The degree of uncertainty is represented anywhere between 0 and 1 within the fuzzy set; such uncertain sets are usually represented by natural languages, so that we can uniformly assign them into system inputs to fuzzy system. "The membership function (also known as Mamdani's fuzzy inference method) of a fuzzy set A in the X discourse universe is defined as μA: X [0, 1], where each element of X is mapped to a value between 0 and 1." The value, also known as membership degree or membership value, measures out the grade membership of the element in X to the fuzzy set A.
- **Execution**: This process shows the execution of these fuzzy inputs to desired output, crisp data, by manipulating it. This comprises two stages:
  a. FL operations are done by Boolean operators such as AND (intersection), OR (union) and NOT (complement) to convert these expressions to fuzzy systems. There are other operators that can be used, such as semantics or linguistics in nature known as mathematical operators.
  b. IF-THEN mapping is used to differentiate between rules depending on input and output.

- **De-fuzzification**: The goal is to use fuzzy truth values to get a continuous variable. This will be simple if the resulting truth values were exactly those produced from the fuzzy rules of a given number in a system [26]. However, since all output truth values are calculated individually, they do not represent such a set of numbers in most cases. Then, one has to decide for a number that best matches the "intention" encoded in the value of truth.

Some of the classifications of fuzzy systems are Tsukamoto fuzzy systems), Takagi-Sugeno-Kang fuzzy systems [22] and Mamdani-Assilan fuzzy systems.

21.3.2 Wind Turbine Fuzzy Model

Due to cost-effectiveness compared with other traditional forms of green power resources, wind source is among the most effective reusable energy options for generating huge electricity. To be the most appropriate renewable energy option for the production of electricity takes a special position. It is not environmentally damaging because it is an abundant resource found in nature. It is known quite well that wind-generated electrical energy is clean, healthy, economic and ecologically responsible (or "green") and the impact is typically lower than that of fossil fuels or nuclear power [27]. Lately, it has been made even more cost effective by emerging technologies in electronics, motor drives and WTs. Wind or PV energy resources are plentiful, and can potentially address all global demands for electricity by tapping only a small portion of it. However, because renewable energy is typically intermittent, sufficient storage and distribution capabilities are necessary for their used economical rates. Wind power may therefore be used by changing it mechanically to electric power using WTs. Over the past two decades, numerous (WT) ideas of renewable wind technologies have rapidly evolved and greatly increased the capacity of wind force. The faster speed activity and direct drive (DD) rotor generators in WTs are new advances in wind energy conversion system technology. Adjustable speed activity has several advantages over the production of stable speeds such as improved power acquisition, maximum power point tracking (MPPT) functions above wide dynamic range of wind speed, higher voltage efficiency, decreased mechanical stress and improved aerodynamic noise system stability. It provides increased performance and decreased mechanical stress in contrast to fixed speed operations. A hybrid wind generation system uses horizontal axis fluctuating speed of WTs which are normally equipped with a shaft of a generator that is without or with a speed-up gear categorizing into gear drive (GD) and DD. A gearbox is used in GD known as squirrel cage induction generator, which is graded as active stall and pitch control WT and works in applications with constant speed. Specifically, in high-electric WTs, the doubly fed induction generator is used instead of alternative speed WTs. The gearless DD and WTs are used as lesser sized WTs using PMSG (permanent magnetic synchronous generator) with high shaft number to eradicate the requirement of gearbox which can be interpreted to a higher efficiency. PMSG becomes increasingly desirable due to the benefits of permanent magnet (PM) equipment than power excited equipment. It shows improved effectiveness, increased power consumption, no extra power source for magnet field operation and better performance owing to the unavailability of mechanical parts such as slip rings [28]. Additionally, the prices of these metallic

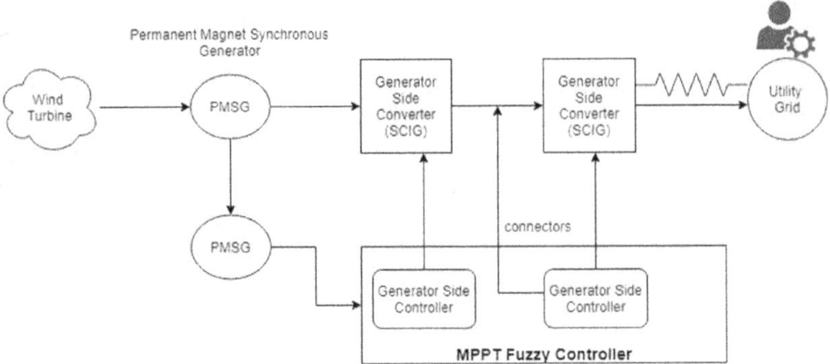

FIGURE 21.5 Wind turbine structure.

magnets are very low but the achievements are very high, thus giving us the advantage to use them in real life. Consequently, the advantages help us to make DD PM WT generator systems more eye-catching in small- and medium-scale WT applications [29]. The generator may be either an induction or a machine of the synchronous type. There are numerous possible configurations of wind generation systems [30]. Robust controller has been designed to trace the highest power which is accessible in the wind itself including tip speed ratios (TSRs), power signal feedbacks, hill climb methods and fuzzy methods. The fuzzy controllers, e.g., MPPT controllers, are used to control the generator and to produce the desired output on the utility grid through FLs. The system has four-quadrant AC drives, which help the machines to work continuously and synchronously without any brakes [31]. The WT structure is shown in Figure 21.5.

21.3.2.1 Wind Turbine Model Description

There are various mathematical expressions which show the real working of WTs, described in the following steps:

- WT transforms wind energy into mechanical energy. The energy is initiated via WT at the generator's shaft [32]. This can be stated as:

$$P_{\mathrm{m}} = 1/2\, C_{\mathrm{p}}\left(\lambda,\ \beta\right) \rho \mathrm{Au}^3 \tag{21.1}$$

where ρ is the air density approximately, β is the pitch angle in degree, A is the area swept by rotor blades in m^2, u is the wind speed in m/s and $C_p(\lambda, \beta)$ is the WT power and coefficient. The turbine power coefficient $C_p(\lambda, \beta)$ defines the WT's power extraction efficiency, and is defined as the ratio between the turbine shaft's mechanical power and available wind power [33]. Based on the generic equation, the characteristics of the turbine modeling equation are shown as follows:

$$C_{\mathrm{p}}\left(\lambda, \beta\right) = 0.5176(116 \times 1/\lambda_i - 0.4\beta - 5)\mathrm{e}^{-21/\lambda i} + 0.0068\lambda \tag{21.2}$$

$$1/\lambda_i = 1/\lambda + 0.08\beta - 0.035/1 + \beta^3$$

where C_p is a nonlinear function of both TSR λ and the blade pitch angle β. λ is the ratio of the turbine tip speed, $w_m \times R$, to the wind speed, u. λ is expressed as:

$$\lambda = \omega_m \times R/u \tag{21.3}$$

where w_m is the rotational speed, R is the turbine blade radius and u is the wind speed independently. For fixed pitch angle β, C_p reverts to nonlinear function of λ only. Depending on equation 21.3, ω_m and λ show a relationship. Therefore, at a certain u, the power is maximized in a certain ω_m, called the optimum rational speed, denoted by ω_{opt} and corresponds to λ_{opt}. Electric torque is produced and calculated in PMSG model, and results finally in equation 21.4:

$$T_e = 3/2\ P\ \psi\ F\ i_{sq} = K_c i_{sq} \tag{21.4}$$

where T_e is electrical torque, i_{sq} is quadrant axis component of the stator in rotor frame and K_c is the torque constant between i_{sq} and T_e.

- Fuzzy logic controller (FLC) can be used to locate the reference velocity and also to track the maximum power peak. The controller device configuration consists of three fuzzy controllers (FLC1, FLC2 and FLC3) with Mamdani's type FIS used in all of them [34]. The power is maximized at a certain wind speed ω with maximum rotational velocity ω_{opt}, fitting the optimum TSR as λ_{opt}. The TSR is stable for all power peaks, so the maximum power at variable wind speed, WT, should always operate at λ_{opt} in speeds below rated speeds. This happens by controlling the rational speed of the WT at the shaft and should be equal to the optimum rational speed. The relationship between ω_{opt} and λ_{opt} for wind speed u and constant R is represented in equation 21.5.

$$\omega_{opt} = \lambda_{opt}/R\ u \tag{21.5}$$

Equation 21.5 depicts the relationship between optimum wind and linear rotational speed. FLC is used to search the reference rotational speed which tracks the peak power at variable wind speeds. The basic diagram of MPPT fuzzy controller is depicted in Figure 21.6.

- The grid-side converter's power flow is controlled to maintain the DC-link voltage at a set point of 600 V. As increasing the output power to DC-link capacitor than the input power causes the DC-link voltage to decrease and vice versa, the output power is regulated to maintain the DC-link voltage approximately constant. The DC-link voltage was maintained, and the reactive power that flows into the grid was regulated at zero. This was achieved using the *d-q* vector control method, by controlling the grid side converter currents. By aligning the *d*-axis and the grid voltage of the reference frame

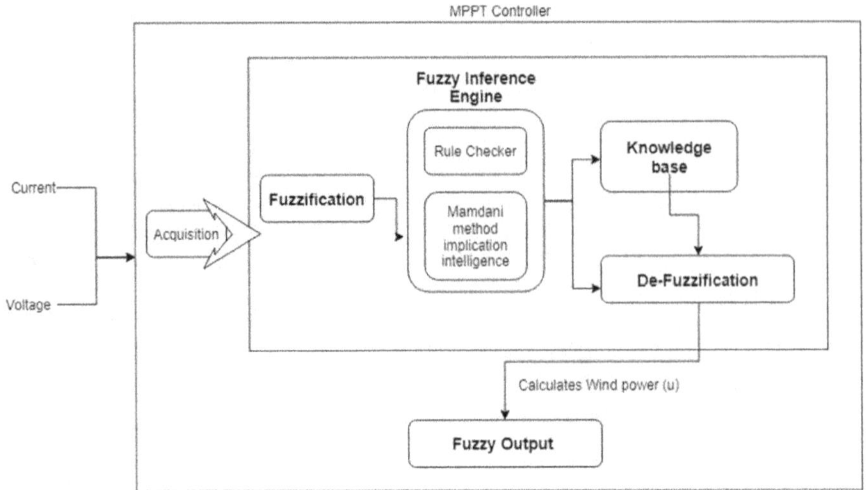

FIGURE 21.6 MPPT controller.

position $dq = 0$, we can obtain the active and reactive power from the following (equation 21.6).

$$P_s = 3/2 \; v_d i_d \tag{21.6}$$

Active and de-active controllers are achieved by controlling *d*- and *q*-axis. For active power control, the *d*-axis current relation is set by an external DC-link voltage control loop. The internal control loop manages the reactive power by setting the current reference *q*-axis to the unit power factor zero value as shown in equation 21.7.

$$Q_s = 3/2 \; v_d i_q \tag{21.7}$$

21.4 NEURAL NETWORK

A modern ANN [35–39] is composed of neural network or connectors of neurons or nodes. Thus, a neural network is either a biological neural network composed of actual biological neurons or an ANN for solving complex AI issues, and the connections of these biological neurons are modeled as weights. Positive weights represent an exciting bond, while negative weights indicate inhibitory bonding. All inputs are adjusted and weighted. The action is called a linear combination. Eventually, the output amplitude is regulated by an activation function. For example, an appropriate performance range of output is normally between 0 and 1, or may be between −1 and 1. Such artificial NNs can be utilized for predictive modeling, adaptive management and applications, where they can be educated and trained via a dataset [40].

Experience-based self-learning can occur within networks which can draw conclusions from a complex and seemingly unrelated collection of knowledge. An NN also known as a neurocomputer is the most common type of AI for individual thinking emulation in comparison with the knowledge rule-based ES and FL that replicate the behavioral nature of person thoughts and resemble the human brain in a similar way by using electrical circuits or computer software programs. It is said that our brain system structure consists of billions of nerve cells [35] or organic neurons. Although we know how our neural system works, the way how they are coordinated together in a group is not known. It helps us solve issues associated with recognition of pattern and graphics processing and manipulation, which are tough to resolve by traditional methods. Stochastic multivariate I/O map-based or pattern identification through individual brain's auto-associative memory properties is the secret to neurocomputing. The property helps us to identify a person when we see his or her profile, e.g., child recognizing alphabet character. This pattern recognition can be acquired through learning and supervised training datasets. A biological neural network [35, 41, 42] is a cluster of neurons in the brain's nervous system which acts as the basic processing element (PE) and accepts and integrates stimuli from other identical neurons through thousands of input paths called dendrites [43]. Each electric input signal that flows through a dendrite passes through a brainstem or synaptic junction with a gap or vent between it. The vent is loaded with a neurotransmitter fluid that either accelerates or slows down the control signals. Afterwards, impulses are gathered into the nucleus changing them in nonlinear way at the output before flowing through axons and different neurons [35]. The neurotransmitter fluid modification of the resistance of the synaptic distance leads to the brain's "memory" or "knowledge." It is believed that, according to the neuron theory, our nervous system has a scattered memory which is associative in nature (i.e., having characteristics purely based on high intelligence) and contributed by the synaptic bond of the cells. It means humans do not have an amalgamate memory similar to a computer.

The ANN is similar to biological NN which acts as the PE, and works similar to human neurons in computer. Each input signal, continuous variable or discrete pulses, flows through a weight called synaptic weight that can be positive integer, negative integer or non-integer. The summing node or hidden layer accumulates all of the input-weighted signals, adds to the weighted bias signal and then passes through the nonlinear, linear activation or transfer function to the output $F(S)$. Some common functions for activating the type are identified as threshold, sigmoid, hyperbolic tan and linear bipolar known as tan sigmoid [44–46]. The structure of ANN is shown in Figure 21.7.

There are three different types of ANN described as follows:

- **Feedforward ANN**: This is the first simplest designed of ANN [47, 48]. As this type of ANN includes node links, the information is moved from the input nodes to the available hidden nodes and forward to the output nodes in one direction, and the desired network is without cycles or loops, i.e., having no shape. Some of the examples of feedforward ANN are: perceptron network, back-propagation model, modular neural network (MNN) [35], etc.

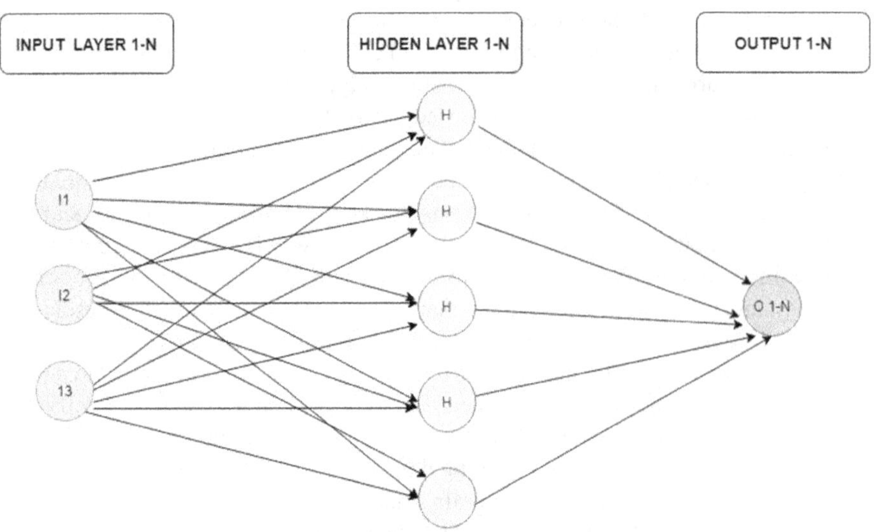

FIGURE 21.7 Structure of artificial neural network.

- **Feed backward ANN or recurrent:** This type of neural network forms loops and cycles taking information from input to output in looped or circular manner. Some of the examples of feed backward ANN are: real-time current NNW, brain state in box and bidirectional associate memory network.
- **Competitive learning ANN**: It combines both feed backward and feedforward ANNs [24, 43, 44]. The input layer is linear, and all the units in the next layer give their outputs. The second-layer outputs can be either linear or nonlinear depending on the requests used in it.

21.4.1 Application of NNWs in Smart Grid Technology Using Renewable Power Resources

ANN (NNW) tends to have the highest ability of all AI techniques consisting of sustainable green power and intelligent electric grid applications [35, 49]. At this moment, our knowledge base is very inadequate in this application, so massive effort is being taken by many scientists to explore the maximum potential of NNW applications in those fields. The novel application of ANN, formally called "adaptive neuro fuzzy inference system" (ANFIS), is mainly in intelligent hybrid fuzzy NNW execution of the inference system. FIS is composed of FL which has an ongoing set of truth values from 0 to 1, decision based IF-THEN fuzzy rules and fuzzy interpretations, i.e., comparisons are based on individual reasoning via small, medium and large lingual variables [23, 50–52]. The IF part of the decision-based fuzzy rules is called antecedent and the THEN part of it is indicated as the final result or consequence. ANFIS frameworks are taking benefits of the neural networks' conscious learning capabilities to automatic adjustments to the fuzzy system membership functions in an acceptable

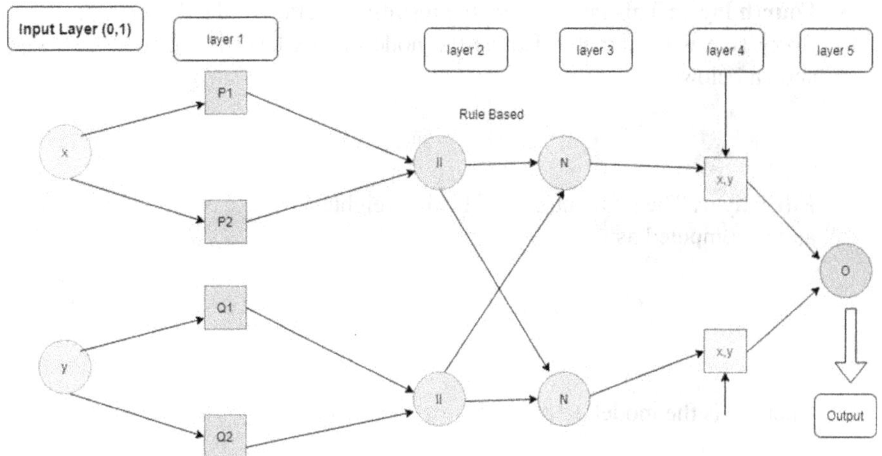

FIGURE 21.8 Structure of ANFIS.

way [53]. The general ANFIS architecture is usually network of five layers working together in mesh as shown in Figure 21.8 [54] and discussed in the following [55–57].

- **First layer**: It is referred to as input plate which projects the fresh inputs into system by applying membership functions to it. The degree of membership function ranges from 0 to 1 nodule, and it stores the parameters of membership function in the form of bell-shaped curve and the equation is indicated below:

$$O_{1,j} = \mu_{p1}(x) \tag{21.8}$$

$$O_{1j} = \mu_{qj2}(y) \tag{21.9}$$

$$\mu(x) = \exp\{-1/2(x - c/\sigma)\} \tag{21.10}$$

  where inputs are taken as i, j and are indicated as i = 1, 2 and j = 3, 4; p and q are the linguistic variable; and c and σ are the mean and SD of the bell-shaped membership function, respectively (also known as premise parameter).
- **Second layer**: Within the preceding rule, each node of this layer performs neural operation "AND" just to evaluate the appropriate and analogous firing strength w_k. Therefore, the node function is written as:

$$O_{2,k} = \mu_{pk}(x) \times \mu_{qk}(y) = w_k \tag{21.11}$$

- **Third layer:** A normalization process to produce the normalized firing strength is performed by the node of this layer given below:

$$O_{3,\,k} = w_k / \Sigma w_k = \overline{w}_k \tag{21.12}$$

- **Fourth layer:** This layer covers the resulting portion of the decision-based fuzzy rule with output, whereas the node of this layer is adaptive and is shown below:

$$O_{4,\,k} = \overline{w}_k z_k \tag{21.13}$$

- **Fifth layer:** The final output will be the weighted average of all rule outputs and is computed as:

$$O_5 = \Sigma_{k=1-2Tw} \overline{w}_k \left(\alpha_k + \beta_k x + \gamma_k y\right) = \psi \tag{21.14}$$

$$\Psi = AP$$

where ψ is the model output and is in matrix form.

There are other options which help ANFIS to succeed greatly and are as follow:

I. It refines bizarre IF-THEN rules for describing the complex system behavior.
II. It does not require human and understands exactly how things happen.
III. Implementation is simple.
IV. It allows for quick and precise learning.
V. It provides the desired data set, the usage of greater choice of membership functions and powerful capacity for generalization.
VI. It is an excellent explanation driven by fuzzy regulations and is easy to embed equally the linguistic and numeric expertise in trouble solving.

21.5 CONCLUSION

Robust AI unlocks fresh platform with RESs for the industrialization of smart power grids. AI in smart grids is an enthusiastic project in nature which will be carried out in numerous years by the consolidated efforts of eminent researchers skilled herein. This helps us to support and achieve the targets of smart grid such as possibility within electrical apparatus, trustworthiness, electric standards, power effectiveness and protection of the electric systems giving best usage of resources and budgeted power. This module also describes the AI used in the field of smart grid, various AI types in it and specific NNW implementations in combination with renewable intelligent power framework. It also includes the ANFIS wind generation system. It is interesting to note that NNWs are now gaining utmost accent for outlook applications among all the AI techniques. Ubiquitous AI systems with the implementation of efficient and economical system-specific NNW microprocessors are anticipated to lead a new sort of industrialization era in the near term.

REFERENCES

1. Bush S. F. (2014). Smart Grids. Piscataway, NJ: IEEE Press/Wiley.
2. Fang X., Misra S., Xue G., and Yang D. (2011). Smart grid — The new and improved power grid: A survey. *IEEE Communication Surveys & Tutorials*, 14(4), 944–980.

3. Simoes M., Roche R., Kyriakides E., Miraoui A., Blunier B., McBee K., Suryanarayanan S., Nguyen P., and Ribeiro P. (2011). Smart grid technologies and progress in Europe and the USA. In: *IEEE Energy Conversion Congress and Exposition (ECCE 2011)*, 383–390.
4. Metz C. (2018). Big bets on AI open a new chip frontier for start up's too. *New York Times*, January 14.
5. Ferreira A., Leitao P., and Vrba P. (2014). Challenges of ICT and artificial intelligence in smart grids. In: *IEEE International Workshop on Intelligent Energy System*, 6–11.
6. Ipakchi A. and Albuyeh F. (2009). Grid of the future. *IEEE Power and Energy Magazine*, 7(2), 52–62.
7. Bose B. K. (2015). Energy environment and power electronics. In: *IEEE Industrial Electronics Society Distinguished Lecture Presentation*, Colorado School of Mines, Denver.
8. Chakraborty A. and Illic M. (2012). Power electronics and smart systems. In: *Control and Optimization Methods for Electric Smart Grids.* New York, NY: Springer.
9. Bojkovic Z. and Bakmaz B. (2012). Smart grid communications architecture: A survey and challenges. In: *Proceedings of the 11th International Conference on Applied Computer and Applied Computational Science (ACACOS)*, 83–89.
10. Gharavi H. and Ghafurian R. (2011). Smart grid: The electric energy system of the future. *Proceedings of the IEEE*, 99(6), 917–921
11. Bose B. K. (2017). Power electronics in smart grid and renewable energy systems. *Proceedings of the IEEE Power Electronics Magazine*, 4(4), 1997–2285.
12. Abu-Rub H., Malinowski M., and Al Haddad K. (2014). *Power Electronics for Renewable Energy Systems, Transportation and Industrial Applications.* Piscataway, NJ: IEEE Press/John Wiley.
13. Makala B. and Bakovic T. (2020). Artificial intelligence in the power sector. EMCompass No. 81. International Finance Corporation, Washington, DC. © International Finance Corporation. https://openknowledge.worldbank.org/handle/10986/34303.
14. Venayagamoorty G. K. (2009). Potentials and promises of computational intelligence for smart grid. In: *IEEE PES General Meeting Conference*, 1–6.
15. Greer C., Wollman D. A., Prochaska D. E., Boynton P. A., Mazer J. A., Nguyen C. T., Fitz Patrick G. J., Nelson T. L., Koepka G. H., Hefner Jr A. R., and Pillitteri V. Y. (2014). NIST framework and roadmap for smart grid interoperability standards. NIST Release 3.0, October 1.
16. Bose B. K. (2019). Artificial intelligence techniques in power electronics and motor drives. In: *IEEE Power Electronics and Industry Application Societies*, Phoenix, Arizona, 625–675.
17. Ramos C. and Liu C. C. (2011). AI in power systems and energy markets. *IEEE Intelligent Systems*, 26(2), 5–8.
18. Bose B. K. (2006). Power Electronics and Motor Drives: Advances and Trends. Burlington, MA: Academic Press Elsevier.
19. Vedder R. G. (1989). PC based expert system shells: some desirable and less desirable characteristics. *Expert System*, 6(1), 28–42.
20. Daoshen C. and Bose B. K (1992). Expert system based automated selection of industrial ac drives. In: *IEEE/IAS Annual Meeting Conference Record*, 387–392.
21. Chhaya S. M. and Bose B. K. (1995). Expert system aided automated design, simulation and controller tuning of ac drive system. *IEEE IECON Conference Record*, 1, 712–718.
22. Takagi T. and Sugeno M. (1985). Fuzzy identification of a system and its applications to modeling and control. *IEEE Transactions on Systems, Man, and Cybernetics*, 1, 116–132.
23. Jang S. R., Sun C. T., and Mizutani E. (1997). *Neuro-Fuzzy and Soft Computing: A Computational Approach to Learning and Machine Intelligence.* Upper Saddle River, NJ: Prentice Hall, 1482–1488.

24. Cirstea M. N., Dinu A., Khor J. G., and Cormick M. M. (2002). Neural and Fuzzy Logic Control of Drives and Power Systems. Burlington, MA: Newnes, Elsevier.
25. Radhakrishnan B. M., Srinivasan D., and Mehta R. (2016). Fuzzy-based multi-agent system for distributed energy management in smart grids. *International Journal of Uncertainty, Fuzziness and Knowledge-Based Systems*, 24(5), 781–803.
26. Munakata Y. and Yashvant J. (1994). Fuzzy system: An overview. *Communication of the ACM*, 37(3), 69–76.
27. Kariniotakis G., Pinson P., Siebert N., Giebal G., Barthelmie R. (2004). The state of the art in short term prediction of wind power - from an offshore perspective. In: *SeaTech Week - Ocean Energy Conference ADEME-IFREMER*, Brest, France
28. Bose B. K. (1998). A high performance inverter-fed drive system of an interior permanent magnet synchronous machine. *IEEE Transactions on Industry Application*, 24(6), 989–997.
29. Xu R. I., Xu X., and Chen M. (2011). The application of genetic neural network on wind power prediction. *ICICA Springer, CCIS 244*, 2, 379–386.
30. Chang W. Y. (2014). A literature review of wind forecasting methods. *Journal of Power and Energy Engineering*, 2, 161–168.
31. Costa A., Crespo A., Navarro J., Lizcano G., Madsen H., and Feitosa E. (2008). A review on the young history of the wind power short term prediction. *Renewable and Sustainable Energy Reviews*, 12(6), 1725–1744.
32. Sideratos G. and Hatziargyriou N. D. (2007). An advanced statistical method for wind power forecasting. *IEEE Transactions on Power System*, 22(1), 258–265.
33. Lei M., Shiyan L., Chuanwen J., Hongling L., and Yan Z. (2009). A review on the forecasting of wind speed and generated power. *Renewable Sustainable Energy Reviews*, 13(4), 915–920.
34. Li L., Wang M. H., Zhu F. F., and Wang C. S. (2009). Wind power forecasting based on time series and neural network. In: *Proceedings of the Second Symposium International Computer Science and Computational Technology*, 293–297.
35. Bose B. K. (1994). Expert system, fuzzy logic, and neural network applications in power electronics and motion control. *Proceedings of the IEEE*, 8(8), 1303–1323.
36. Principe C., Eulianoand N. R., and Lefebvre W. C. (2000). *Neural and Adaptive Systems: Fundamentals through Simulations*, Vol. 672. New York, NY: Wiley.
37. Haykin S. (1994). *Neural Networks: A Comprehensive Foundation*. New York, NY: Macmillan Publishing.
38. Bose B. K. (2007). Neural network applications in power electronics and motor drives-An introduction and perspective. *IEEE Transactions on Industrial Electronics*, 54(1), 14–33.
39. Ciabattoni L., Ippoliti G., Longhi G., and Ceralietti W. M. (2013). Online tuned neural networks for fuzzy supervisory control of PV-battery system. In: *IEEE PES Innovative Smart Grid Technologies Conference (ISGT)*, Washington, DC, 1–6.
40. Gomes P. and Castro R. (2012). Wind speed and wind power forecasting using statistical models: AutoRegressive moving average (ARMA) and artificial neural network (ANN). *International Journal of Sustainable Energy Development*. 10.20533/ijsed.2046.3707.2012.0007.
41. Lee J., Park G. L., Kim E. H., Kim Y., and Lee I. W. (2012). Wind speed modeling based on artificial neural networks for Jeju area. *International Journal of Control and Automation*, 5(2), 81–88.
42. Catalao J. B. S., Pousinho H. M. I., and Mendes V. M. F. (2009). An artificial neural network approach for short term wind power forecasting in Portugal. In: *International Conference on Intelligent System Applications to Power System*, 1–5.
43. Deshmukh M. K. and Moorthy C. B. (2010). Application of genetic algorithm to neural network model for estimation of wind power potential. *Journal of Engineering, Science and Management Education*, 2, 42–48.

44. Nair S. V., Kothari P., and Lodha K. (2014). ANN and statistical theory based forecasting and analysis of power system variables. *International Journal of Emerging Technology and Advanced Engineering*, 4(7), 753–758.
45. Bose B. K. (2007). Neural network applications in power electronics and motor drives-an introduction and perspective. *IEEE Transactions on Industrial Electronics*, 54(1), 12–13.
46. Swami A., Mendel J. M., and Nikias C. L. (1998). Higher order spectral analysis toolbox. Math Works Inc. Neural Networks Toolbox User's Guide.
47. Catalao J. P. S., Mariano S. J. P. S., Mendes V. M. F., and Ferreira L. A. F. M. (2007). An artificial neural network approach for short-term electricity prices forecasting. In: *International Conference on Intelligent Systems Applications to Power Systems*, 15–23.
48. Zhang W. T., Xu F. Y., and Zhou L. (2010). Artificial neural network for load forecasting in smart grid. In: *International Conference Machine Learning and Cybernetics*, 3200–3206.
49. Bose B. K. (2017). Artificial intelligence techniques in smart grid and renewable energy systems: Some example applications. *Proceedings of the IEEE*, 105(11), 2262–2273.
50. Jang J. S. R. (1993). ANFIS: Adaptive network-based fuzzy inference systems. *IEEE Transactions on Systems, Man, and Cybernetics*, 23(3), 665–685.
51. Kassa Y., Zhang J. H., Zheng D. H., and Wei D. (2016). Short term wind power prediction using ANFIS. In: *IEEE International Conference on Power and Renewable Energy*, 388–393.
52. Catalao P. S., Pousinho H. M. I., and Mendes V. M. F. (2011). Hybrid wavelet-PSO-ANFIS approach for short-term wind power forecasting in Portugal. *IEEE Transactions on Sustainable Energy*, 2(1), 50–59.
53. Hmouz A., Shen J., Hmouz R., and Jun Y. (2012). Modeling and simulation of an adaptive neuro-fuzzy inference system (ANFIS) for mobile learning. *IEEE Transactions on Learning Technologies*, 5(3), 226–237.
54. Chen B., Matthews P. C., and Tavner P. J. (2013). Wind turbine pitch faults prognosis using a-prior knowledge-based ANFIS. *Expert Systems with Applications*, 40(17), 6863–6876.
55. Mohandes M., Rehman S., and Rahman S.M. (2012). Estimation of wind speed profile using adaptive neusssro fuzzy inference system ANFIS. *Applied Energy*, 88(11), 4024–4032.
56. Catalao P. S., Pousinho H. M. I., and Mendes V. M. F. (2011). Hybrid wavelet-PSO-ANFIS approach for short-term wind power forecasting in Portugal. *IEEE Transactions on Sustainable Energy*, 2(1), 50–59.
57. Castellanos F., and James N. (2009). Average hourly wind speed forecasting with ANFIS II. *In:* 11th *Americas Conference on Wind Engineering.*

22 Parameter Identification of a New Reverse Two-Diode Model by Moth Flame Optimizer

Saumyadip Hazra[1], Souvik Ganguli[1], and Suman Lata Tripathi[2]
[1]Department of Electrical & Instrumentation Engineering, Thapar Institute of Engineering & Technology, Patiala, Punjab, India
[2]School of Electronics and Electrical Engineering, Lovely Professional University, Jalandhar, Punjab, India

CONTENTS

22.1 INTRODUCTION

Environmental pollution has been one of the popular topics of discussion and research for the past few decades. Several researchers throughout the world have contributed much to the development of new technologies for reducing pollution [1]. Alongside this, the growing energy demand, changes in climate, global warming, and the nearly ending fossil fuels presented further challenges in front of the society to work with an alternative source of energy [2]. The existing methods for the production of energy need to be replaced by the use of renewable energy resources. The significance of renewable energy resources can be explained based on the fact that they provide more flexibility in their installations, they can be replenished, and, most importantly, they do not pollute the environment [3]. The primary renewable sources in use these days are nuclear, biomass, wind, solar, geothermal, etc. The solar power belonging to this category has the highest potential for the generation of electricity due to its unlimited availability. It is available to most of the countries in the world [4]. A considerable amount of flexibility is in the hands of the user for its installation and usage. Solar power has provided a significant contribution toward the growth of the distributed generation, and low maintenance, low-cost electricity, and long-lasting nature are some of the added advantages.

Most importantly, its eco-friendly nature has contributed much toward its growth as clean energy is produced by it [5]. The solar plants may range from small to huge plants and require a hefty amount of money to invest in their installation. Reduced efficiency and low utility factor are some of the other disadvantages faced by solar power. Every year huge amount of e-waste is generated from the equipment used for solar power, which needs to be disposed of properly [6]. The demerits of solar power have been a new topic for research and development in recent years, and much progress has been recorded. For instance, a significant amount of growth has been recorded in increasing the efficiency of the cell [7]. There are various types of solar panels available in the market. These include monocrystalline, polycrystalline, organic, and thin-film solar panels. Monocrystalline solar panels use bars made of silicon drawn into wafers and use single crystal silicon. Polycrystalline solar panels use various silicon bars melted down together to form a single wafer. Organic solar panels use carbon-based technology for manufacturing the semiconductor material and can also be referred to as plastic solar cells. Thin-film solar panels, which are of particular interest in this chapter, consist of fragile sheets of semiconductor material and are much thinner than the conventional silicon-based panel. They are the lightest solar panels available and can be of many types and provide great flexibility in use. They significantly lack the traditional panels in terms of efficiency as it ranges between 6% and 10% [8–10].

For the proper estimation of the performance of the solar panels, it becomes essential to gain knowledge about the exact behavior of the panel and its characteristics based on the exact mathematical model. It can be achieved by plotting the accurate voltage-current plots for the cell. Also, to control the whole system, the knowledge of the behavior of the cell is required. The performance parameters and the efficiency of the cell largely depend on the model chosen and can be determined accurately with the help of intrinsic parameters [11, 12]. With passing time, there may be slight changes in them due to the operating conditions, which include different types of faults or maloperations encountered. Unfortunately, these parameters for the evaluation of the panel are not provided by the manufacturer. Therefore, a task lies to extract the parameters from the cell and do further calculations. It is quite a well-known fact that the power output of the panel depends on the irradiance level. This fact has been used for the determination of the intrinsic parameters from the datasheet information provided by the manufacturers. Hence, choosing the proper mathematical model and carrying out appropriate calculations can serve the purpose of performance evaluation of the panel [13, 14].

Several mathematical modeling methods have been presented in the literature for the parameter estimation of photovoltaic (PV) cells. Some of the most widely used models are single-diode model (SDM), double-diode model (DDM), modified double-diode model (MDDM), and three-diode model (TDM). In this chapter, reverse two-diode model (RTDM) for the parameter estimation of the thin-film cells has been presented. This model has been used in the literature for the estimation of parameters of organic solar cells and gave satisfactory results. RTDM can be seen as the modification of the SDM and the DDM. In this model, the second diode accounts for the bends in the I-V and P-V curves of the cell. The conventional methods used for the modeling of the solar cells are not suitable for the modeling of the organic and thin-film cells, and hence the modeling of these cells required some modifications in it. According to Pillai *et al.* [15], who estimated the parameters of organic solar cells using this method, the model failed to give excellent convergence characteristics. Still, it improvised after adding a diode in the circuit with a resistor placed parallel to it under normal irradiance levels. However, in this chapter, it has been shown that the estimation of parameters through this method gave outstanding convergence characteristics along with smooth P-V and I-V curves. The computational complexity increased largely due to the position of the diode and a resistor parallel to the diode in an unconventional place. RTDM method has been specifically designed for the organic cells, and this chapter proves that the same can be applied as well for the extraction of parameters of thin-film solar cells [15].

The solution to the complex implicit equations is indeed a challenging task. The methods incorporated for solving analytical and numerical methods. Analytical methods are the least effective methods for solving them since they may need the values at all the data points of the characteristic P-V and I-V curves for solar cells [16]. Besides this, analytical methods may produce inferior results. Moreover, considerable time is consumed in solving them through analytic methods since the equations involved are non-linear. The other way is the numerical method, which may be used for the solution, but again, they also provide poor results in many cases. Their solutions largely depend on the initial value taken, and the convergence is also slow [17]. In the case of problems which include multiple maxima or minima, they may get struck locally in any of them. Due to the presence of multiple maxima and minima, they can also provide false results due to their inability to find the global maxima [18]. Hence, a subset of numerical methods, known as metaheuristic algorithms, is used, which also acts as the optimization process. These are the population-based methods where the population are called as search agents and are initialized and updated during the whole course of iteration stochastically. The values are gradually improved during the iterations, and also the target is updated based on calculated values. The iterations end by identifying the fittest agent and assuming that either it has reached to the optimal value or it is nearest to the optimal value of the problem. The advantage of these algorithms lies in the fact that any dimension problem can be tackled using them [19, 20].

Artificial intelligence (AI) is the intelligence shown by agents other than humans for performing various operations on a set of problems. The 'intelligence' is depicted by showcasing different capabilities, out of which the most prominent one is decision-making. The optimization process employing any smart device such as computers falls under the category of AI. Many engineering problems that require optimization for their solution can rarely be solved with the help of exhaustive searches.

The solution to these problems lies in the use of heuristics, where the agents are randomized initially and then gradually reach up to the optimum point. Advancements made in these heuristic algorithms, known as metaheuristic algorithms, have extensively been used for the engineering problems, and extraction of PV cell parameters is one of those. These AI-based algorithms are equipped with capabilities for decision-making and to seek the optimal solution for the problem without getting stuck in the local maxima.

In this chapter, firstly, the parameters are extracted for the newly developed RTDM using moth flame optimization (MFO) [26]. Other algorithms such as equilibrium optimizer (EO) [21] and Harris hawk optimization (HHO) [22], developed in the year 2020 and 2019 respectively, are used for comparison. Even MFO has proved to give excellent results earlier [23], where the parameters for DDM, TDM, and MDDM were extracted and hence the choice to find the parameters of the RTDM was employed in this chapter. Moreover, other algorithms such as salp swarm algorithm (SSA) also gave good results, as presented by Abbasi *et al.* [24]. Statistical analysis has been performed on the results and the convergence curve of them has been shown along with their P-V and I-V curves.

The rest of the chapter is organized as follows. Section 22.2 is the problem formulation section where the mathematical model of the diode model has been described. Section 22.3 is the methodology section where a brief introduction has been presented on the metaheuristic algorithms used for this study. Section 22.4 presents the results obtained after the calculations are performed on the developed equations. Finally, Section 22.5 presents the conclusion of the whole chapter.

22.2 PROBLEM FORMULATION

The newly explored RTDM consists of a current source, a series resistance, two parallel resistances, and two diodes. The model is shown in Figure 22.1, and for the mathematically modeling it, first Kirchhoff's current law (KCL) has to be applied in it, which results in:

$$I_{PV} = I_{D1} + I_{R_{P1}} + I \tag{22.1}$$

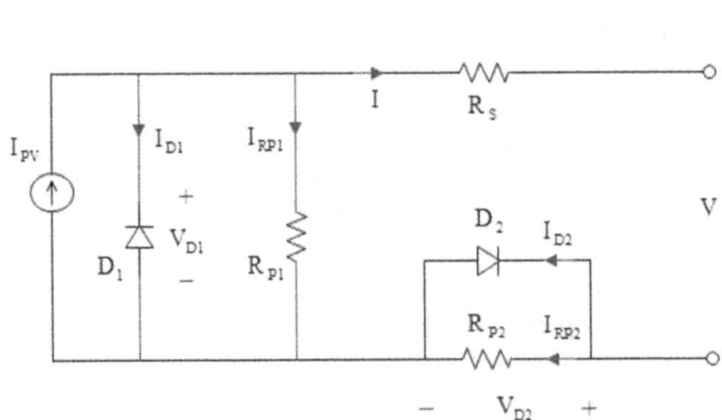

FIGURE 22.1 Mathematical model of RTDM.

where I_{PV} represents the PV generated by the cell, I_{D1} is the current flowing through the diode 1, $I_{R_{P1}}$ is the current flowing through the first parallel resistance, and I depicts the load current flowing through the circuit. The diode current equation 22.1 can be substituted from Shockley's equation of diode and can be written as:

$$I_{D1} = I_{01}\left[\exp\left(\frac{q(V+V_{D2}+IR_s)}{a_1KT}\right)-1\right] \tag{22.2}$$

In the above equation, a_1 is the ideality factor for the first diode. The current flowing through the parallel resistance $I_{R_{P1}}$ in equation 22.1 can be calculated from the following equation:

$$I = \frac{V+V_{D2}+IR_s}{R_{P1}}. \tag{22.3}$$

It should be noted that in equations 22.2 and 22.3, the term V_{D2} is the voltage across the second diode present in the circuit and its value can be calculated using the equation:

$$V_{D2} = \left[I - I_{02}\left[\exp\left(\frac{q(V+V_{D2}+IR_s)}{a_2KT}\right)-1\right]\right]\times R_{P2} \tag{22.4}$$

It is interesting to note at this point that the equation of V_{D2} forms an implicit equation as there exists a term of V_{D2}. For solving implicit equations, direct solution method cannot be applied. Hence, first, the constraints are generated by using the datasheet and using the values present in it by imposing various conditions to the final equation which is given by substitution of equations 22.2 and 22.3 in equation 22.1:

$$I_{PV} = I_{01}\left[\exp\left(\frac{q(V+V_{D2}+IR_s)}{a_1KT}\right)-1\right]+\frac{V+V_{D2}+IR_s}{R_{P1}}+I \tag{22.5}$$

The first condition applied is the open-circuit condition in which the open-circuit voltage is considered when the load current through the circuit becomes zero:

$$V_{D2} = -I_{02}\left[\exp\left(\frac{q(V_{OC}+V_{D2})}{a_2KT}\right)-1\right]\times R_{P2} \tag{22.6}$$

Hence, equation 22.5 can be written as:

$$I_{PV} = I_{01}\left[\exp\left(\frac{q(V_{OC}+V_{D2})}{a_1KT}\right)-1\right]+\frac{V_{OC}+V_{D2}}{R_{P1}} \tag{22.7}$$

The second condition is the short-circuit condition in which the load terminal is short circuited and the short-circuit current is considered from datasheet. Then, equations 22.4 and 22.5 can be modified as:

$$V_{D2} = \left[I_{SC} - I_{02} \left[\exp\left(\frac{q(V_{D2} + I_{SC}R_s)}{a_2 KT} \right) - 1 \right] \right] \times R_{P2} \tag{22.8}$$

$$I_{PV} = I_{01} \left[\exp\left(\frac{q(V_{D2} + I_{SC}R_s)}{a_1 KT} \right) - 1 \right] + \frac{V_{D2} + I_{SC}R_s}{R_{P1}} + I_{SC}. \tag{22.9}$$

The third condition is derived by using the maximum power point voltage and current from the datasheet and applying the same to equations 22.4 and 22.5, which can be modified as:

$$V_{D2} = \left[I_{MP} - I_{02} \left[\exp\left(\frac{q(V_{MP} + V_{D2} + I_{MP}R_s)}{a_2 KT} \right) - 1 \right] \right] \times R_{P2} \tag{22.10}$$

$$I_{PV} = I_{01} \left[\exp\left(\frac{q(V_{MP} + V_{D2} + I_{MP}R_s)}{a_1 KT} \right) - 1 \right] + \frac{V_{MP} + V_{D2} + I_{MP}R_s}{R_{P1}} + I_{MP} \tag{22.11}$$

Based on all the constraints generated, the errors obtained from each of the equation are used for the formulation of the objective function. The errors are then clubbed together by taking their squares. The concept of squared errors has been used here because of its ability to reduce the overall error significantly. The error equations of each of the single equations can be written as:

$$Err_{OC} = I_{PV} - I_{01} \left[\exp\left(\frac{q(V_{OC} + V_{D2})}{a_1 KT} \right) - 1 \right] - \frac{V_{OC} + V_{D2}}{R_{P1}} \tag{22.12}$$

$$Err_{SC} = I_{PV} - I_{01} \left[\exp\left(\frac{q(V_{D2} + I_{SC}R_s)}{a_1 KT} \right) - 1 \right] - \frac{V_{D2} + I_{SC}R_s}{R_{P1}} - I_{SC} \tag{22.13}$$

$$Err_{MP} = I_{PV} - I_{01} \left[\exp\left(\frac{q(V_{MP} + V_{D2} + I_{MP}R_s)}{a_1 KT} \right) - 1 \right] - \frac{V_{MP} + V_{D2} + I_{MP}R_s}{R_{P1}} - I_{MP} \tag{22.14}$$

The final error equation is given as follows in the lines as [25]:

$$E = E_{OC}^2 + E_{SC}^2 + E_{MP}^2 \tag{22.15}$$

22.3 PROPOSED TECHNIQUE

22.3.1 Moth Flame Optimization (MFO)

Moths are small insects that are known for their traveling during nights. They use a method known as the transverse orientation for navigation to travel by taking the help of the moon. They achieve it by maintaining a special angle with the moon, but the problem arises for them when they encounter any artificial light in their path, which makes them travel in spiral lines. In the MFO algorithm, the moths are considered as the search agents, and the flame is considered the best position acquired by the moths during their search. The spiral movement of the moths is considered where the moth determines the starting point, and the position of the flame determines the final point. When the distance between the moth and flame decreases, the frequency of updating the position of the moth increases. The position of the moth is updated by assuming a hyperellipse around the flame, and when the moth moves around it, the exploitation phase occurs. Since there are a number of moths considered, there are several flames as well, and during each iteration, the fitness of the moth and the flame is calculated. Then the best flame is assigned to the moth having the best fitness value and is done to prevent the sticking around near the local maxima. The algorithm continues until the number of iterations is not reached [26]. Some of the algorithms, with which comparison is carried out, are discussed in the next subsections.

22.3.2 Equilibrium Optimizer

The equilibrium optimizer (EO) is based on the balancing of mass in a controlled volume where the mass can flow till the equilibrium condition is reached. It also obeys the laws of physics where the amount of mass left is equal to the amount of mass entered and follows the first-order ordinary differential equation. The concentration and solution are considered in EO, where the solution is the same as a particle, and concentration is the same as the particle's position in the particle swarm optimization (PSO) algorithm. The position of the particle is updated based on the three main terms in EO. The first is the concentration at the equilibrium; the second is the difference between the position of particle and equilibrium; and the third is related to generation rate, which contributes mainly during the exploitation phase of the algorithm. EO algorithm starts by randomly assuming the positions of the particles uniformly distributed over the concentration constructed based on the number of particles and dimensions. Then to choose the equilibrium concentration, the four best solutions obtained so far are saved, and then their average is taken to update the positions of the particle scattered throughout [21].

22.3.3 Harris Hawk Optimization (HHO)

Hawks are intelligent birds that are known for their hunting techniques. Harris hawks are predator birds that are found in Arizona, the United States. The technique used by Harris hawks is called the technique of 'surprise pounce' in which multiple hawks attack from different locations simultaneously so that the rabbit can be attacked as soon as it comes out of the cover. A variety of techniques for the hunting of the rabbit are depicted by intelligent hawks, which perpetually depend on the circumstances which depend on several factors. The detection of the prey requires the great eyes of the hawks, or sometimes it may involve the watching of the sites by many hawks, which may last up to many hours. The rabbit is considered as the optimum solution and the hawk near it or which catches it is considered as the best hawk. The exploration phase involves the hawks searching the rabbit all over the search space with some random initial values. The transition from exploration to exploitation depends on the energy of the rabbit. The energy of the rabbit decreases throughout iterations, and it is to be detected by the hawks when to attack the prey and when to keep looking for it. The rabbit uses every method it possesses to escape from the hunt and tries to escape even from very dangerous situations. The exploitation phase includes the process of making the rabbit tired so that it can be hunted down and usually takes several minutes; after that, the movement of the rabbit decreases. After that, depending on the final movements of the rabbit, soft or hard besiege or soft besiege with rapid dives starts until the hawks finally hunt down the rabbit [22].

22.3.4 Whale Optimization Algorithm (WOA)

Whales are one of the largest living mammals and are marine animals. They have a huge body and come near the surface of the water after a short duration to breathe. They are considered as one of the smart animals, and there are many species of whales. They may live alone or in a small group or may live along with their family. One of the types of the whale is the humpback whale, which has an interesting hunting technique for the hunting of krill and small fishes, known as bubble net method. This method involves the spiral movement of the whale by releasing bubbles in between and has been used for the optimization process. The whales can identify the locations of the prey and update themselves accordingly. Initially, the position of the optimal design is not known in the search space, so the one close to the prey is considered as the best. When the agent with the best position is defined, then other search agents are made to update their positions based on the position of the best agent. The bubble net method includes updating of the position of the whale based on the shrinking spiral technique where it moves in a spiral shape, and the shape shrinks continuously. The exploration phase of the algorithm is the one in which the whale searches for the presence of the prey. The humpback whales search for their prey randomly over the entire search space based on the positions of the other whales. The algorithms start by randomly assuming the solutions over the entire search space. At the end of each iteration, the search agents apprise their positions based on the randomly selected agent or the best search agent obtained till then. Then the whales perform circular motion or the shrinking spiral motion based on the prey by ejecting bubbles and finally reach out to the point where they eat their prey. This algorithm acts as a global optimization algorithm [27].

22.3.5 Sine Cosine Algorithm (SCA)

As the name clarifies itself, it uses the sine and cosine functions for obtaining the optimum value of the optimization problem. It is also a population-based method. The sine and cosine functions are oscillating functions, and the use of these functions allows the repositioning of a solution around the other solutions. It adds up to improve the exploitation of the search space. For the exploration of the search space, the functions also look up outside their search space. The exploration phase is supported by changing the ranges of sine and cosine functions whenever required. When the range of either sine or cosine function is changed, then the solution updates itself, whether it is inside or outside the search space. The random location is defined using a variable that has a range as that of the sine or cosine function, which guarantees the exploration and exploitation of the algorithm. The algorithm starts by initializing the positions of the solutions randomly, and the one with the best position is set to be the destination. The positions of the solutions are updated based on the destination. Besides, the ranges of the sine and cosine functions are changed accordingly to ensure the exploitation phase of the algorithm. This process continues until the maximum number of iterations is not reached. The algorithm can store the best solution obtained so far, and with each iteration, there is a possibility of a change of the solution [28].

22.3.6 Salp Swarm Algorithm (SSA)

Salps are marine animals that are similar to jellyfish and have a transparent body. The salps can form a swarm and move together in the search for food. It is known that the salps form a chain of salps and move together. The salp at the starting of the chain is assumed to be the leader salp that is responsible for guiding the members to the optimum food location. The positions of salps are represented by a matrix of order equal to the number of variables in the problem. The food source is considered as the optimum solution to the problem, but it is unknown. The leader salp updates his/her position at the end of each iteration based on the position of the food. Hence, the best position obtained by the chain is assumed as the optimum solution and also as the position of the food source. The algorithm starts by randomizing the positions of the salps all over the search space, and then it calculates the fitness of each salp present. The best salp with the best position obtained is then assumed to be the food source, which is then to be chased by the chain. At the end of each iteration, the positions of all the salps are updated, and then again, the fitness of each salp is calculated. During the iterations, if it is found that any of the salps have gone off the boundary or gone off the chain, it is brought back into boundary or chain. This process is continuously performed until the desired result is not found. In every iteration, the position of food particle is updated because, during the movements of the salps, they may encounter a better position, or the salps may become fitter, or their potential to find the maxima may increase [29].

22.4 SIMULATION RESULTS AND DISCUSSIONS

Newly developed RTDM was used for the study where the parameters of the two commercially available thin-film solar panels, Astronergy CHSM5001T 110W and Astronergy CHSM5011T 110W, are extracted based on the datasheet information given in Table 22.1.

TABLE 22.1
Datasheet Information of the Thin-Film Solar Panels

	Company	
	Astronergy CHSM5001T	**Astronergy CHSM5011T**
Maximum power (W)	110	110
Open-circuit voltage (V)	129.1	167
Short-circuit current (A)	1.519	1.08
MPP voltage (V)	89.46	0.905
MPP current (A)	1.230	121.5

As described earlier, very less work has been done in the field of parameter extraction of thin-film panels and the complexity of the equations is high, which demanded the proper choice of the variable ranges and the same is reported in Table 22.2.

The calculation is carried out assuming the standard test conditions (STC) for the panels and eight parameters are estimated. Some of the well-known and recent meta-heuristic algorithms which have proved to give good results as per the literature for different types of models have been used considering 20 search agents and overall 200 iterations.

After obtaining the results, statistical analysis is done considering independent 30 runs for each of the six algorithms used for the study. The statistical analysis includes the minimum, maximum, average, and standard deviation. The statistical analysis tells us about how good results have been given by the algorithms used and whether they have provided satisfactory results or not, and gives the quantitative analysis. The result of the statistical analysis is presented in Table 22.3, and according to the results, MFO outshines all the other algorithms by its best results, where it has produced around 10^{16} times better results for minimum value of error when compared against SCA algorithm for the first solar panel. For the second panel considered for the study, the best results are again produced by the MFO algorithm, and SCA gave the worst results. In this case, the least error of MFO is around 100 times better than the least error of the SCA algorithm. From the maximum, minimum, and standard deviation of the results, it can be concluded that the variation of the results produced by the algorithm was very less; hence, it is evident that the algorithms can produce stable results.

TABLE 22.2
Upper and Lower Bounds of the Eight Decision Variables

Parameters	Upper Bound	Lower Bound
Photovoltaic current (I_{pv})	0	5
Ideality factors, a_1, a_2	0.5	2
Series resistance, R_s	0	0.5
Parallel resistances, R_{P1}, R_{P2}	0.01	500
Reverse saturation currents, I_{01}, I_{02}	0	1e-06

TABLE 22.3
Statistical Analysis for the Results Obtained after 30 Independent Runs

Test Models	Methods	Minimum	Maximum	Average	Std. dev.
Astronergy CHSM5001T	MFO	**1.0457e-17**	**8.6309e-17**	**3.1387e-17**	**2.0247e-17**
	EO	1.0100e-05	9.9400e-05	3.4833e-05	2.5930e-05
	HHO	0.0113	0.0967	0.0507	0.0274
	WOA	0.1863	0.4208	0.3432	0.0752
	SCA	0.3662	0.3698	0.3674	0.0011
	SSA	0.3657	0.3719	0.3692	0.0021
Astronergy CHSM5011T	MFO	**0.0017**	**0.0022**	**0.0020**	**2.0440e-04**
	EO	0.0185	0.0497	0.0222	0.0053
	HHO	0.0117	0.0948	0.0536	0.0298
	WOA	0.1030	0.5087	0.2404	0.0957
	SCA	0.1048	0.2354	0.2191	0.0392
	SSA	0.1041	0.2845	0.2247	0.0454

The bold letters represent the best values amongst the algorithms compared.

To further verify the accurateness and validity of the results obtained, Kruskal-Wallis non-parametric test [30] is performed, which tests whether the samples have been generated from the same distribution or not. This test is generally performed on two or more samples having the same or a different number of sample sizes. This is a stochastic test that is done to check the stochastic dominance of one sample over another. Based on the Kruskal-Wallis test, the mean rank graphs have been produced for both the type of cells considered. Figure 22.2 presents the mean rank

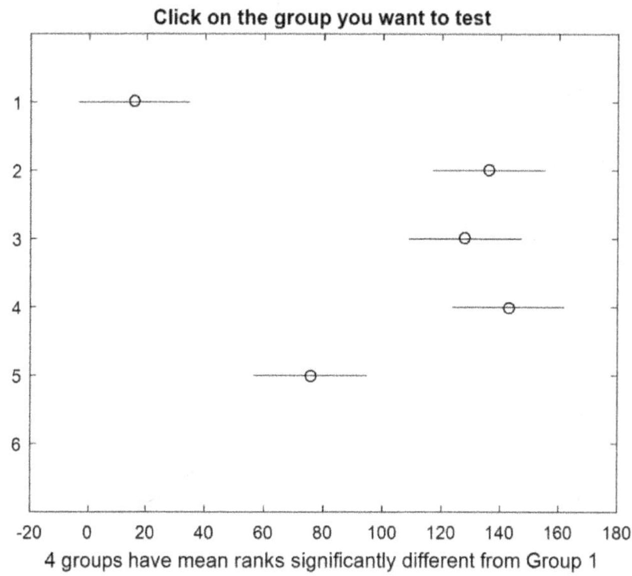

FIGURE 22.2 Kruskal-Wallis test diagram for CHSM5001T model.

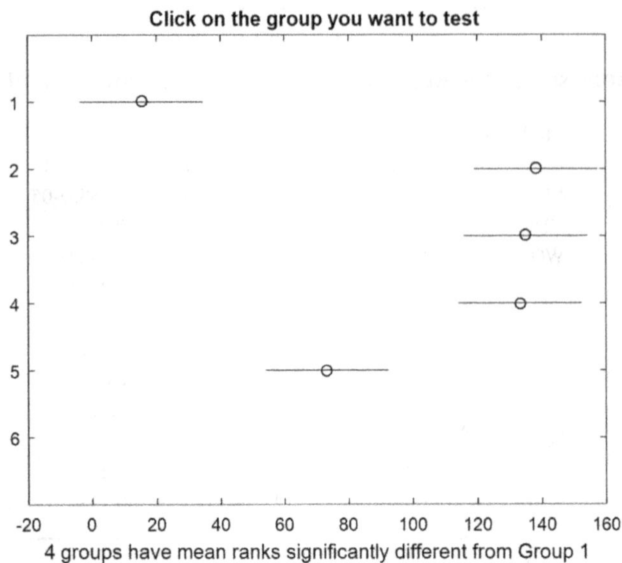

FIGURE 22.3 Kruskal-Wallis test diagram for CHSM5011T model.

graph of CHSM5001T, it can be seen very clearly that the mean rank of MFO has a significant difference from the four other algorithms, proving the superiority of the algorithm. In the same manner, Figure 22.3 represents the mean rank graph for the CHSM5011T panel, and the same result has been obtained where the MFO has a different mean rank from the four other algorithms. Only the EO algorithm reaches close to the MFO rank and is represented by number 6 along the y-axis. Therefore, MFO supersedes four algorithms, viz., SCA, SSA, WOA, and HHO in terms of Kruskal-Wallis test methodology.

Wilcoxon rank-sum test [31] is another non-parametric statistical test that is performed on two or more groups where the difference between the set of pairs is calculated, and these are further used for the determination of the significance of the results. In simpler words, this statistical test is performed to determine whether the result obtained is significant or not. The same has been performed in this study concerning the MFO algorithm, and the p-values have been calculated for all the other algorithms. The results are shown in Table 22.4. If the p-value is under 0.05,

TABLE 22.4
***P*-Values for Wilcoxon Rank-Sum Test**

Test System	Method	WOA	SCA	SSA	HHO	EO
CHSM5001T model	MFO	3.3170e-08	1.5616e-10	1.7501e-08	3.3170e-08	1.0132e-10
CHSM5011T model	MFO	8.4236e-09	8.4236e-09	2.6172e-06	1.5477e-08	2.7634e-11

TABLE 22.5
Holm-Bonferroni Test Results for *p*-Value Corrections

Test System	Method	Amended *p*-Values	*H*-Value
Test 1	MFO	$10^{-7}\times$ [0.6634 0.0062 0.5250 0.6634 0.0051]	[1 1 1 1 1]
Test 2	MFO	$10^{-5}\times$ [0.0034 0.0034 0.2617 0.0031 0.0000]	[1 1 1 1 1]

then the result is considered insignificant statistically. However, in the table, it can be seen that all of them have their values much lower than 0.05; hence, all the obtained results are significant.

Further, the same Wilcoxon test is performed with the help of corrections made by the Holm-Bonferroni test [32], who made a correction when several statistical tests are made with dependent and independent variables, and the results are presented in Table 22.5. It is again seen that the *p*-values are much less than 0.05, which again verifies that the results obtained are correct.

As mentioned earlier, 20 search agents were considered for 200 iterations, and based on that, a comparative convergence characteristic of all the six metaheuristic algorithms for the least error values obtained by them is shown in Figures 22.4 and 22.5. Figure 22.4 represents to the first solar panel considered where it can be clearly seen that the best results are shown by MFO, which is followed by the EO algorithm. Other than these, the algorithms gave moderate results, and almost converging results are obtained for WOA, SCA, and SSA. The texture of the curve is smooth.

Figure 22.5 shows the convergence characteristics for the second solar panel, and in this case as well, the best result is obtained by the MFO algorithm, followed by the

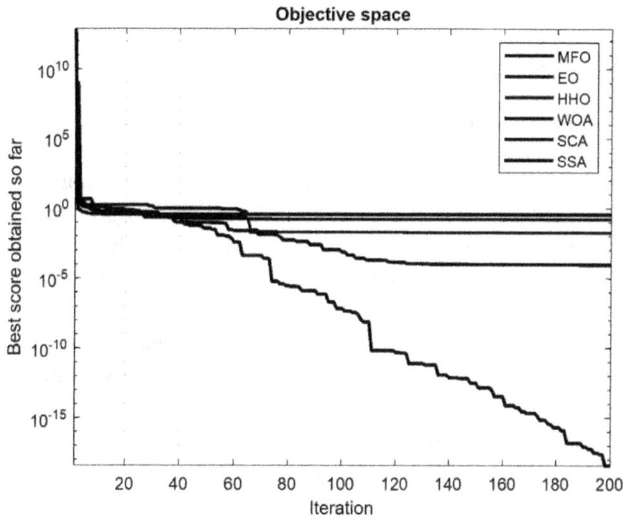

FIGURE 22.4 Convergence characteristics of algorithms for CHSM5001T panel.

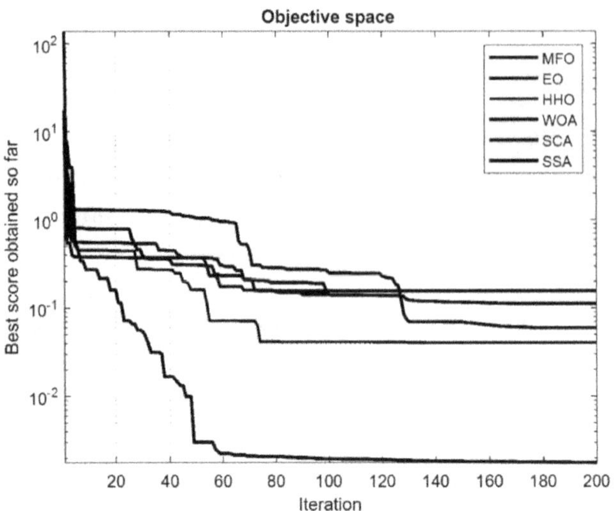

FIGURE 22.5 Convergence curves for different algorithms for the CHSM5011T panel.

HHO algorithm. The convergence characteristics are almost overlapping for WOA, SCA, and SSA. The texture of the curves is smooth. As described in the preceding sections, neither the convergence speed nor the characteristics are affected for the algorithms considered for this study.

A new formulation approach has been adopted to construct a new set of equations. The unknown parameters have been determined using the MFO technique. As many as five algorithms are used for comparison. Statistical analysis has been conducted to validate the significance of the results. Thus, the MFO algorithm proves superior to the other algorithms under study to measure the eight parameters of the RTDM.

22.5 CONCLUSION

In this chapter, the parameter estimation of the newly developed reverse diode model of a thin-film solar panel for the two commercially available panels has been done, although the model was formulated earlier to estimate the parameters of organic solar cells. Since the conventional models were not able to give accurate values of the parameters, this new model was developed. MFO algorithm has been applied for parameter identification of these solar panels. Five, two new and three recently developed, heuristic methods are used to compare the MFO algorithm. Due to the non-conventional nature of the model, the complexity increased largely, and the results obtained proved that still valid results are produced from the algorithms. To validate the statement, various statistical analyses have been done on the result, including the Kruskal-Wallis non-parametric test, which is done to determine the stochastic dominance of one algorithm over the other, and the mean ranks obtained are presented graphically which proves that MFO algorithm has significantly different mean rank than the other algorithms. Wilcoxon rank-sum test has also been performed, which is again a non-parametric test, and the significance of the obtained results has been tested statistically with respect to MFO algorithm and all the *p*-values came out to be much lower than 0.05, indicating their significance. The Wilcoxon rank-sum test has also been performed by considering the Holm-Bonferroni corrections in the Wilcoxon rank-sum test and the *p*-values were much less than 0.05, indicating their significance. Finally, the comparative convergence curves for both the panels have been presented, which prove that the best result has been provided by the MFO algorithm, followed by EO and HHO algorithm. The curves also prove that the convergence speed and accuracy of the algorithms have not been affected.

REFERENCES

1. Wu, L., Chen, Z., Long, C., Cheng, S., Lin, P., Chen, Y., & Chen, H. (2018). Parameter extraction of photovoltaic models from measured IV characteristics curves using a hybrid trust-region reflective algorithm. *Applied Energy, 232*, 36–53.
2. Chellaswamy, C., & Ramesh, R. (2016). Parameter extraction of solar cell models based on adaptive differential evolution algorithm. *Renewable Energy, 97*, 823–837.
3. Jordehi, A. R. (2016). Parameter estimation of solar photovoltaic (PV) cells: A review. *Renewable and Sustainable Energy Reviews, 61*, 354–371.
4. Abbassi, R., Abbassi, A., Jemli, M., & Chebbi, S. (2018). Identification of unknown parameters of solar cell models: A comprehensive overview of available approaches. *Renewable and Sustainable Energy Reviews, 90*, 453–474.
5. Mehta, H. K., Warke, H., Kukadiya, K., & Panchal, A. K. (2019). Accurate expressions for single-diode-model solar cell parameterization. *IEEE Journal of Photovoltaics, 9*(3), 803–810.
6. Qais, M. H., Hasanien, H. M., Alghuwainem, S., & Nouh, A. S. (2019). Coyote optimization algorithm for parameters extraction of three-diode photovoltaic models of photovoltaic modules. *Energy, 187*, 116001.
7. Yousri, D., Thanikanti, S. B., Allam, D., Ramachandaramurthy, V. K., & Eteiba, M. B. (2020). Fractional chaotic ensemble particle swarm optimizer for identifying the single, double, and three diode photovoltaic models' parameters. *Energy, 195*, 116979.
8. Nogueira, C. E. C., Bedin, J., Niedzialkoski, R. K., de Souza, S. N. M., & das Neves, J. C. M. (2015). Performance of monocrystalline and polycrystalline solar panels in a water pumping system in Brazil. *Renewable and Sustainable Energy Reviews, 51*, 1610–1616.
9. Zhao, J., Li, Y., Yang, G., Jiang, K., Lin, H., Ade, H., ... & Yan, H. (2016). Efficient organic solar cells processed from hydrocarbon solvents. *Nature Energy, 1*(2), 1–7.
10. Han, G., Zhang, S., Boix, P. P., Wong, L. H., Sun, L., & Lien, S. Y. (2017). Towards high efficiency thin film solar cells. *Progress in Materials Science, 87*, 246–291.
11. Qais, M. H., Hasanien, H. M., & Alghuwainem, S. (2020). Parameters extraction of three-diode photovoltaic model using computation and Harris Hawks optimization. *Energy, 195*, 117040.
12. Jordehi, A. R. (2018). Enhanced leader particle swarm optimisation (ELPSO): An efficient algorithm for parameter estimation of photovoltaic (PV) cells and modules. *Solar Energy, 159*, 78–87.
13. Elaziz, M. A., & Oliva, D. (2018). Parameter estimation of solar cells diode models by an improved opposition-based whale optimization algorithm. *Energy Conversion and Management, 171*, 1843–1859.
14. Cuevas, E., Gálvez, J., & Avalos, O. (2020). Comparison of solar cells parameters estimation using several optimization algorithms. In: *Recent Metaheuristics Algorithms for Parameter Identification* (pp. 51–95). Springer, Cham.

15. Pillai, D. S., Sahoo, B., Ram, J. P., Laudani, A., Rajasekar, N., & Sudhakar, N. (2017). Modelling of organic photovoltaic cells based on an improved reverse double diode model. *Energy Procedia*, *117*, 1054–1061.
16. Muangkote, N., Sunat, K., Chiewchanwattana, S., & Kaiwinit, S. (2019). An advanced onlooker-ranking-based adaptive differential evolution to extract the parameters of solar cell models. *Renewable Energy*, *134*, 1129–1147.
17. Yu, K., Qu, B., Yue, C., Ge, S., Chen, X., & Liang, J. (2019). A performance-guided JAYA algorithm for parameters identification of photovoltaic cell and module. *Applied Energy*, *237*, 241–257.
18. Yu, K., Liang, J. J., Qu, B. Y., Chen, X., & Wang, H. (2017). Parameters identification of photovoltaic models using an improved JAYA optimization algorithm. *Energy Conversion and Management*, *150*, 742–753.
19. Derick, M., Rani, C., Rajesh, M., Farrag, M. E., Wang, Y., & Busawon, K. (2017). An improved optimization technique for estimation of solar photovoltaic parameters. *Solar Energy*, *157*, 116–124.
20. Beigi, A. M., & Maroosi, A. (2018). Parameter identification for solar cells and module using a hybrid firefly and pattern search algorithms. *Solar Energy*, *171*, 435–446.
21. Faramarzi, A., Heidarinejad, M., Stephens, B., & Mirjalili, S. (2020). Equilibrium optimizer: A novel optimization algorithm. *Knowledge-Based Systems*, *191*, 105190.
22. Heidari, A. A., Mirjalili, S., Faris, H., Aljarah, I., Mafarja, M., & Chen, H. (2019). Harris hawks optimization: Algorithm and applications. *Future Generation Computer Systems*, *97*, 849–872.
23. Allam, D., Yousri, D. A., & Eteiba, M. B. (2016). Parameters extraction of the three diode model for the multi-crystalline solar cell/module using moth-flame optimization algorithm. *Energy Conversion and Management*, *123*, 535–548.
24. Abbassi, R., Abbassi, A., Heidari, A. A., & Mirjalili, S. (2019). An efficient salp swarm-inspired algorithm for parameters identification of photovoltaic cell models. *Energy Conversion and Management*, *179*, 362–372.
25. Biswas, P. P., Suganthan, P. N., Wu, G., & Amaratunga, G. A. (2019). Parameter estimation of solar cells using datasheet information with the application of an adaptive differential evolution algorithm. *Renewable Energy*, *132*, 425–438.
26. Mirjalili, S. (2015). Moth-flame optimization algorithm: A novel nature-inspired heuristic paradigm. *Knowledge-Based Systems*, *89*, 228–249.
27. Mirjalili, S., & Lewis, A. (2016). The whale optimization algorithm. *Advances in Engineering Software*, *95*, 51–67.
28. Mirjalili, S. (2016). SCA: A sine cosine algorithm for solving optimization problems. *Knowledge-Based Systems*, *96*, 120–133.
29. Mirjalili, S., Gandomi, A. H., Mirjalili, S. Z., Saremi, S., Faris, H., & Mirjalili, S. M. (2017). Salp swarm algorithm: A bio-inspired optimizer for engineering design problems. *Advances in Engineering Software*, *114*, 163–191.
30. Breslow, N. (1970). A generalized Kruskal-Wallis test for comparing K samples subject to unequal patterns of censorship. *Biometrika*, *57*(3), 579–594.
31. Wilcoxon, F., Katti, S. K., & Wilcox, R. A. (1970). Critical values and probability levels for the Wilcoxon rank sum test and the Wilcoxon signed rank test. In: *Selected Tables in Mathematical Statistics*, Vol. 1 (pp. 171–259). American Mathematical Society, Providence, RI.
32. Hochberg, Y. (1988). A sharper Bonferroni procedure for multiple tests of significance. *Biometrika*, *75*(4), 800–802.

23 Time Series Energy Prediction and Improved Decision-Making

Iram Naim[1] *and Tripti Mahara*[2]
[1]Department of Computer Science and Information Technology, MJP Rohilkhand University, Bareilly, Uttar Pradesh, India
[2]School of Business and Management, Christ University, Bangalore, Karnataka, India

CONTENTS

23.1 INTRODUCTION

In a manufacturing organization, in order to ensure efficient production their operational planning is based on the availability of resources. As energy is a prime mover in a manufacturing setup [1, 2], the procurement strategy of this resource should include a forecasting methodology for optimum cost-benefit balance. Natural gas is the cleanest fossil fuel [3] among the available fossil fuels and is one of the most widely used energy resources in manufacturing organizations in India. The country's overall gas production increased by 0.69% in 2018–2019. In terms of absolute data, it stood at 32,873 units against 32,649 MMSCM (million metric standard cubic meter) of 2017–2018. It is worth mentioning that the output of 2017–2018 was also 2.35% higher than its preceding year, changing the previous 6 years' output that depicted a downward trend [4]. It is used in both industrial and domestic fronts of the country. In industry, it is mainly used as a fuel to generate power, as transportation fuel for vehicles, for heating purpose, in steel manufacturing industries and in petrochemical. It is also utilized for cooking in domestic households [5]. India currently consumes 166 million standard cubic meters per day (MSCMD) of gas and this demand is expected to double by 2030 [6]. In order to meet their requirement, manufacturing companies purchase natural gas from local distribution companies. Procurement of natural gas is associated with many variable factors, making the cost of procurement highly volatile and fluctuating in the market. To overcome these types of volatilities, manufacturing companies generally keep multiple contracts like short-term (spot), medium-term (up to 18 months) and long-term (from 18 months to 15-20 years) [7] to procure the gas.

Observing these fluctuations in the price of the natural gas, it is essential to predict the quantity of natural gas to be procured based on the historical data. If any mathematical model predicts future data taking historical data as input, it is termed as time series forecasting. Analyzing this type of data has become a recent area of focus in artificial intelligence (AI), as accurate forecasting is becoming increasingly vital across all kinds of industries in order to make more informed decisions. Essentially, applying AI to time series analysis allows us to better uncover the meaning of hidden patterns in the data. There are many statistical methods and machine learning and deep learning algorithms to perform time series predictions, but classical methods such as ETS (error, trend, seasonal) and ARIMA (auto-regressive integrated moving average) outperform machine learning and deep learning methods for one-step and multi-step forecasting on univariate datasets [8] Thus, the study here used traditional time series forecasting to achieve the following objectives:

- To come up with a framework for natural gas procurement.
- To make effective use of natural gas data to mitigate price risk and allow more accurate natural gas budgeting and forecasting.
- Development of a personalized strategy that integrates various factors that are important for a manufacturing organization.

The remainder of this chapter is framed in various sections. Related work has been compiled in Section 23.2. Section 23.3 describes the proposed framework with its various components. Section 23.4 discusses the managerial inferences, followed by conclusion in Section 23.5.

23.2 RELATED WORK

Literature review is divided into two parts: the first part provides details about existing forecasting methods for natural gas prediction and the second part discusses procurement strategies for natural gas in the manufacturing domain.

Recently, the use of forecasting can be seen in various fields of resource planning such as human resource planning [9–11], enterprise resource planning [12–14] and in many other fields. Time series forecasting techniques are common methods used for data related to time. From the literature, it is evident that ARIMA is a good method for predicting the output on short-term basis. Time series-based daily gas requirement of a town has been predicted using the degree-day concept in [15]. A time series forecasting model to identify the parameters affecting natural gas demand has been depicted in [16] using multivariable regression analysis. Another work based on degree-day method of forecasting is shown in [17], where the researchers did monthly forecast and partitioned years based on season of heating and nonheating. Ediger and Akar [18] used the ARIMA and seasonal ARIMA (SARIMA) methods to estimate the future primary energy demand of Turkey from 2005 to 2020. ARIMA-based forecasting has been done in [19], in which the past values have been obtained in terms of present and lagged values of white noise. ARIMA model of forecast is also used to predict the monthly natural gas consumption of Turkey [20, 21]. Forecasting for the daily trend is presented using ARIMA forecast and artificial neural network in [22] for local distribution companies. Zhao et al. [23] used ARIMA with historical fuel cost data to develop a three-step-ahead fuel cost distribution prediction model. Karabiber and Xydis [24] predicted the consumption of natural gas in four subnets of Denmark.

There are various research works on natural gas procurement and its improvement. Studies have been done for improvement of internal procurement processes, as discussed in [25] and [26]. Research work is also done for observing the behavior of suppliers [27] and integration of buyer and supplier [28]. There are papers for different network charges on natural gas, as discussed in [29]. Supplier selection [30] is also an important criterion for procuring any commodity. A study in Colombia [31] defined a model that captures the structure of the gas network and ensures gas supply based on market rules. Some of the studies [32–34] used mixed-integer programming model for natural gas purchase and related constraints. Kaming et al. [35] developed a model for engineering procurement construction contract for natural gas project in Indonesia. Shahrukh et al. [36] found out the benefits of spot purchase of natural gas. Sillanpää et al. [37] defined procurement decisions with nonlinear costs and multi-period commitments.

Generally, the procurement strategies are available for local distribution companies of natural gas. There are few research studies available that represent procurement suggestion at organization level. One of the important studies that focuses on the similar problem has been performed in the country Slovenia [38]. This study focuses on analyzing the use of natural gas at particular establishment level and suggests an efficient procurement on the basis of analysis.

23.3 METHODOLOGY: PROPOSED NATURAL GAS PROCUREMENT STRATEGY IN MANUFACTURING INDUSTRY

To formulate the procurement strategy for natural gas, fuel resource planning is necessary. It is possible when the consumption of natural gas at organization level is known. As depicted in Figure 23.1, along with the consumption data, there are internal and external factors that contribute to the development of procurement strategy. The proposed framework for natural gas procurement used in manufacturing industry is presented in Figure 23.1. It constitutes natural gas consumption forecast, fuel resource planning, internal factors and external factors.

This framework helps in providing the important input for big modelling systems required for companies involved in predicting the prices of natural gas over long period of time depending on demand and supply constraints. It is important that the accurate demand forecast for natural gas is done to increase effectiveness of procurement strategy. The details of each variable affecting procurement strategy are discussed in the section ahead.

23.3.1 Natural Gas Consumption Forecast

Forecasting of natural gas consumption is generally stored based on certain time duration. Thus, it reflects time series data. To forecast this type of data, two important steps are analysis of the time series data and selection of appropriate model.

23.3.1.1 Analysis of Time Series Data

Time series data comprise four constituents. These are seasonal component, trend component, cyclic component and random component. Seasonal elements repeat themselves over a specific duration, i.e., repetition of pattern on a weekly basis or monthly basis, etc. Trend shifts in either upward or downward direction in a predictable pattern. Cyclic elements consist of long duration cycles in comparison to seasonal elements, whereas a random element does not carry any fixed pattern [39].

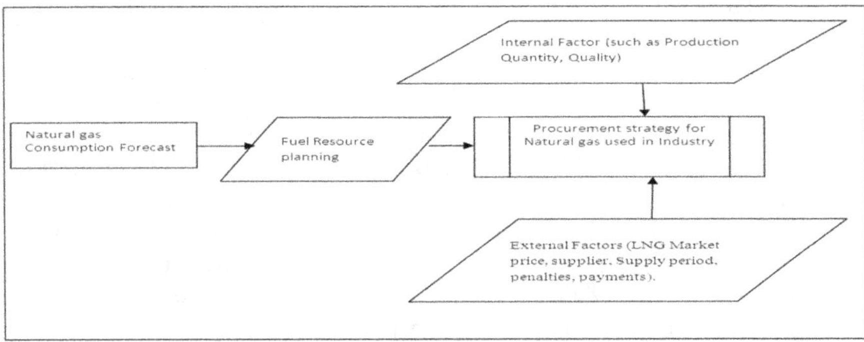

FIGURE 23.1 Framework for natural gas procurement within a manufacturing plant.

To analyze these components in a time series data, the following plots [40] can be useful: a run sequence plot, a seasonal plot, multiple box plots, autocorrelation function (ACF) plot and partial autocorrelation function (PACF) plot.

23.3.1.2 Selection of Forecasting Method

Selecting the most appropriate forecasting methodology is crucial. ARIMA is one of the common methods used for short-term forecasting. Future data for a known time series, X_t, can be depicted by using ARMA model [41]. As the name suggests, ARMA contains two sections: the first is "AR" – autoregressive component and the second one is "MA" – moving average component. In the AR part, the variable is regressed based on its own lagged values. The AR component of the model provides that the future values of X are weighted averages of current and past realizations. In the MA part, error term is modelled by treating it as linear combination of error terms. ARIMA model [42] is a generalization of ARMA model. Time series is made stationary in ARIMA model by performing the differencing.

- For nil differencing, $x_t = X_t$.
- For single differencing, $x_t = X_t - X_{t-1}$.

The general expression is depicted as:

$$X_t = c + \sum_{i=1}^{p} \varphi_i x_{t-i} - \sum_{i=1}^{q} \theta_i \varepsilon_{t-i}, \tag{23.1}$$

where moving average component becomes nonpositive as each differencing minimizes the error.

The common approaches to ARIMA forecasting are the Box-Jenkins ARIMA [43] and Hyndman-Khandakar ARIMA [44]. In this chapter, the Hyndman-Khandakar algorithm is used to select the most suitable ARIMA (p, d, q) model for dataset under consideration. An ARIMA (p, d, q) equation is given by:

$$\varphi(B)(1 - Bd)Xt = c + \theta(B)\varepsilon t \tag{23.2}$$

where εt is a white noise, B is the indication of backshift operator, whereas $\varphi(z)$ is the polynomial having order p and $\theta(z)$ is polynomial of order. Select p, coefficient of AR component q, coefficient of MA component and c, constant by minimizing AICc. The criterion for the selection of suitable model is Akaike information criterion (AIC) [45]. The seasonal ARIMA model is a combination of seasonal term with ARIMA model [20]:

$$\text{SARIMA}(p,d,q)(P,D,Q)_s, \tag{23.3}$$

in which S signifies count for yearly observations, whereas seasonal part is represented in capital letters and nonseasonal parts are depicted in lowercase notations.

23.3.2 Fuel Resource Planning

Before commencing fuel resource planning, the planning horizon is to be identified. It is defined as the time frame for which the planning is done. A planning horizon should be based on long- and short-term goals of production, amount and quality of resources required, and other relevant conditions. It is necessary for all the departments to project their fuel requirement for a time frame such as weekly and monthly. One of the important steps to define planning horizon is based on the results of forecasting. Depending on the forecasting horizon with accurate forecasted value having minimum error, fuel procurement time duration can be chosen.

23.3.3 Internal Factors

Product quantity and product quality are two important factors that contribute in developing procurement strategy. It is prudent to identify the production target of various departments as well as the overall plant. The short-term and yearly production figures directly affect the fuel consumption in the plant. In any manufacturing plant, the production targets depend on the entire range of products.

23.3.4 External Factors

External factors such as liquefied natural gas (LNG) market price, supplier, supply period, penalties and payments have significant contribution in procurement strategy. Each of them are explained in the following.

23.3.4.1 LNG Market Price

The market price of natural gas is dependent on international market and month and location of supply. Adequate knowledge of market trends for prices of natural gas should be collected before finalizing the procurement strategy. Good strategy and analysis of the market will save investment cost and help in economical procurement of natural gas.

23.3.4.2 Supplier

In India, natural gas industry started in 1960 from Assam and Gujarat. In the 1970s, Oil and Natural Gas Corporation (ONGC) of India discovered the South basin fields. Later on, Gujarat, KG basin, Cauvery basin, Tripura, Assam, etc., were identified as major potential areas for extraction of natural gas in the country. In the year 2004, LNG terminal of capacity 4 MMTPA (million metric tonnes per annum) was set up at Dahej and import of LNG was started from Qatar [46]. Gas producers, gas suppliers and gas transmitters are the three entities involved in production and transmission of gas. In India, ONGC, Oil India Limited (OIL), British Gas (BG) India and Gujarat State Petroleum Corporation Ltd (GSPC) are main gas producers and Petronet LNG Ltd (PLL), Shell Oil Company, ONGC, Gas Authority of India Limited (GAIL) are gas suppliers. GAIL, GSPC, ONGC, Reliance Industries

Limited (RIL), OIL and Assam Gas Company Limited (AGCL) are the companies involved in providing gas transport facility from seller's premises to buyers' premises through pipelines

23.3.4.3 Supply Period

The supply period is the entire contract period in which gas is received by the buyer through identified suppliers. As the prices of natural gas are volatile, the supply period is to be finalized with due care as there may be a chance of costly procurement of natural gas for the selected supply period.

23.3.4.4 Penalties

There are penalty clauses [47] associated with contracts of gas agreement. Some of the common penalties are take or pay, shortfall quantity, properly nominated contract quantity (PNCQ), unauthorized over-drawl, ship or pay and payments.

23.4 VALIDATION OF PROPOSED FRAMEWORK: CASE STUDY APPROACH

To validate the proposed framework, a case of a manufacturing organization involved in production of steel is considered. Being a consumer of natural gas, prediction of natural gas demand for the plant is very helpful in fuel resource planning. Use of a good forecasting technique will suggest precise amounts of fuel to be procured by the organization.

A monthly dataset of natural gas consumption data has been obtained from this manufacturing organization from April 2010 to March 2017. This time series is taken only for validation of proposed work. The measurement for gas consumption during the relevant period was in standard cubic meter (SCM). This data of natural gas consumption is for the entire plant covering all the processes performed in steel manufacturing. To identify the nature of data and its trend over time, a plot of consumption vis-à-vis time is depicted in Figure 23.2.

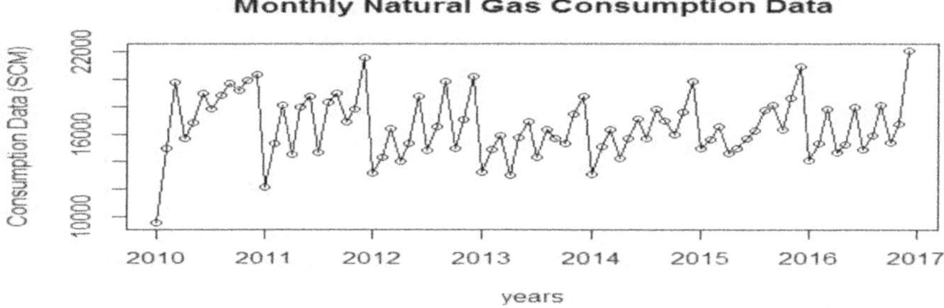

FIGURE 23.2 Monthly data for natural gas consumption.

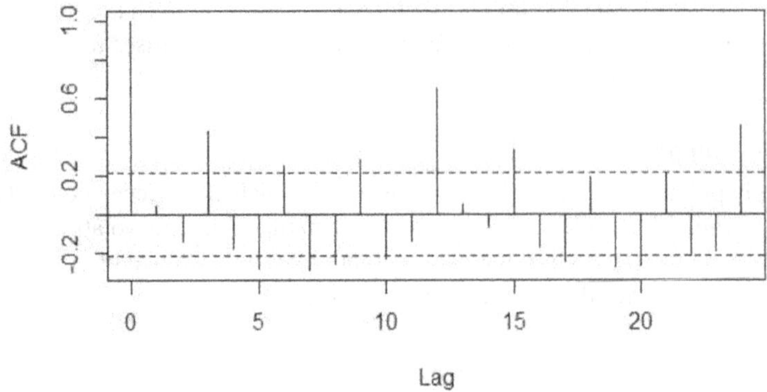

FIGURE 23.3 ACF for dataset.

23.4.1 Natural Gas Consumption Forecast

23.4.1.1 Analysis of Time Series Data

In order to check the randomness of the data, ACF plot and PACF plot have been drawn and are depicted in Figures 23.3 and 23.4, respectively. These plots are helpful in recognizing the relation of lagged values in the series. As is clear from the plots, there exists significant value at lag 12, which represents seasonal periodic behavior.

23.4.1.2 Selection of Forecasting Model

Validation of forecasting can be effectively performed by segregating the available data into two sections, i.e., training and test data. Training data, also termed as in-sample period, is a set of data that helps in recognizing its interconnection or relationship, whereas test data, also known as out-sample period, is used for

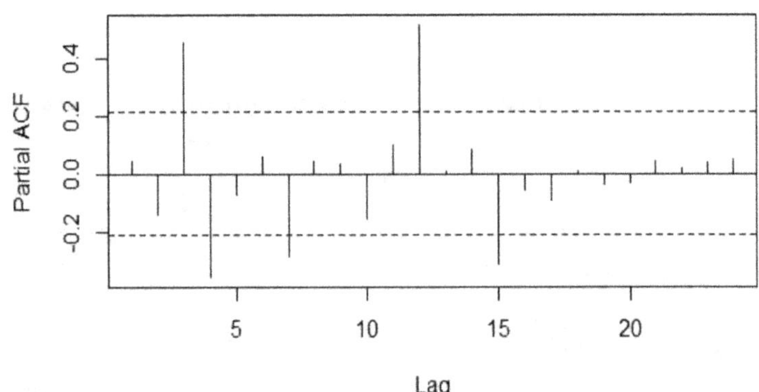

FIGURE 23.4 PACF for dataset.

validating the accuracy of forecasting. Training set, i.e., the data from the 1st month to 80th month, is utilized for developing the model, and the remainder of the data from the 81st to 84th month is used as test set for training purpose. Time series cross-validation and error analysis are used for validation purpose. Different training datasets are being used, each containing one additional observation over its preceding one. Eighty months' data is used for 1-month forecast, and prediction is done for the 81st month. This step is repeated after incrementing the training set data by one more month data, i.e., now 81 observations will be considered for training and forecasting will be done for the 82nd month. On the similar increment, the other steps will be performed.

Hyndman-Khandakar algorithm is utilized in ARIMA forecasting to optimize the AIC values and reach a suitable model with p, d and q parameters. The values of p and q are incremented and reduced by a factor of unity. The identified 1-month forecast model is of order ARIMA (0,1,1)(0,1,1) [12]. This is a seasonal ARIMA model. Where $p = 0$, it represents no autoregressive (AR) component of order 1. Order of differencing is depicted through value of parameter d; here, $d = 1$ represents stationarity in time series and is autocorrelated. Also, $q = 1$ shows presence of moving average term. Similarly, all the methods are being applied to remaining training set for performing 1-month forecast. Figure 23.5(a)–(d) represents forecasted values for training set 1, training set 2, training set 3, and training set 4, respectively, for 1-month ahead. Table 23.1 gives the summary for achieved values of forecasting in different steps.

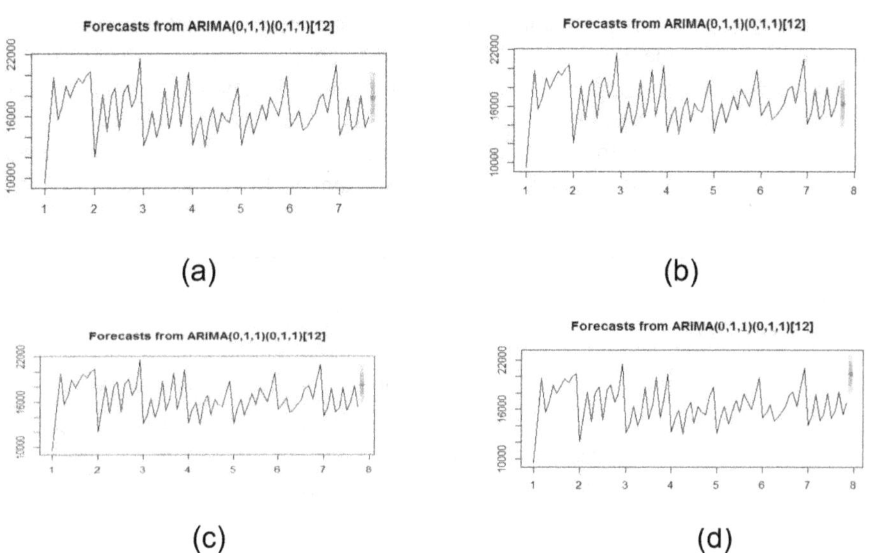

FIGURE 23.5 Result of 1-month forecasting from ARIMA model for different training sets.

TABLE 23.1
One-Month Forecast Values of Various Models

Type	Training Set	Test Set	ARIMA (0,1,1)(0,1,1) [12]
1-month forecast	Training set 1	18120	17789.41
	Training set 2	15667	16272.86
	Training set 3	17780	18312.13
	Training set 4	21091	20637.05

23.4.2 Fuel Resource Planning

To decide the time duration for natural gas procurement, we will take into account the results obtained from forecasting in Section 23.4.1. The forecasting results obtained from ARIMA (0,1,1)(0,1,1) [12] are very close to the test dataset. The accuracy measures are calculated and the errors are averaged across all test sets. Error components that occur during forecasting are listed in Table 23.2. Hence, the results obtained by ARIMA are used as an input for fuel resource planning.

From the error analysis, it can be suggested that the procurement practice for this organization to procure natural gas from the market for supply can be a period of 3 months as it is an efficient short-term forecasting algorithm.

23.4.3 Internal Factors

The case company has production targets related to steel manufacturing. Usually, the companies decide their annual target for the financial year starting from April to March. The case company divides its annual production target into smaller subparts that are based on the four quarters, i.e., April to June, July to September, October to December and January to March. It can also be seen in Figure 23.6 that at the end of each quarter, i.e., in the months of June, September, December and March, the

TABLE 23.2
Analysis of Error

		Errors for 1-Month Forecast				
Method	**Errors**	**Tr1**	**Tr2**	**Tr3**	**Tr4**	**Average**
ARIMA (0,1,1)	RMSE	330.59	605.862	532.128	453.949	489.5267
(0,1,1)	MAE	330.59	605.862	532.128	453.949	480.6323
[12]	MAPE	1.82445	3.86712	2.99285	2.15234	2.709188
	MASE	0.1404	0.25745	0.226	0.19304	0.204222

Note: RMSE (root mean square error), MAE (mean absolute error), MAPE (mean absolute percentage error), MASE (mean absolute scaled error).

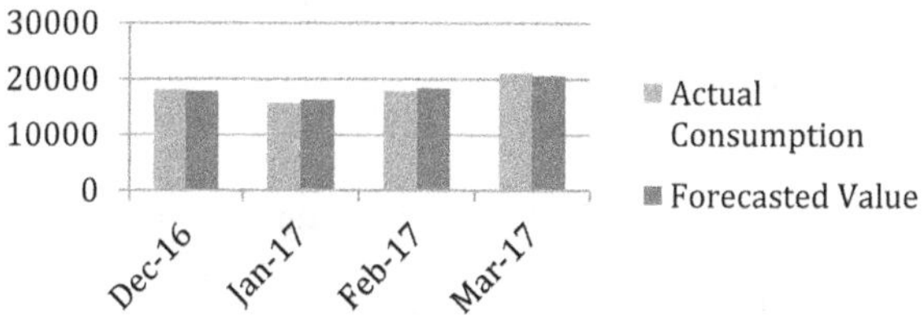

FIGURE 23.6 Actual consumption vs. forecasted value.

consumption of natural gas is high in quantity. This also made a valid point with Section 23.4.2, where we have suggested the period of 3 months for the procurement in accordance with quarterly production targets.

23.4.4 External Factors

The market prices of natural gas are highly fluctuating and change each fortnightly depending on the international market. The market price of LNG is decided by the government. Payments are to be done for the minimum contracted quantity, even in case consumption is less than minimum contracted quantity. The results of monthly natural gas forecasting (Figure 23.6) is very close to the actual consumption. Thus, it is prudent to make procurement of natural gas on quarterly frequency. However, the case company can protect these penalties as they can use a flexible range of ± 5% under procurement contracts because all the forecasted values lie within this range.

23.5 CONCLUSIONS

In this chapter, we propose a framework for fuel resource planning and related procurement strategy that can be used by any manufacturing organization. This procurement strategy incorporates short-term forecasting results along with inputs from various internal and external factors. The main focus of the research is to depict the importance of forecasting in effective procurement. Thus, ARIMA is consiered one of the most suitable techniques for short term forecasting. It is used to suggest the procurement quantity, supply period and range for procurement. Fuel resource planning can further be improved by considering other internal and external factors. Due to unavailability of specific data for external factors for the case organization, they have not been included in the study.

REFERENCES

1. T. A. Napp, A. Gambhir, T. P. Hills, N. Florin, and P. S. Fennell, "A review of the technologies, economics and policy instruments for decarbonising energy-intensive manufacturing industries," *Renewable and Sustainable Energy Reviews*, vol. 30, pp. 616–640, 2014.
2. P. Thollander, M. Danestig, and P. Rohdin, "Energy policies for increased industrial energy efficiency: Evaluation of a local energy programme for manufacturing SMEs," *Energy Policy*, vol. 35, no. 11, pp. 5774–5783, 2007.

3. A. H. Kakaee, A. Paykani, and M. Ghajar, "The influence of fuel composition on the combustion and emission characteristics of natural gas fueled engines," *Renewable and Sustainable Energy Reviews*, vol. 38, pp. 64–78, 2014.
4. "India natural gas output grew 0.69% in FY19," 2020. [Online]. Available: https://www.indoasiancommodities.com/2019/05/01/india-natural-gas-output-grew-0-69-in-fy19/ [accessed: 18 June 2020].
5. ET EnergyWorld, "India's natural gas production grew for second consecutive year in 2018-2019," 2020. [Online]. Available: https://energy.economictimes.indiatimes.com/news/oil-and-gas/indias-natural-gas-production-grew-for-second-consecutive-year-in-2018-2019/69108091 [accessed: 28 May 2020].
6. Reuters, "Global oil majors see surge in Indian demand for natural gas," 2019. [Online]. Available: https://www.reuters.com/article/us-energy-india-gas/global-oil-majors-see-surge-in-indian-demand-for-natural-gas-idUSKBN1WT1QI [accessed: 18 June 2020].
7. K. Neuhoff and C. Von Hirschhausen, "Long-term vs. Short-term Contracts: A European Perspective on Natural Gas," Working Papers EPRG 0505, Energy Policy Research Group, Cambridge Judge Business School, University of Cambridge, 2006.
8. S. Makridakis, E. Spiliotis, and V. Assimakopoulos, "Statistical and Machine Learning forecasting methods: Concerns and ways forward," *PLoS ONE*, vol. 13, no. 3, pp. 1–26, 2018.
9. Y. Zhang, F. Li, and B. Liu, "Research on the forecasting model of total human resource demand of large central enterprise group based on C-D production function," in: *DEStech Transactions on Computer Science and Engineering*, 2017. DOI: 10.12783/dtcse/itme2017/8008
10. D. Y. Wang and M. Lu, "Discussion on human resources management issues for SMEs — A case study of Qinzhouhuang Millet Co., Ltd.," in: *Proceedings of the International Conference on Humanity and Social Science (ICHSS2016)*, pp. 22–27, 2017.
11. C. D. Ittner and J. Michels, "Risk-based forecasting and planning and management earnings forecasts," *Review of Accounting Studies*, vol. 22, no. 3, pp. 1005–1047, 2017.
12. V. A. Mabert, A. Soni, and M. A. Venkataramanan, "Enterprise resource planning: Managing the implementation process," *European Journal of Operational Research*, vol. 146, no. 2, pp. 302–314, 2003.
13. V. C. Sugiarto, R. Sarno, and D. Sunaryono, "Sales forecasting using Holt-Winters in enterprise resource planning at sales and distribution module," in: *Proceedings of the 2016 International Conference on Information and Communication Technology and Systems (ICTS 2016)*, pp. 8–13, 2017.
14. P. M. Catt, R. H. Barbour, and D. J. Robb, "Assessing forecast model performance in an ERP environment," *Industrial Management & Data Systems*, vol. 108, no. 5, pp. 677–697, 2008.
15. F. Gümrah, D. Katircioglu, Y. Aykan, S. Okumus, and N. Kilinçer, "Modeling of gas demand using degree-day concept: Case study for Ankara," *Energy Sources*, vol. 23, no. 2, pp. 101–114, 2001.
16. F. B. Gorucu and F. Gumrah, "Evaluation and forecasting of gas consumption by statistical analysis," *Energy Sources*, vol. 26, no. 3, pp. 267–276, 2004.
17. H. Aras and N. Aras, "Forecasting residential natural gas demand," *Energy Sources*, vol. 26, no. 5, pp. 463–472, 2004.
18. V. Ş. Ediger and S. Akar, "ARIMA forecasting of primary energy demand by fuel in Turkey," *Energy Policy*, vol. 35, no. 3, pp. 1701–1708, 2007.
19. F. Faisal, "Time series ARIMA forecasting of natural gas consumption in Bangladesh's power sector," *Elixir*, vol. 49, pp. 9985–9990, 2012.
20. M. Akpinar and N. Yumusak, "Year ahead demand forecast of city natural gas using seasonal time series methods," *Energies*, vol. 9, no. 9, p. 727, 2016.

21. M. Akpinar and N. Yumusak, "Forecasting household natural gas consumption with ARIMA model: A case study of removing cycle," in: *AICT 2013 – 7th International Conference on Application of Information and Communication Technologies, Conference Proceedings*, 2013.
22. C. V. Cardoso and G. L. Cruz, "Forecasting natural gas consumption using ARIMA models and artificial neural networks," *IEEE Latin America Transactions*, vol. 14, no. 5, pp. 2233–2238, 2016.
23. Z. Zhao, C. Fu, C. Wang, and C. J. Miller, "Improvement to the prediction of fuel cost distributions using ARIMA model," in: *IEEE Power and Energy Society General Meeting*, 2018.
24. O. A. Karabiber and G. Xydis, "Forecasting day-ahead natural gas demand in Denmark," *Journal of Natural Gas Science and Engineering*, vol. 76, p. 103193, 2020.
25. S. Chaturvedi and D. Chakrabarti, "Internal purchase process : A study for setting improvement target," *International Journal of Procurement Management*, vol. 8, no. 6, pp. 753–768, 2015.
26. J. J. Sikorski, O. R. Inderwildi, M. Q. Lim, S. S. Garud, J. Neukäufer, and M. Kraft, "Enhanced procurement and production strategies for chemical plants: Utilizing real-time financial data and advanced algorithms," *Industrial & Engineering Chemistry Research*, vol. 58, no. 8, pp. 3072–3081, 2019.
27. F. Rocha, "Procurement as innovation policy and its distinguishing effects on innovative efforts of the Brazilian oil and gas suppliers," *Economics of Innovation and New Technology*, vol. 27, no. 9, pp. 1–20, 2017.
28. G. Salema and A. Buvik, "Buyer-supplier integration and logistics performance in healthcare facilities in Tanzania : The moderating effect of centralised decision control," *International Journal of Procurement Management*, vol. 11, no. 2, pp. 250–265, 2018.
29. C. Mosácula, J. P. Chaves-ávila, and J. Reneses, "Designing natural gas network charges: A proposed methodology and critical review of the Spanish case," *Utilities Policy*, vol. 54, pp. 22–36, 2018.
30. M. Silva and P. S. Figueiredo, "Supplier selection: A proposed framework for decision making," *International Journal of Procurement Management*, vol. 11, no. 2, pp. 233–249, 2018.
31. J. Villada and Y. Olaya, "A simulation approach for analysis of short-term security of natural gas supply in Colombia," *Energy Policy*, vol. 53, pp. 11–26, 2013.
32. L. Contesse, J. C. Ferrer, and S. Maturana, "A mixed-integer programming model for gas purchase and transportation," *Annals of Operations Research*, vol. 139, no. 1, pp. 39–63, 2005.
33. T. Li, M. Eremia, and M. Shahidehpour, "Interdependency of natural gas network and power system security," *IEEE Transactions on Power Systems*, vol. 23, no. 4, pp. 1817–1824, 2008.
34. C. M. Correa-Posada and P. Sanchez-Martin, "Integrated power and natural gas model for energy adequacy in short-term operation," *IEEE Transactions on Power Systems*, vol. 30, no. 6, pp. 3347–3355, 2015.
35. P. F. Kaming, A. Koesmargono, and B. W. Aji, "Delay model for Engineering Procurement Construction (EPC): A case of Liquefied Natural Gas (LNG) projects in Indonesia," in: *The 2nd Conference for Civil Engineering Research Networks (ConCERN-2 2018), MATEC Web Conference*, vol. 270, 2019.
36. M. Shahrukh, R. Srinivasan, and I. A. Karimi, "Evaluating the benefits of LNG procurement through spot market purchase," in: *Computer Aided Chemical Engineering, Vol. 46* (pp. 1723–1728). Elsevier B.V., Amsterdam, 2019.
37. V. Sillanpää, J. Liesiö, and A. Käki, "Procurement decisions over multiple periods under piecewise-linear shortage costs and fixed capacity commitments," *Omega (United Kingdom)*,, vol. 100, p. 102207, 2021.

38. M. Kovačič and B. Šarler, "Genetic programming prediction of the natural gas consumption in a steel plant," *Energy*, vol. 66, pp. 273–284, 2014.
39. A. S. Weigend, *Time Series Prediction*. Routledge, London, 2018.
40. R. J. Hyndman and G. Athanasopoulos, *Forecasting: Principles and Practice*, 2nd edition. OTexts, Melbourne, Australia, 2018.
41. Liu, Y. and Tajbakhsh, S.D., 2020. Fitting ARMA Time Series Models without Identification: A Proximal Approach. arXiv preprint arXiv:2002.06777.
42. M. Gerolimetto, "ARIMA and SARIMA Models," Ca' Foscari University of Venice, Italy, 2010.
43. G. Box, G. Jenkins, G. Reinsel, and G. Ljung, *Time Series Analysis: Forecasting and Control*. John Wiley & Sons, Hoboken, NJ, 2008.
44. R. Hyndman and Y. Khandakar, "Automatic time series for forecasting: The forecast package for R," *Journal of Statistical Software*, vol. 27, no. 3, 2007.
45. K. Aho, D. Derryberry, and T. Peterson, "Model selection for ecologists: The worldviews of AIC and BIC," *Ecology*, vol. 95, no. 3, pp. 631–636, 2014.
46. Ministry of Petroleum and Natural Gas, Government of India, "LPG-policies and guidelines." [Online]. Available: http://petroleum.nic.in/marketing/policies-and-guidelines/lpg-policies-and-guidelines [accessed: 19 May 2020].
47. "RLNG, B I D Agreement, Gail," 2016. [Online]. Available: https://powermin.gov.in/sites/default/files/uploads/E-bid_RLNG_Agreement-for_2nd-phase.pdf

24 Machine Learning-Enabled Cyber Security in Smart Grids

Anand Sharma
Department of Computer Science, Mody University of Science and Technology, Sikar, Rajasthan, India

Manish Kumar
Department of Nuclear Engineering, School of Technology, Pandit Deendayal Energy University, Gandhinagar, Gujarat, India

Nitai Pal
Department of Electrical Engineering, Indian Institute of Technology (ISM), Dhanbad, Jharkhand, India

CONTENTS

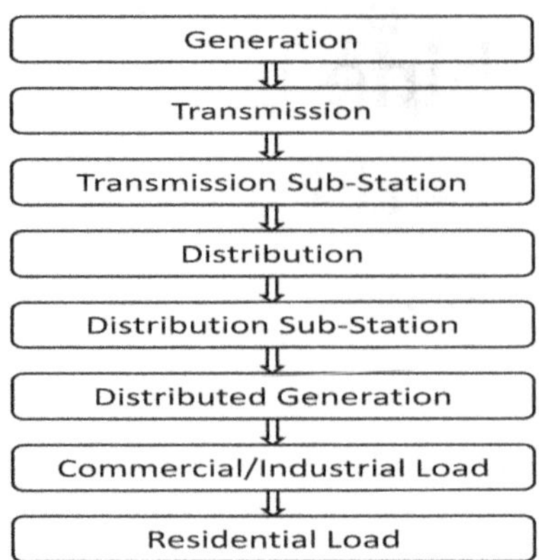

FIGURE 24.1 Conventional power system.

24.1 INTRODUCTION

The need for electricity is growing day by day due to industrialization, urbanization, and growing populations. Fossil fuels are the main contributor in the power generation today. We need to minimize the losses to reduce resources consumption while delivering more power to users matching with growing demand. The world is paying more attention toward renewable energy as it is a clean and green source of energy. Photovoltaic (PV) installations are either grid connected or stand-alone with hybrid power systems [1]. PV and wind energy are used widely for enhancing efficiency, reliability, availability, communication capacity, etc. Renewable energy sources with battery-based storage are necessary to meet the load requirements. A fuel cell converts chemical energy into electricity as long as fuel and oxidant are supplied. Gaseous oxygen is the most common oxidant as it easily available from the air and is easily stored. Electrons can flow from the positive terminal to the negative electrode via the external circuit due to non-electronic conductor of electrolyte. Oxygen reacts with electrons and hydrogen ions to produce water at the cathode [2]. Battery is the most common electrochemical device. The conventional model is shown in Figure 24.1.

To achieve a revolutionized energy sector, it required the multi-direction information flow between different units of the grids. It provides the required finished data such as substation metering, generation, transmission and distributions, security, controlling and monitoring of services, and assets [3]. The smart grid is defined by the National Institute of Standards and Technology (NIST) as the integration of conventional grid with information and communication technology (ICT) [4, 5]. Later, IEEE modified this representation and explained the multidirectional interaction for improvements. Such expansions and enhancements motivated the ICT experts and users to activate the grid proficiently by deploying renewable resources also [6, 7]. The deployments of smart grids along with networking capabilities and distributed intelligence have improved the competence and dependability, but have increased the risk of cyberattacks [8]. In the smart grid, two-way communication [9] enables it to access and report the conditions to its neighboring agents via the communication path and port with processors [10, 11]. The conventional systems are upgraded

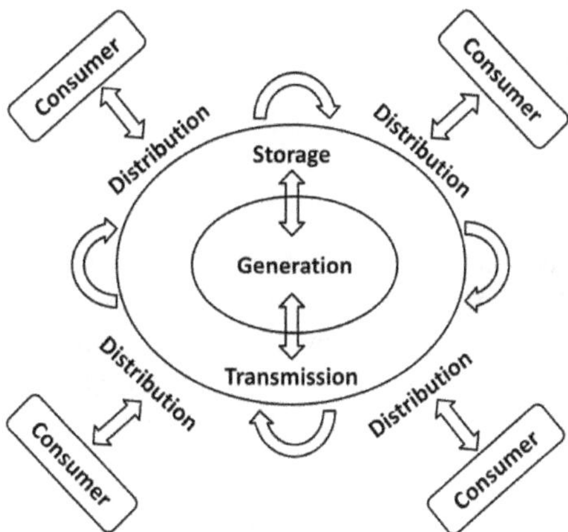

FIGURE 24.2 Smart grid infrastructure.

globally to provide all the working shown in Figure 24.2. All the stakeholders of the grid utilize network layers and software to handle the grid proficiently [12, 13].

24.1.1 Integration of Generation and Storage

Distributed generation (DG) includes renewable generation, wind source, fossil-based, micro-turbines, and small-scale power sources, such as private electric generating units. Energy storage devices include batteries, flywheels, compressed air, super capacitors, and many alternatives that are based on new or emerging technologies. DG with energy storage is used for increasing the efficiency of the transportation sector, reducing the carbon intensity of grid-supplied electricity, and meeting the peak power requirements with more resilient energy supply. Energy storage, geographic information system, community energy storage, outage management system, rooftop PV panels, and micro-grids are important challenges associated with integrating distributed energy resources with the grid [14].

24.1.2 Communication Infrastructure

The communication channels transfer signals from the center to the corresponding equipment and make the power system secure and stable. Supervisory control and data acquisition (SCADA) are required for interfacing the different control and protection system. Human-machine interface (HMI) is used for interaction between the associated systems and human. It gives the operator minimum input to gain the desired output and minimizes undesired outputs on the basis of real-time data [15]. Monitoring and control of the power system changed from manual to computer-based using high-speed hardware, quick processor software, and fast communication system.

24.1.3 Renewable Energy Integration with Convention System

Different renewable sources such as solar, wind, geothermal, and hydropower large-scale grid are integrated with the existing system. Few challenges arise during integration of these non-traditional and renewable power sources onto the grid. The changeable nature of the output is taken into consideration for optimization of the operation of the power system.

24.1.4 Data Acquisition and Controlling

SCADA software is used for remote location data collection in real time to control equipment and conditions. SCADA is also used for process control and data analysis. It is a collection of equipment which can provide remote location handling to operator. The operator can also get information related to the exact condition of process or device. SCADA system helps in capturing the values and equipment data accurately using automation system [16].

Data supervisory system continuously monitors the equipment very effectively, so there is no need of extra person for maintenance. Therefore, SCADA system also decreases operation and maintenance cost. The major components of SCADA system are HMI, programmable logic controllers (PLCs), communication infrastructure, remote terminal units (RTUs), and supervisory computers. Communications between the devices are made using RTUs and PLCs, including HMI software and field connection controllers. Several HMIs are hosted on client sites for data acquisition with multiple servers, and distributed software applications may use in larger SCADA systems [17].

24.2 CYBER SECURITY

The security issue in smart grid is due to large data processing and communication system of SCADA and attachment of smart meter in a large quantity in every utility. The SCADA system and smart meter are hubs of hacking for malevolent hackers. A smart grid has both wireless and wired communication network. Sensors, meters, Phasor Measurement Units and other high measurement and monitoring devices also help in increasing the system reliability. The accurate and wide range of measurement devices are the backbone of effective functioning of smart grid [18]. During the transformation of information, there can be denial-of-service by the operator which may interrupt, obstruct or corrupt the information transmission, resulting in unavailability of the information to the nodes. At this point, the hacker may use legitimate method to intentionally delay these time-critical messages. This way the hacker can monitor and control the devices and create a bootless flux. It is easy for the attacker to launch DoS attacks against smart grids, especially in the wireless network of power system. Privacy in smart grid means not to disclose the private data given to energy supplier [19]. It should not be accessed by unauthorized access. The private data includes the identification of number of the smart meter and the power used by the consumer. These types of privacy issues should be addressed appropriately using proper protocol and security system to reduce customers' worries related to information leakages. Also, we should prevent any virus from attacking the system. Figure 24.3 shows the multilayer security system in the smart grid.

Transmission lines face both temporary and permanent faults because of large physical dimension and environment. Temporary faults are generally self-cleared and permanent faults may be mitigated after detection with available conventional protective relay equipment.

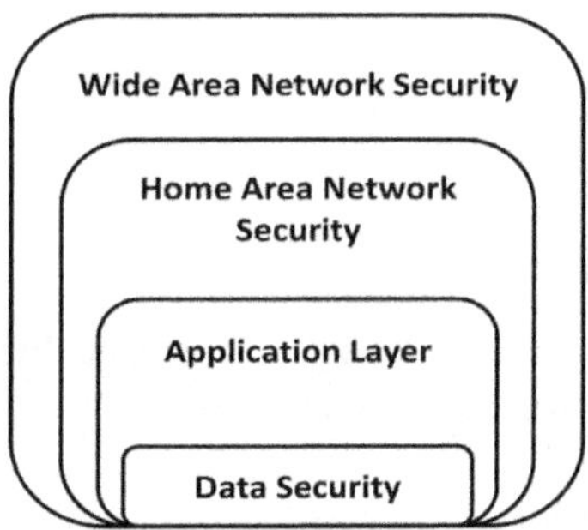

FIGURE 24.3 Data security with multilayer system.

The fault location should be estimated with better accuracy for fast and reliable fault detection. A good fault detection, classification, direction estimation, and location scheme save both money and time for the inspection and repair work with faster and better service. The convention fault location methods estimate fault location with 10–20% error. This fault location algorithm utilizes synchronized current and voltage phasors of fast and secure operation of power grid with better security.

Previously, the simplest method to assault the electrical grid would have been to truly get to and obliterate segments. In any case, the power framework and ICT are exceptionally coupled, which presented new security concerns [20]. It is imperative to take note that present security approaches are inappropriate, not suitable, deficiently adaptable, inconsistent or basically lacking which should be supplanted by new and progressed procedures to guarantee the security of the exceptionally complex and massive smart grid.

24.2.1 Objectives

The interdependency and interconnectivity of the framework for reliable and versatile power supply, is vital and must be very much organized for a protected and productive energy delivery [21]. Notwithstanding an advancing threat and diligent auxiliary difficulties, compelling cyber security strategy advancement is particularly significant. What's more, it must have significant level of discernibleness, accessibility, and controllability and must encourage extended arrangement of sustainable power sources for improved framework which provides decrease in cost of tasks and upkeep cost [22].

24.2.2 Requirement

Smart grid design and framework are confronted with heaps of security threats and difficulties extending from thefts, cyber assaults, fear-based oppression, catastrophic events, and so on. The smart grid security issues include the following:

- Availability
- Integrity
- Authentication

- Authorization
- Confidentiality
- Non-repudiation

24.2.3 Key Challenges

Smart grid is vulnerable to different dangers and difficulties. Following are different digital challenges for which security is necessary.

- Connectivity: Countless devices of smart grid that interoperate in the network.
- Trust: Consumers are dependable on the plan choice and will choose according to his understanding and requirements.
- Consumer's privacy: Customer's security is a significant angle in any framework.
- Vulnerabilities of software: Software experience a wide assortment of malwares.

24.3 CYBER SECURITY THREATS AND ATTACKS

Cyberattacks are apparently the most talked about assaults on smart grids because of the vulnerabilities of the foundation to cyber assaults. Having distinguished different threats and difficulties faced by grid, secrecy, authentication, and security of information are basic for grid productivity that must be ensured to forestall unapproved alterations through the infrastructure [23]. An attacker may bargain a portion of the correspondence hardware, for example, multiplexers influencing an immediate harm or utilizing it as a secondary passage to dispatch future attacks [24]. As smart grid utilizes TCP/IP, it attracts DoS attack. It may endeavor to block, delay or degenerate data transmission so as to make smart grid assets inaccessible. Appropriated cyber security frameworks are along these lines intended to monitor the design in keeping up information integrity [25].

On the off chance that the underneath database is not appropriately arranged, an attacker may access the database, and afterward utilize his/her aptitudes to misuse the framework [26]. Google has launched the Google Power Meter [27] software for tracking everyday energy utilization of consumers, so the consumers re-examine their energy utilization information which may additionally uncover their exercises to looming foes. An attacker may create malware to contaminate smart energy meters or organization servers. It can be utilized to supplant or adjoin any capacity to a device or a framework, for example, transfer of sensitive data [28].

24.4 PRESENT SECURITY SYSTEMS

Cyber security for the smart grid is a critical matter that pulls in the consideration of analysts and industry experts. The different ICT fundamentally received in these proposed security structures by different analysts incorporate the web, WiFi, ZigBee, WiMax, 4G, and Bluetooth. Additionally, new rising innovations, for example, micro-grids, virtual power plants, appropriated insight procedures, smart metering framework and sustainable power source assets, have been thought of to make the grid matrix increasingly dispersed, adaptable for activity and versatile to different security threats [29] since flexibility is consistently a key thought for basic foundations like power grids [30].

The next way of designing a smart grid is by giving information insurance and item authentication. Concerning confirmation, certain necessities must be met. These prerequisites satisfy high effectiveness, tolerance to attacks, and the multicast support [31]. An enhancement model for an insignificant cost obtainment plan for back-start hotspots for helpful self-recuperating was introduced in [32]. In [33], improvement of grid strength in the use of micro-grid to reestablish basic burdens (e.g., clinics, road lights, web, and some other correspondence equipment, etc.) was investigated.

A consistence check is done by means of computerized tool that performs checks over all parts of framework to guarantee that designs of every segment are perfect alleviation and insurance [34]. The device can likewise bring up shortcomings that need consideration. This is significant on the grounds that in a basic framework, for example, the smart grid, a shortcoming in one segment can cause an immense security breach [35].

24.5 PROPOSED MACHINE LEARNING SYSTEM

Security arrangements produced for conventional IT systems are not compelling in grid network due to the significant contrasts between them. Their objective of securities is distinctive, for example, security of IT systems intends to maintain the confidentiality, integrity, and availability only, while security of grid network intends to give overall cyber security consisting of hardware and electrical line security and data transmission security. Additionally, the security design of IT systems is not quite the same as grid since security of IT systems is accomplished by giving more assurance at the middle of the system, while the assurance in grid is done at every edge of network.

Machine learning is genuinely a new idea on the innovation map. We realize that typical computers, telephones, and mobile phones are created remembering the idea of machine learning in the most recent decade or somewhere in the vicinity. In the event that this machine learning can likewise be actualized on a grid system, it will absolutely improve the performance with the security and the speed of counter-assault once an assault is distinguished. On the off chance that the framework would itself be able to gain from the assaults that it has experienced before and afterward, track it and gain from the past encounters, then it can perform countermeasures all alone and furthermore keep the human association least so the people/administrators can think ahead and concoct their own countermeasures. Figure 24.4 shows the deployment of machine learning along with the infrastructure in smart grid for cyber security.

During the generation, transmission, and distribution phase, machine learning algorithm is implemented with feature extraction, normalization, analysis, and anomaly detection. Here, in this smart grid the artificial neural network algorithm is implemented.

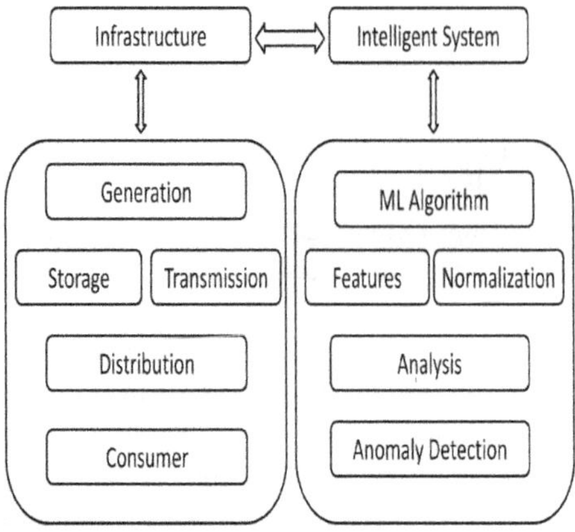

FIGURE 24.4 Deployment of machine learning in smart grid for cyber security.

24.5.1 Implementation and Result

Machine learning algorithm is studied with three conditions: attack 0%, attack 15%, and attack 25%. The precision, recall, and F_1 score are measured and classified accordingly.

24.5.1.1 Attack 0%

Identifying attacks is easier when the error is higher compared to the normal situation. Table 24.1 represents the classification report for attack 0%. It is observed that the F_1 score is 0.99.

24.5.1.2 Attack 15%

Table 24.2 represents the classification report for attack 15%. An F_1 score of 0.97 is observed.

24.5.1.3 Attack 25%

Table 24.3 represents the classification report for attack 15%. An F_1 score of 0.80 is observed.

The overall results are better in the attack 25% despite the fact that F_1 score is lower. It is because the attacks are higher as the time increases.

TABLE 24.1
Classification Report: Attack 0%

	Precision	Recall	F_1 Score	Support
0	1.00	0.99	0.99	65071
1	0.98	1.00	0.99	42762
Micro avg	0.99	0.99	0.99	107,268
Macro avg	0.99	0.99	0.99	107,268
Weighted avg	0.99	0.99	0.99	107,268

TABLE 24.2
Classification Report: Attack 15%

	Precision	Recall	F_1 Score	Support
0	0.97	0.98	0.98	66167
1	0.97	0.96	0.96	44563
Micro avg	0.97	0.97	0.97	116,256
Macro avg	0.97	0.97	0.97	116,256
Weighted avg	0.97	0.97	0.97	116,256

TABLE 24.3
Classification Report: Attack 25%

	Precision	Recall	F_1 Score	Support
0	0.89	0.98	0.94	89562
1	0.89	0.53	0.66	23542
Micro avg	0.89	0.89	0.89	116,256
Macro avg	0.89	0.76	0.80	116,256
Weighted avg	0.89	0.89	0.88	116,256

24.6 ADVANTAGES AND FEATURES OF PROPOSED ML SECURITY SYSTEM

It is appropriate to build up a continuous and developing intelligent security arrangement of all the associated parts for improved security and strength. The proposed security system depends on the wellsprings of threats for distinguishing and recognized the threats. This system is totally focused on utilizing the procedure of threat-tracing to the source, for an extensive, however confounded, yet an increasingly proficient method for fighting the perils presented by the recognized dangers.

In the proposed framework, identity is confirmed through strong authentication components. This shows the execution of a certain deny strategy to such an extent that entrance to the system is conceded distinctly through unequivocal access consents. A vigorous authentication mechanism is conveyed while imparting grid parties. The mechanism is worked progressively with certain limitations, for example, least computational expense, low communication overhead and power to assaults, particularly DoS attack.

The maker has installed in its items a safe storage that contains sensitive information for programming approval. By using keys, framework can approve any recently downloaded program before running. For the smooth conduction of structure, defenselessness evaluation is performed in any event every year to ensure that components that interface with the edge are secure.

The arrangement of proposed framework will give the accompanying key ideas with respect to digital security in shrewd networks.

- Data quality and integrity
- Transparency
- Data minimization
- Accountability and auditing
- Security

Interruption Detection System advancements are enlarged host-based safeguards to shield the framework from inside and outside attacks. Since the existence pattern of the keen lattice is longer than that of the IT frameworks included, all IT innovations are being able to be redesigned.

24.7 CONCLUSION

As of late, the quantities of cyberattacks are expanding quickly. The insightful cyberpsychological militants with detail and propelled power framework information might have the option to make confidentiality, integrity or accessibility attack on the system. Smart grid cyber security is a basic issue that has gained the attention of analysts and cyber experts. Security of smart grid from cyberattack is not just a worry of the architects and specialists; it is additionally the obligation of the legislature to guarantee the security of this national basic foundation. The shifting from convention grid to machine learning based smart grid provides ceaseless advancement which enhance the reliability, security and effectiveness of the grid.

The advancement of computational frameworks has made the way for a universe of chances for machine learning applications through calculations or hybrid techniques that improve proficiency and are getting continuously progressively incredible and equipped for preparing a lot of data.

This chapter has addressed different smart grid security attacks and difficulties with its solution by proposing modified grid structure which provides better security and strength to the grid. It additionally summed up the security necessities and difficulties of the smart grid and existing security arrangements. For improved smart grid security and reliability, machine learning algorithms ought to be put all through framework.

Providing security in smart grid cyber space is still under examination and requires more examination to defeat the vulnerabilities. Additionally, wide research exploration is needed by government offices, scholarly world, organizations, and proficient and pertinent bodies for the assessment of smart grid security and protection issues to improve consumers' trust and mindfulness.

REFERENCES

1. J. M. Carrasco, L. G. Franquelo, J. T. Bialasiewicz, E. Galván, R. C. Portillo, M. A. Martín Prats, J. E. León, and N. Moreno-Alfonso, "Power-electronic "systems for the grid integration of renewable energy sources: A survey," *IEEE Transaction on Industrial Electronics*, vol. 53, no. 4, pp. 1002–1016, 2006.
2. R Gupta, S. N. Singh, and S. K. Singal, "Automation of small hydropower station," in: International Conference on Small Hydropower - Hydro Sri Lanka, October 22–24, 2007.
3. L. M. Camarinha-Matos, "Collaborative smart grids – a survey on trends," *Renewable and Sustainable Energy Reviews,* vol. 65, pp. 283–294, 2016.
4. NIST, Introduction to NISTIR 7628 Guidelines for Smart Grid Cyber Security. http://www.nist.gov/smartgrid/upload/nistir-7628_total.pdf
5. R. Apel, "Smart grid architecture model: Methodology and practical application," in: Workshop of Electrical Power Control Centers, 2013.
6. M. Miller, M. Johns, E. Sortomme, and S. Venkata, "Advanced integration of distributed energy resources" in: Power and Energy Society General Meeting, pp. 1–2, July 2012.
7. R. Morales Gonzalez, B. Asare-Bediako, J. Cobben, W. Kling, G. Scharrenberg, and D. Dijkstra, "Distributed energy resources for a zero-energy neighborhood," in: 3rd IEEE PES International Conf. and Exhibition on Innovative Smart Grid Technologies, pp.1–8, 2012.
8. A. Sanjab, W. Saad, I. Guvenc, A. Sarwat, and S. Biswas, "Smart grid security: Threats, challenges, and solutions." arXiv:1606.06992. 2016
9. W. Wang and Z. Lu, "Cyber security in the smart grid: Survey and challenges," *Computer Networks*, vol. 57, no. 7, pp. 1344–1371, 2013.
10. E. De Santis, A. Rizzi, and A. Sadeghian, "A learning intelligent system for classification and characterization of localized faults in smart grids," paper presented at the 2017 IEEE Congress on Evolutionary Computation (CEC), pp. 2669–2676, 2017.
11. T. M. Lawrence, R. T. Watson, M.-C. Boudreau, and J. Mohammadpour, "Data flow requirements for integrating smart buildings and a smart grid through model predictive control," *Procedia Engineering*, vol. 180, pp. 1402–1412, 2017.
12. M. Moretti, S. N. Djomo, H. Azadi, K. May, K. De Vos, S. Van Passel, and N. Witters, "A systematic review of environmental and economic impacts of smart grids," *Renewable and Sustainable Energy Reviews*, vol. 68, pp. 888–898, 2017.
13. T. Abdulrahman, O. Isiwekpeni, N. Surajudeen-Bakinde, and A. Otuoze, "Design, specification and implementation of a distributed home automation system," *Procedia Computer Science*, vol. 94, pp. 473–478, 2016.
14. L. Zhou, Y. Chen, K. Guo, and F. Jia, "New approach for MPPT control of photovoltaic system with mutative-scale dual-carrier chaotic search," *IEEE Transactions on Power Electronics*, Vol. 26, no. 4, April 2011.
15. J. Hieb, "Security hardened remote terminal units for SCADA networks," Electronic Theses and Dissertations, Paper 615, University of Louisville, 2008.
16. G. Lu, D. De, and W.-Z. Song, "Smart grid lab: A laboratory-based smart grid testbed," in: Proceedings of the IEEE Conference on Smart Grid Communications, 2010.
17. M. S. Thomas and J. D. Mc Donald, *Power System SCADA and Smart Grids.* Oxfordshire: Taylor & Francis Group, LLC; 2015
18. The Smart Grid Interoperability Panel - Smart Grid Cybersecurity Working Group, "Smart grid cyber security strategy and requirements," NIST IR-7628, February 2010.
19. S. Tan, D. De, W.-Z. Song, J. Yang, and S. K. Das, "Survey of security advances in smart grid: A data driven approach," *IEEE Communications Surveys and Tutorials,* vol. 19, no. 1, pp. 397–422, 2017.

20. S. Paul and Z. Ni, "Vulnerability analysis for simultaneous attack in smart grid security," in: Eighth Conference on Innovative Smart Grid Technologies, pp. 1–5, April 2017.
21. S. Ruj and A. Pal, "Analyzing cascading failures in smart grids under random and targeted attacks," paper presented at the 28th International Conference on Advanced Information Networking and Applications (AINA), pp. 226–233, 2014
22. X. Yu and Y. Xue, "Smart grids: A cyber-physical systems perspective," *Proceedings of the IEEE*, vol. 104, no. 5, pp. 1058–1070, 2016.
23. K. Moslehi and R. Kumar, "A reliability perspective of the smart grid," *IEEE Transactions on Smart Grid*, vol. 1, no. 1, pp. 57–64, 2010.
24. K. Wang et al., "Wireless big data computing in smart grid," *IEEE Wireless Communications*, vol. 24, no. 2, pp. 58–64, 2017.
25. I. Esnaola, S.M. Perlaza, H.V. Poor, and O. Kosut, "Maximum distortion attacks in electricity grids," *IEEE Transactions on Smart Grid*, vol. 7, no. 4, pp. 2007–2015, 2016.
26. V. Aravinthan, V. Namboodiri, S. Sunku, and W. Jewell, "Wireless AMI application and security for controlled home area networks," in: Proceedings of the IEEE Power and Energy Society General Meeting, pp. 1–8, July 2011.
27. A. Singh, J. Fernandes, M. Yadav, N. Patil and M. M. Charadva, "Wireless multiuser real time LCD display using short message service basedon GSM technology," *International Journal of Advanced Foundation and Research in Science and Engineering*, vol. 2, pp. 173–177, 2016
28. D. Wei et al. "Protecting smart grid automation systems against cyberattacks," *IEEE Transactions on Smart Grid*, vol. 2, no. 4, pp. 782–795, 2011.
29. Y. Xu, C. C. Liu, K. Schneider, F. Tuffner, and D. Ton, "Microgrids for service restoration to critical load in a resilient distribution system," *IEEE Transactions Smart Grid*, vol. 1, no. 99, 2016.
30. Y. Wang, M. M. Amin, J. Fu, and H. B. Moussa, "A novel data analytical approach for false data injection cyber-physical attack mitigation in smart grids," *IEEE Access*, vol. 5, pp. 26022–26033, 2017.
31. M. B. Line, I. A. Tondel, and M. G. Jaatun, "Cyber security challenges in smart grids," in: 2nd IEEE PES International Conference and Exhibition, Innovative Smart Grid Technologies (ISGT Europe), Manchester, 2011.
32. F. Qiu, J. Wang, C. Chen, and J. Tong, "Optimal black start resource allocation," *IEEE Transactions on Power Systems*, vol. 31, no. 3, pp. 2493–2494, 2016.
33. X. Liu, M. Shahidehpour, Z. Li, X. Liu, Y. Cao, and Z. Bie, "Microgrids for enhancing the power grid resilience in extreme conditions," *IEEE Transactions on Smart Grid*, vol. 8, no. 2, no. 3, pp. 589–597, 2016. http://dx.doi.org/10.1109/TSG.2016.2579999.
34. M. Kammerstetter, "Architecture-driven smart grid security management," in: ACM Workshop on Information Hiding and Multimedia Security, 2014.
35. S. Ahmed, Y. Lee, S. Hyun, and I. Koo, "Feature selection–based detection of covert cyber deception assaults in smart grid communications networks using machine learning," *IEEE Access*, vol. 6, pp. 27518–27529, 2018.

Index

N

O

P

R

S

T

U

V

W

X

For Product Safety Concerns and Information please contact our EU representative GPSR@taylorandfrancis.com
Taylor & Francis Verlag GmbH, Kaufingerstraße 24, 80331 München, Germany

www.ingramcontent.com/pod-product-compliance
Lightning Source LLC
Chambersburg PA
CBHW060748220326
41598CB00022B/2367